Student Solutions Manual

Elementary Technical Mathematics

ELEVENTH EDITION

Dale Ewen
Parkland Community College

C. Robert Nelson
Champaign Centennial High School

Prepared by

James Lapp

CENGAGE
Learning·

Australia • Brazil • Mexico • Singapore • United Kingdom • United States

For product information and technology assistance, contact us at **Cengage Learning Customer & Sales Support, 1-800-354-9706**.

For permission to use material from this text or product, submit all requests online at **www.cengage.com/permissions**
Further permissions questions can be emailed to **permissionrequest@cengage.com**.

ISBN-13: 978-1-285-19927-6
ISBN-10: 1-285-19927-8

Cengage Learning
200 First Stamford Place, 4th Floor
Stamford, CT 06902
USA

Cengage Learning is a leading provider of customized learning solutions with office locations around the globe, including Singapore, the United Kingdom, Australia, Mexico, Brazil, and Japan. Locate your local office at: **www.cengage.com/global**.

Cengage Learning products are represented in Canada by Nelson Education, Ltd.

To learn more about Cengage Learning Solutions, visit **www.cengage.com**.

Purchase any of our products at your local college store or at our preferred online store **www.cengagebrain.com**.

Printed in the United States of America
1 2 3 4 5 6 7 17 16 15 14 13

Table of Contents

Chapter 1: Basic Concepts

Section 1.1: Review of Basic Operations

1. 3255

3. 1454

5. 795,776

7. 5164

9. 26,008

11. 2820

13. 4195 Ω

15. The sum of the lengths is 224 feet, so 224 studs are required.

17. 39 ft

19. Input: 1925 cm^3 ; Output: 1425 cm^3

1925 cm^3 − 1425 cm^3 = 500 cm^3

21. 27,216

23. 18,172,065

25. 35,360,000

27. 1809

29. 389

31. 844 r 40

33. 31 mi/gal × 16 gal = 496 mi

35. 1300 cm^3 ÷ 4 = 325 cm^3

37. 2340 km ÷ 180 L = 13 km/L

39. $516 ÷ 6 h = $86/h

51. a.

$$\frac{856\text{ lb} + 754\text{ lb} + 1044\text{ lb} + 928\text{ lb} + 888\text{ lb} + 734\text{ lb} + 953\text{ lb} + 891\text{ lb}}{8} = \frac{7048\text{ lb}}{8\text{ days}} = 881\text{ lb/day}$$

b. $\dfrac{4320\text{ lb}}{36\text{ days}} = 120$ lb/day ; $\dfrac{120\text{ lb/day}}{8\text{ steers}} = 115$ lb/day/steer

53.

$$\frac{92,480\text{ lb}}{32\text{ lb/bu}} = 2890\text{ bu}$$

$$\frac{2890\text{ bu}}{34\text{ acre}} = 85\text{ lb/acre}$$

55.

$175,000 − $300 = $172,000

$$\frac{\$172,000}{10} = \$17,200$$

59. $I = \dfrac{E}{R} = \dfrac{220}{44} = 5$ A

41. 125 mi/h × 4 h = 500 mi

43.

5 × 18 ft = 90 ft

42 × 15 ft = 630 ft

158 × 12 ft = 1896 ft

105 × 10 ft = 1050 ft

79 × 8 ft = 632 ft

87 × 6 ft = <u>522 ft</u>

Total = 4820 ft

45.

First draftperson:

8 × 30 × 80 = 19,200 drawings

Second draftperson:

8 × 30 × 120 = 28,800 drawings

Difference:

28,800 − 19,200 = 9600 drawings

47.

17 ft 5 in. = 17 ft × 12 in./ft + 5 in.

= 209 in.

209 in. − 75 in. = 134 in.

134 in. ÷ 2 = 67 in. from either corner

49. $\dfrac{6864\text{ bu}}{156\text{ acre}} = 44$ bu/acre

57.

30 ft × 12 in./ft = 360 in.

360 in. − 2 × 5 in. = 350 in.

350 in. ÷ 10 in. = 35

One additional daylily is required at the end of the planting so 35 + 1 = 36 daylilies are needed in total.

61. $E = IR = (2)(12) = 24$ V

63. 220 × 4 oz = 880 oz

1

65. $800 \text{ mg} \div 800 \text{ mg} = 4$ tablets

67.

$$14 \text{ ft } 6 \text{ in.} - 4 \times (2 \text{ ft } 6 \text{ in.}) - 3 \times (1 \text{ ft})$$
$$= 14 \text{ ft } 6 \text{ in.} - 10 \text{ ft} - 3 \text{ ft}$$
$$= 1 \text{ ft } 6 \text{ in.}$$
$$(1 \text{ ft } 6 \text{ in.}) \div 2$$
$$= 18 \text{ in.} \div 2$$
$$= 9 \text{ in.}$$

69.

$$8 \text{ ft} - 3 \times (10 \text{ in.}) - 2 \times (1 \text{ ft } 2 \text{ in.})$$
$$= 96 \text{ in.} - 3 \times 10 \text{ in.} - 2 \times 14 \text{ in.}$$
$$= 96 \text{ in.} - 30 \text{ in.} - 28 \text{ in.}$$
$$= 38 \text{ in.}$$
$$38 \text{ in.} \div 2$$
$$= 19 \text{ in.}$$

71.

$$2500 \div 1000 = 2.5$$
$$2.5 \times 8540 \text{ bd ft} = 213{,}500 \text{ bd ft}$$

73. $50 + 125 + 110 + 35 = 320$ seats

75. $2 \times 90 + 3 \times 4 + 2 \times 4 = 180 + 12 + 8 = 200$ items.

77. a. $\$131 + \$152 + \$128 = \$411$

    b. $\$411 \div 3 = \$137$

Section 1.2: Order of Operations

1.

$$8 - 3(4 - 2)$$
$$= 8 - 3(2)$$
$$= 8 - 6$$
$$= 2$$

3.

$$(8 + 6) - (7 - 3)$$
$$= 14 - 4$$
$$= 10$$

5.

$$2(9 + 5) - 6 \times (13 + 2) \div 9$$
$$= 2(14) - 6 \times 15 \div 9$$
$$= 28 - 90 \div 9$$
$$= 28 - 10$$
$$= 18$$

7.

$$27 + 13 \times (7 - 3)(12 + 6) \div 9$$
$$= 27 + 13 \times (4)(18) \div 9$$
$$= 27 + 52(18) \div 9$$
$$= 27 + 936 \div 9$$
$$= 27 + 104$$
$$= 131$$

9.

$$16 + 4(7 + 8) - 3$$
$$= 16 + 4(15) - 3$$
$$= 16 + 60 - 3$$
$$= 73$$

11.

$$9 - 2(17 - 15) + 18$$
$$= 9 - 2(2) + 18$$
$$= 9 - 4 + 18$$
$$= 23$$

13.

$$(39 - 18) - (23 - 18)$$
$$= 21 - 5$$
$$= 16$$

15.

$$3(8 + 6) - 7(13 + 3) \div 14$$
$$= 3(14) - 7(16) \div 14$$
$$= 42 - 112 \div 14$$
$$= 42 - 8$$
$$= 34$$

17.

$$42 + 12(9-3)(12+13) \div 30$$
$$= 42 + 12(6)(25) \div 30$$
$$= 42 + 72(25) \div 30$$
$$= 42 + 1800 \div 30$$
$$= 42 + 60$$
$$= 102$$

19.

$$38 + 9 \times (8+4) - 3(5-2)$$
$$= 38 + 9 \times 12 - 3(3)$$
$$= 38 + 108 - 9$$
$$= 137$$

21.

$$27 - 2 \times (18-9) - 3 + 8(43-15)$$
$$= 27 - 2 \times 9 - 3 + 8(28)$$
$$= 27 - 18 - 3 + 224$$
$$= 230$$

23.

$$12 \times 9 \div 18 \times 64 \div 8 + 7$$
$$= 108 \div 18 \times 64 \div 8 + 7$$
$$= 6 \times 64 \div 8 + 7$$
$$= 384 \div 8 + 7$$
$$= 48 + 7$$
$$= 55$$

25.

$$7 + 6(3+2) - 7 - 5(4+2)$$
$$= 7 + 6(5) - 7 - 5(6)$$
$$= 7 + 30 - 7 - 30$$
$$= 0$$

27.

$$3 + 17(2 \times 2) - 67$$
$$= 3 + 17(4) - 67$$
$$= 3 + 68 - 67$$
$$= 4$$

29.

$$28 - 4(2 \times 3) + 4 - (16 \times 8) \div (4 \times 4)$$
$$= 28 - 4(6) + 4 - 128 \div 16$$
$$= 28 - 24 + 4 - 8$$
$$= 0$$

31.

$$24/(6-2) + 4 \times 3 - 15/3$$
$$= 24/4 + 12 - 5$$
$$= 6 + 12 - 5$$
$$= 13$$

33.

$$3 \times 15 \div 9 + (13-5)/2 \times 4 - 2$$
$$= 45 \div 9 + 8/2 \times 4 - 2$$
$$= 5 + 4 \times 4 - 2$$
$$= 5 + 16 - 2$$
$$19$$

35.

$$10 + 4^2$$
$$= 10 + 16$$
$$= 26$$

37.

$$\frac{20 + (2 \cdot 3)^2}{7 \cdot 2^3}$$
$$= \frac{20 + 6^2}{7 \cdot 8}$$
$$= \frac{20 + 36}{56}$$
$$= \frac{56}{56}$$
$$= 1$$

39.

$$6[3 + 2(2+5)]$$
$$= 6[3 + 2(7)]$$
$$= 6[3 + 14]$$
$$= 6[17]$$
$$= 102$$

41.

$$5 \times 2 + 3[2(5-3) + 4(4+2) - 3]$$
$$= 10 + 3[2(2) + 4(6) - 3]$$
$$= 10 + 3[4 + 24 - 3]$$
$$= 10 + 3[25]$$
$$= 10 + 75$$
$$= 85$$

Section 1.3: Area and Volume

1.

$A = l \times w$

$A = 12 \text{ yd} \times 8 \text{ yd}$

$\quad = 96 \text{ yd}^2$

5.

$A = l \times w$

$A = 191 \text{ in.} \times 73 \text{ in.}$

$\quad = 13,943 \text{ in}^2$

3.

$A = l \times w$

$A = 4100 \text{ ft} \times 75 \text{ ft}$

$\quad = 307,500 \text{ ft}^2$

7.

Area of outer rectangle: $9 \text{ cm} \times 12 \text{ cm} = 108 \text{ cm}^2$

Area of inner rectangle: $6 \text{ cm} \times 4 \text{ cm} \quad = \underline{24 \text{ cm}^2}$

Total area: $\qquad\qquad\qquad = 84 \text{ cm}^2$

9.

Area of left rectangle: $\quad 8 \text{ in.} \times 3 \text{ in.} = 24 \text{ in}^2$

Area of middle rectangle: $2 \text{ in.} \times 6 \text{ in.} = 12 \text{ in}^2$

Area of right rectangle: $\quad 3 \text{ in.} \times 4 \text{ in.} = \underline{12 \text{ in}^2}$

Total area: $\qquad\qquad\qquad = 48 \text{ in}^2$

11.

Area of upper rectangle: $3 \text{ in.} \times 6 \text{ in.} = 24 \text{ in}^2$

Area of lower rectangle: $7 \text{ in.} \times 4 \text{ in.} = \underline{28 \text{ in}^2}$

Total area: $\qquad\qquad\qquad = 52 \text{ in}^2$

13. $\dfrac{48 \text{ in.} \times 36 \text{ in.}}{4 \text{ in.} \times 4 \text{ in.}} = \dfrac{1728 \text{ in}^2}{16 \text{ in}^2} = 108$ tiles are needed.

15.

Area of ceiling: $\qquad\qquad 12 \text{ ft} \times 16 \text{ ft} = 192 \text{ ft}^2$

Area of left/right walls: $\quad 2 \times 8 \text{ ft} \times 12 \text{ ft} = 192 \text{ ft}^2$

Area of front/back walls: $2 \times 8 \text{ ft} \times 16 \text{ ft} = \underline{256 \text{ ft}^2}$

Total area: $\qquad\qquad\qquad\qquad = 640 \text{ ft}^2$

Twenty rooms will be $20 \times 640 \text{ ft}^2 = 12,800 \text{ ft}^2$ so $12,800 \text{ ft}^2 \div 640 \text{ ft}^2 = 32$ gallons of paint will be needed.

17. a.

$A = l \times w$

$A = 24 \text{ ft} \times 45 \text{ ft}$

$\quad = 1080 \text{ ft}^2$

Value $= 1080 \text{ ft}^2 \times \$110/\text{ft}^2$

$\qquad = \$118,800$

17. (continued)

    b.

      Area of upper rectangle: $24 \text{ ft} \times 85 \text{ ft} = 2040 \text{ ft}^2$

      Area of lower rectangle: $19 \text{ ft} \times 16 \text{ ft} = \underline{304 \text{ ft}^2}$

      Total area: $ = 2344 \text{ ft}^2$

      Value $= 2344 \text{ ft}^2 \times \$110/\text{ft}^2 = \$257{,}840$

19.

$$V = l \times w \times h$$
$$V = 3 \text{ m} \times 4 \text{ m} \times 8 \text{ m}$$
$$= 96 \text{ m}^3$$

21.

    Volume of upper box: $6 \text{ cm} \times 4 \text{ cm} \times 5 \text{ cm} = 120 \text{ cm}^3$

    Volume of lower box: $6 \text{ cm} \times 20 \text{ cm} \times 5 \text{ cm} = \underline{600 \text{ cm}^3}$

    Total Volume: $ = 720 \text{ cm}^3$

23.

    Volume of left box: $5 \text{ in.} \times 6 \text{ in.} \times 40 \text{ in.} = 1200 \text{ in}^3$

    Volume of middle box: $25 \text{ in.} \times 6 \text{ in.} \times 10 \text{ in.} = 1500 \text{ in}^3$

    Volume of right box: $5 \text{ in.} \times 6 \text{ in.} \times 40 \text{ in.} = \underline{1200 \text{ in}^3}$

    Total Volume: $ = 3900 \text{ in}^3$

25.

$$V = l \times w \times h$$
$$V = 10 \text{ cm} \times 12 \text{ cm} \times 5 \text{ cm}$$
$$= 600 \text{ cm}^3$$

27.

$$V = l \times w \times h$$
$$V = 8 \text{ in.} \times 20 \text{ in.} \times 72 \text{ in.}$$
$$= 11{,}520 \text{ in}^3$$

29.

$$V = l \times w \times h$$
$$V = 3 \text{ ft} \times 5 \text{ ft} \times 2 \text{ ft}$$
$$= 30 \text{ ft}^3$$

31.

$$V = l \times w \times h$$
$$V = 15 \text{ ft} \times 12 \text{ ft} \times 2 \text{ ft}$$
$$= 360 \text{ ft}^3$$

So, the cement will weigh
$360 \text{ ft}^3 \times 193 \text{ lb/ft}^3 = 69{,}480 \text{ lb.}$

33.

$$V = l \times w \times h$$
$$V = 8 \text{ ft} \times 5 \text{ ft} \times 6 \text{ ft}$$
$$= 240 \text{ ft}^3$$

So, the water will weigh
$240 \text{ ft}^3 \times 62 \text{ lb/ft}^3 = 14{,}880 \text{ lb.}$

35.

$$V = l \times w \times h$$
$$V = 100 \text{ ft} \times 50 \text{ ft} \times 10 \text{ ft}$$
$$= 50{,}000 \text{ ft}^3$$

So, the cost of heating the space will be
$50{,}000 \text{ ft}^3 \div 1000 \text{ ft}^3 \times \$55 = \$2750.$

37. The height of the cardboard sheet would be $16 \text{ in.} + 9 \text{ in.} = 25 \text{ in.}$ and the width would be $4 \times 9 \text{ in.} + 1 \text{ in.} = 37 \text{ in.}$

39.

$$V = l \times w \times h$$
$$V = 4 \text{ ft} \times 4 \text{ ft} \times 8 \text{ ft}$$
$$= 128 \text{ ft}^3$$

41.

$$8 \text{ ft} \times 12 \text{ in./ft} = 96 \text{ in.}$$

$$24 \text{ ft} \times 12 \text{ in./ft} = 288 \text{ in.}$$

$$V = l \times w \times h$$

$$= 96 \text{ in.} \times 288 \text{ in.} \times 3 \text{ in.}$$

$$= 82944 \text{ in}^3$$

$1 \text{ ft}^3 = 1728 \text{ in}^3$, so $\dfrac{82944 \text{ in}^3}{1728 \text{ in}^3} = 48 \text{ ft}^3$ of mulch are needed.

Section 1.4: Formulas

1.

$$W = fd$$

$$W = (30)(20)$$

$$= 600$$

3.

$$W = fd$$

$$W = (1125)(10)$$

$$= 11,250$$

5.

$$W = fd$$

$$W = (176)(326)$$

$$= 57,376$$

7.

$$f = ma$$

$$f = (1600)(24)$$

$$= 38,400$$

9.

$$I = \frac{E}{R}$$

$$I = \frac{120}{15}$$

$$= 8$$

11.

$$P = IE$$

$$P = (29)(173)$$

$$= 5017$$

13.

$$A = \frac{1}{2}bh$$

$$A = \frac{1}{2}(10 \text{ in.})(8 \text{ in.})$$

$$= 40 \text{ in}^2$$

15.

$$A = \frac{1}{2}bh$$

$$A = \frac{1}{2}(54 \text{ ft})(30 \text{ ft})$$

$$= 810 \text{ ft}^2$$

17.

$$A = lw$$

$$A = (8 \text{ m})(7 \text{ m})$$

$$= 56 \text{ m}^2$$

19.

$$A = lw$$

$$A = (36 \text{ ft})(18 \text{ ft})$$

$$= 648 \text{ ft}^2$$

21.

$$A = \left(\frac{a+b}{2}\right)h$$

$$A = \left(\frac{7 \text{ ft} + 9 \text{ ft}}{2}\right)(4 \text{ ft})$$

$$= \left(\frac{16 \text{ ft}}{2}\right)(4 \text{ ft})$$

$$= (8 \text{ ft})(4 \text{ ft})$$

$$= 32 \text{ ft}^2$$

23.

$$A = \left(\frac{a+b}{2}\right)h$$

$$A = \left(\frac{96 \text{ cm} + 24 \text{ cm}}{2}\right)(30 \text{ cm})$$

$$= \left(\frac{120 \text{ cm}}{2}\right)(30 \text{ cm})$$

$$= (60 \text{ cm})(30 \text{ cm})$$

$$= 1800 \text{ cm}^2$$

25.

$$V = lwh$$

$$V = (25 \text{ cm})(15 \text{ cm})(12 \text{ cm})$$

$$= 4500 \text{ cm}^3$$

27.

$$v = v_0 + gt$$

$$v = 12 + (32)(5)$$

$$= 172$$

29.

$$I = \frac{E}{Z}$$

$$I = \frac{240}{15}$$

$$= 16$$

31.

$$P = cd^2 SN$$

$$P = (0.7853)(3)^2(4)(4)$$

$$= 113.1$$

33.

Area of left rectangle: $55 \text{ ft} \times 120 \text{ ft} = 6600 \text{ ft}^2$

Area of middle rectangle: $160 \text{ ft} \times 60 \text{ ft} = 9600 \text{ ft}^2$

Area of right rectangle: $260 \text{ ft} \times 60 \text{ ft} = \underline{21,600 \text{ ft}^2}$

Total area: $= 31,800 \text{ ft}^2$

Area in tsf $= 31,800 \text{ ft}^2 \div 1000 = 31.8$ tsf

Section 1.5: Prime Factorization

1. a. $1+5=6$ is divisible by 3, so 15 is divisible by 3.

b. 15 is not divisible by 4.

3. a. $9+6=15$ is divisible by 3, so 96 is divisible by 3.

b. 96 is divisible by 4.

5. a. $7+8=15$ is divisible by 3, so 78 is divisible by 3.

b. 78 is not divisible by 4.

7. 53 is prime

9. $93 = 3 \cdot 31$ is not prime

11. $16 = 2 \cdot 2 \cdot 2 \cdot 2$ is not prime

13. $39 = 3 \cdot 13$ is not prime

15. 458 is even, so it is divisible by 2.

17. $315,817$ is odd, so it is not divisible by 2.

19. 1367 is odd, so it is not divisible by 2.

21. $3+8+7=18$ is divisible by 3, so 387 is divisible by 3.

23. $4+5+3+1+2+8=23$ is not divisible by 3, so $453,128$ is not divisible by 3.

25. $2+1+8+7+4+5=27$ is divisible by 3, so $218,745$ is divisible by 3.

27. 70 ends in 0, so it is divisible by 5.

29. 366 does not end in 0 or 5, so it is not divisible by 5.

31. $63,227$ does not end in 0 or 5, so it is not divisible by 5.

33. 56 is even, so it is divisible by 2.

35. $2+1+8=11$ is not divisible by 3, so 218 is not divisible by 3.

37. 528 does not end in 0 or 5, so it is not divisible by 5.

39. $1+9+8=18$ is divisible by 3, so 198 is divisible by 3.

41. $1,820,670$ is even, so it is divisible by 2.

43. $7,215,720$ ends in 0, so it is divisible by 5.

45. $2 \cdot 2 \cdot 5$

47. $2 \cdot 3 \cdot 11$

49. $2 \cdot 2 \cdot 3 \cdot 3$

51. $3 \cdot 3 \cdot 3$

53. $51 = 3 \cdot 17$

55. $42 = 2 \cdot 3 \cdot 7$

57. $120 = 2 \cdot 2 \cdot 2 \cdot 3 \cdot 5$

59. $171 = 3 \cdot 3 \cdot 19$

61. $105 = 3 \cdot 5 \cdot 7$

63. $252 = 2 \cdot 2 \cdot 3 \cdot 3 \cdot 7$

Unit 1A Review

1. 241

2. 1795

3. $2,711,279$

4. 620

5.

$$3 \times 12 \text{ ft} = 36 \text{ ft}$$
$$8 \times 8 \text{ ft} = 64 \text{ ft}$$
$$9 \times 10 \text{ ft} = 90 \text{ ft}$$
$$\underline{12 \times 6 \text{ ft} = 72 \text{ ft}}$$
$$\text{Total} = 262 \text{ ft}$$

6. $14,244 \text{ lb} \div 56 \text{ lb} = 254 \text{ bu}$

7.

$$6 + 2(5 \times 4 - 2)$$
$$= 6 + 2(20 - 2)$$
$$= 6 + 2(18)$$
$$= 6 + 23$$
$$= 42$$

8.

$$3^2 + 12 \div 3 - 2 \times 3$$
$$= 9 + 4 - 6$$
$$= 7$$

9.

$$12 + 2\left[3(8-2) - 2(3+1)\right]$$
$$= 12 + 2\left[3(6) - 2(4)\right]$$
$$= 12 + 2\left[18 - 8\right]$$
$$= 12 + 2\left[10\right]$$
$$= 12 + 20$$
$$= 32$$

10.

$$\text{Area of left rectangle:} \quad 24 \text{ in.} \times 11 \text{ in.} = 264 \text{ in}^2$$
$$\text{Area of middle rectangle:} \ 15 \text{ in.} \times 11 \text{ in.} = 165 \text{ in}^2$$
$$\text{Area of right rectangle:} \quad 10 \text{ in.} \times 7 \text{ in.} = \underline{70 \ \text{in}^2}$$
$$\text{Total area:} \qquad\qquad\qquad\qquad = 499 \text{ in}^2$$

11.

$$V = lwh$$
$$V = (15 \text{ ft})(8 \text{ ft})(6 \text{ ft})$$
$$= 720 \text{ ft}^3$$

12.

$$d = vt$$
$$d = (45)(4)$$
$$= 180$$

13.

$$I = \frac{E}{R}$$
$$I = \frac{120}{12}$$
$$= 10$$

14.

$$A = \frac{1}{2}bh$$
$$A = \frac{1}{2}(40)(15)$$
$$= 300$$

15. $51 = 3 \cdot 17$ is not prime.

16. 47 is prime.

17. $1 + 9 + 5 = 15$ is divisible by 3, so 195 is not divisible by 3.

18. 821 does not end in 0 or 5, so it is not divisible by 5.

19. $40 = 2 \cdot 2 \cdot 2 \cdot 5$

20. $135 = 3 \cdot 3 \cdot 3 \cdot 5$

Section 1.6: Introduction to Fractions

1. $\dfrac{12}{28} = \dfrac{2 \cdot 2 \cdot 3}{2 \cdot 2 \cdot 7} = \dfrac{3}{7}$

3. $\dfrac{36}{42} = \dfrac{2 \cdot 2 \cdot 3 \cdot 3}{2 \cdot 3 \cdot 7} = \dfrac{6}{7}$

5. $\dfrac{9}{48} = \dfrac{3 \cdot 3}{2 \cdot 2 \cdot 2 \cdot 2 \cdot 3} = \dfrac{3}{16}$

7. $\dfrac{13}{39} = \dfrac{13}{3 \cdot 13} = \dfrac{1}{3}$

9. $\dfrac{48}{60} = \dfrac{2 \cdot 2 \cdot 2 \cdot 2 \cdot 3}{2 \cdot 2 \cdot 3 \cdot 5} = \dfrac{4}{5}$

11. $\dfrac{9}{9} = 1$

13. $\dfrac{0}{8} = 0$

15. $\dfrac{9}{0}$ is undefined

17. $\dfrac{14}{16} = \dfrac{2 \cdot 7}{2 \cdot 2 \cdot 2 \cdot 2} = \dfrac{7}{8}$

19. $\dfrac{27}{36} = \dfrac{3 \cdot 3 \cdot 3}{2 \cdot 2 \cdot 3 \cdot 3} = \dfrac{3}{4}$

21. $\dfrac{12}{16} = \dfrac{2 \cdot 2 \cdot 3}{2 \cdot 2 \cdot 2 \cdot 2} = \dfrac{3}{4}$

23. $\dfrac{20}{25} = \dfrac{2 \cdot 2 \cdot 5}{5 \cdot 5} = \dfrac{4}{5}$

25. $\dfrac{12}{40} = \dfrac{2 \cdot 2 \cdot 3}{2 \cdot 2 \cdot 2 \cdot 5} = \dfrac{3}{10}$

27. $\dfrac{112}{128} = \dfrac{2 \cdot 2 \cdot 2 \cdot 2 \cdot 7}{2 \cdot 2 \cdot 2 \cdot 2 \cdot 2 \cdot 2 \cdot 2} = \dfrac{7}{8}$

29. $\dfrac{112}{144} = \dfrac{2 \cdot 2 \cdot 2 \cdot 2 \cdot 7}{2 \cdot 2 \cdot 2 \cdot 2 \cdot 3 \cdot 3} = \dfrac{7}{9}$

31. $\dfrac{78}{5} = 15 \text{ r } 3 = 15\dfrac{3}{5}$

33. $\dfrac{28}{3} = 9 \text{ r } 1 = 9\dfrac{1}{3}$

35. $\dfrac{45}{36} = \dfrac{5}{4} = 1 \text{ r } 1 = 1\dfrac{1}{4}$

37. $\dfrac{57}{6} = \dfrac{19}{2} = 9 \text{ r } 1 = 9\dfrac{1}{2}$

39. $5\dfrac{15}{12} = 5\dfrac{5}{4} = 5 + \left(1\dfrac{1}{4}\right) = 6\dfrac{1}{4}$

41. $3\dfrac{5}{6} = \dfrac{(3 \times 6) + 5}{6} = \dfrac{23}{6}$

43. $2\dfrac{1}{8} = \dfrac{(2 \times 8) + 1}{8} = \dfrac{17}{8}$

45. $1\dfrac{7}{16} = \dfrac{(1 \times 16) + 7}{16} = \dfrac{23}{16}$

47. $6\dfrac{7}{8} = \dfrac{(6 \times 8) + 7}{8} = \dfrac{55}{8}$

49. $10\dfrac{3}{5} = \dfrac{(10 \times 5) + 3}{5} = \dfrac{53}{5}$

51. $\dfrac{28}{6} = \dfrac{14}{3} = 4 \text{ r } 2 = 4\dfrac{2}{3}$ pies

Section 1.7: Addition and Subtraction of Fractions

1. 16

3. 210

5. 48

7. $\dfrac{2}{3} + \dfrac{1}{6} = \dfrac{4}{6} + \dfrac{1}{6} = \dfrac{5}{6}$

9. $\dfrac{1}{16} + \dfrac{3}{32} = \dfrac{2}{32} + \dfrac{3}{32} = \dfrac{5}{32}$

11. $\dfrac{2}{7} + \dfrac{3}{28} = \dfrac{8}{28} + \dfrac{3}{28} = \dfrac{11}{28}$

13. $\dfrac{3}{8} + \dfrac{5}{64} = \dfrac{24}{64} + \dfrac{5}{64} = \dfrac{29}{64}$

15. $\dfrac{1}{5} + \dfrac{3}{20} = \dfrac{4}{20} + \dfrac{3}{20} = \dfrac{7}{20}$

17. $\dfrac{4}{5} + \dfrac{1}{2} = \dfrac{8}{10} + \dfrac{5}{10} = \dfrac{13}{10} = 1\dfrac{3}{10}$

19. $\dfrac{1}{3} + \dfrac{1}{6} + \dfrac{3}{16} + \dfrac{1}{12} = \dfrac{16}{48} + \dfrac{8}{48} + \dfrac{9}{48} + \dfrac{4}{48} = \dfrac{37}{48}$

21. $\dfrac{1}{20} + \dfrac{1}{30} + \dfrac{1}{40} = \dfrac{6}{120} + \dfrac{4}{120} + \dfrac{3}{120} = \dfrac{13}{120}$

23.
$$\frac{3}{10}+\frac{1}{14}+\frac{4}{15}=\frac{63}{210}+\frac{15}{210}+\frac{56}{210}$$
$$=\frac{134}{210}$$
$$=\frac{67}{105}$$

25. $\dfrac{7}{8}-\dfrac{3}{4}=\dfrac{7}{8}-\dfrac{6}{8}=\dfrac{1}{8}$

27. $\dfrac{4}{5}-\dfrac{3}{10}=\dfrac{8}{10}-\dfrac{3}{10}=\dfrac{5}{10}=\dfrac{1}{2}$

29. $\dfrac{9}{14}-\dfrac{3}{42}=\dfrac{27}{42}-\dfrac{3}{42}=\dfrac{24}{42}=\dfrac{4}{7}$

31. $\dfrac{9}{16}-\dfrac{13}{32}-\dfrac{1}{8}=\dfrac{18}{32}-\dfrac{13}{32}-\dfrac{4}{32}=\dfrac{1}{32}$

33.
$$2\frac{1}{2}=2\frac{2}{4}$$
$$\underline{4\frac{3}{4}=4\frac{3}{4}}$$
$$6\frac{5}{4}=7\frac{1}{4}$$

35.
$$3=2\frac{8}{8}$$
$$\underline{\frac{3}{8}=\frac{3}{8}}$$
$$2\frac{5}{8}$$

37.
$$8\frac{3}{16}=7\frac{19}{16}$$
$$\underline{3\frac{7}{16}=3\frac{7}{16}}$$
$$4\frac{12}{16}=4\frac{3}{4}$$

39.
$$7\frac{3}{16}=6\frac{19}{16}$$
$$\underline{4\frac{7}{8}=4\frac{14}{16}}$$
$$2\frac{5}{16}$$

41.
$$3\frac{4}{5}=3\frac{36}{45}$$
$$\underline{9\frac{8}{9}=9\frac{49}{45}}$$
$$12\frac{86}{45}=13\frac{41}{45}$$

43.
$$3\frac{9}{16}+4\frac{7}{12}+3\frac{1}{6}$$
$$=3\frac{27}{48}+4\frac{28}{48}+3\frac{8}{48}$$
$$=10\frac{63}{48}=10\frac{21}{16}=11\frac{5}{16}$$

45.
$$16\frac{5}{8}-4\frac{7}{12}-2\frac{1}{2}$$
$$=16\frac{15}{24}-4\frac{14}{24}-2\frac{12}{24}$$
$$=15\frac{39}{24}-4\frac{14}{24}-2\frac{12}{24}$$
$$=9\frac{13}{24}$$

47.
$$712\frac{3}{4}\text{ ft}+563\text{ ft}+961\frac{1}{2}\text{ ft}$$
$$=712\frac{3}{4}\text{ ft}+563\text{ ft}+961\frac{2}{4}\text{ ft}$$
$$=2236\frac{5}{4}\text{ ft}=2237\frac{1}{4}\text{ ft}$$

49. a.
$$2\frac{3}{8}\text{ ft}+3\frac{7}{8}\text{ ft}$$
$$=5\frac{10}{8}\text{ ft}=6\frac{2}{8}\text{ ft}=6\frac{1}{4}\text{ ft}$$

b.
$$6\frac{1}{4}\text{ ft}-4\frac{3}{4}\text{ ft}$$
$$=5\frac{5}{4}\text{ ft}-4\frac{3}{4}\text{ ft}$$
$$=1\frac{2}{4}\text{ ft}=1\frac{1}{2}\text{ ft}$$

51.

$$13\frac{3}{4}\text{ gal} + 11\frac{2}{5}\text{ gal} + 10\frac{2}{5}\text{ gal}$$

$$= 13\frac{15}{20}\text{ gal} + 11\frac{8}{20}\text{ gal} + 10\frac{8}{20}\text{ gal}$$

$$= 34\frac{31}{20}\text{ gal} = 35\frac{11}{20}\text{ gal}$$

53.

$$25\frac{1}{4}\text{ gal} - 23\frac{3}{4}\text{ gal}$$

$$= 24\frac{5}{4}\text{ gal} - 23\frac{3}{4}\text{ gal}$$

$$= 1\frac{2}{4}\text{ gal} = 1\frac{1}{2}\text{ gal}$$

55.

$$\frac{1}{3}\text{ h} + \frac{1}{4}\text{ h} + \frac{1}{4}\text{ h}$$

$$= \frac{4}{12}\text{ h} + \frac{3}{12}\text{ h} + \frac{3}{12}\text{ h}$$

$$= \frac{10}{12}\text{ h} = \frac{5}{6}\text{ h}$$

57.

$$\frac{1}{3}\text{ ton} + \frac{3}{4}\text{ ton} + \frac{9}{16}\text{ ton}$$

$$= \frac{16}{48}\text{ ton} + \frac{36}{48}\text{ ton} + \frac{27}{48}\text{ ton}$$

$$= \frac{79}{48}\text{ ton} = 1\frac{31}{48}\text{ ton}$$

59.

$$10\text{ in.} - \frac{3}{4}\text{ in.} - \frac{3}{4}\text{ in.} - \frac{1}{8}\text{ in.} - \frac{1}{8}\text{ in.}$$

$$= 10\text{ in.} - \frac{6}{8}\text{ in.} - \frac{6}{8}\text{ in.} - \frac{1}{8}\text{ in.} - \frac{1}{8}\text{ in.}$$

$$= 10\text{ in.} - \frac{14}{8}\text{ in.}$$

$$= 9\frac{4}{4}\text{ in.} - 1\frac{3}{4}\text{ in.}$$

$$= 8\frac{1}{4}\text{ in.}$$

61. a.

$$5\frac{9}{16}\text{ in.} - 1\frac{1}{8}\text{ in.} - 1\frac{1}{8}\text{ in.}$$

$$= 5\frac{9}{16}\text{ in.} - 1\frac{2}{16}\text{ in.} - 1\frac{2}{16}\text{ in.}$$

$$= 3\frac{5}{16}\text{ in.}$$

b.

$$1\frac{1}{8}\text{ in.} + 2\frac{5}{32}\text{ in.} + 3\frac{5}{16}\text{ in.} + 2\frac{5}{32}\text{ in.} + 1\frac{1}{8}\text{ in.} + 7\frac{11}{16}\text{ in.} + 2\frac{1}{16}\text{ in.} + 4\frac{3}{8}\text{ in.} + 5\frac{1}{16}\text{ in.}$$

$$= 1\frac{4}{32}\text{ in.} + 2\frac{5}{32}\text{ in.} + 3\frac{10}{32}\text{ in.} + 2\frac{5}{32}\text{ in.} + 1\frac{4}{32}\text{ in.} + 7\frac{22}{32}\text{ in.} + 2\frac{2}{32}\text{ in.} + 4\frac{12}{32}\text{ in.} + 5\frac{2}{32}\text{ in.}$$

$$= 27\frac{66}{32}\text{ in.} = 29\frac{2}{32}\text{ in.} = 29\frac{1}{16}\text{ in.}$$

63. a.

$$3\frac{1}{4}\text{ in.} - 1\frac{3}{8}\text{ in.} - 1\frac{5}{8}\text{ in.}$$

$$= 3\frac{1}{4}\text{ in.} - 2\frac{8}{8}\text{ in.}$$

$$= 3\frac{1}{4}\text{ in.} - 3\text{ in.}$$

$$= \frac{1}{4}\text{ in.}$$

63. (continued)

b.

$$3\frac{1}{4}\ \text{in.}+\frac{15}{16}\ \text{in.}+\frac{15}{16}\ \text{in.}+1\frac{7}{8}\ \text{in.}+1\frac{1}{4}\ \text{in.}+\frac{13}{16}\ \text{in.}+1\frac{3}{8}\ \text{in.}+1\frac{7}{8}\ \text{in.}$$

$$=3\frac{4}{16}\ \text{in.}+\frac{15}{16}\ \text{in.}+\frac{15}{16}\ \text{in.}+1\frac{14}{16}\ \text{in.}+1\frac{4}{16}\ \text{in.}+\frac{13}{16}\ \text{in.}+1\frac{6}{16}\ \text{in.}+1\frac{14}{16}\ \text{in.}$$

$$=7\frac{85}{16}\ \text{in.}=12\frac{5}{16}\ \text{in.}$$

65.

$$1\frac{3}{4}\ \text{A}+1\frac{1}{2}\ \text{A}$$

$$=1\frac{3}{4}\ \text{A}+1\frac{2}{4}\ \text{A}$$

$$=2\frac{5}{4}\ \text{A}=3\frac{1}{4}\ \text{A}$$

67.

$$\frac{1}{16}\ \text{A}+\frac{1}{12}\ \text{A}+1\frac{3}{4}\ \text{A}$$

$$=\frac{3}{48}\ \text{A}+\frac{4}{48}\ \text{A}+1\frac{36}{48}\ \text{A}$$

$$=1\frac{43}{48}\ \text{A}$$

69.

$$6\frac{3}{4}\ \text{in.}+2\frac{7}{8}\ \text{in.}$$

$$=6\frac{6}{8}\ \text{in.}+2\frac{7}{8}\ \text{in.}$$

$$=8\frac{13}{8}\ \text{in.}=9\frac{5}{8}\ \text{in.}$$

71. a.

$$6\frac{7}{8}\ \text{in.}+1\frac{3}{8}\ \text{in.}+2\frac{1}{4}\ \text{in.}$$

$$=6\frac{7}{8}\ \text{in.}+1\frac{3}{8}\ \text{in.}+2\frac{2}{8}\ \text{in.}$$

$$=9\frac{12}{8}\ \text{in.}=10\frac{4}{8}\ \text{in.}=10\frac{1}{2}\ \text{in.}$$

b.

$$1\frac{5}{8}\ \text{in.}-\frac{7}{16}\ \text{in.}-\frac{7}{16}\ \text{in.}$$

$$=1\frac{5}{8}\ \text{in.}-\frac{14}{16}\ \text{in.}$$

$$=1\frac{5}{8}\ \text{in.}-\frac{7}{8}\ \text{in.}$$

$$=\frac{13}{8}\ \text{in.}-\frac{7}{8}\ \text{in.}$$

$$=\frac{6}{8}\ \text{in.}=\frac{3}{4}\ \text{in.}$$

73.

a.

$$5\frac{1}{8}\ \text{in.}+5\ \text{in.}+7\frac{5}{8}\ \text{in.}+4\frac{1}{16}\ \text{in.}$$

$$=5\frac{2}{16}\ \text{in.}+5\ \text{in.}+7\frac{10}{16}\ \text{in.}+4\frac{1}{16}\ \text{in.}$$

$$=21\frac{13}{16}\ \text{in.}$$

b.

$$7\frac{1}{4}\ \text{in.}-3\frac{3}{16}\ \text{in.}-3\frac{3}{16}\ \text{in.}$$

$$=7\frac{1}{4}\ \text{in.}-6\frac{6}{16}\ \text{in.}$$

$$=7\frac{1}{4}\ \text{in.}-6\frac{3}{8}\ \text{in.}$$

$$=7\frac{2}{8}\ \text{in.}-6\frac{3}{8}\ \text{in.}$$

$$=6\frac{10}{8}\ \text{in.}-6\frac{3}{8}\ \text{in.}$$

$$=\frac{7}{8}\ \text{in.}$$

75.

$$16 \text{ in.} - 1\frac{5}{8} \text{ in.} = 15\frac{8}{8} \text{ in.} - 1\frac{5}{8} \text{ in.}$$

$$= 14\frac{3}{8} \text{ in.}$$

77.

$$\frac{7}{8} \text{ in.} - \frac{51}{64} \text{ in.} = \frac{56}{64} \text{ in.} - \frac{51}{64} \text{ in.}$$

$$= \frac{5}{64} \text{ in.}$$

79.

One cut:

$$1\frac{7}{8} \text{ in.} - \frac{3}{32} \text{ in.}$$

$$= 1\frac{28}{32} \text{ in.} - \frac{3}{32} \text{ in.}$$

$$= 1\frac{25}{32} \text{ in.}$$

Three cuts:

$$1\frac{7}{8} \text{ in.} - \frac{3}{32} \text{ in.} - \frac{3}{32} \text{ in.} - \frac{3}{32} \text{ in.}$$

$$= 1\frac{28}{32} \text{ in.} - \frac{3}{32} \text{ in.} - \frac{3}{32} \text{ in.} - \frac{3}{32} \text{ in.}$$

$$= 1\frac{19}{32} \text{ in.}$$

81.

Length:

$$\frac{7}{32} \text{ in.} + 3\frac{5}{16} \text{ in.} + \frac{7}{32} \text{ in.} + 3\frac{5}{16} \text{ in.} + \frac{7}{32} \text{ in.} + 3\frac{5}{16} \text{ in.} + \frac{7}{32} \text{ in.}$$

$$= \frac{7}{32} \text{ in.} + 3\frac{10}{32} \text{ in.} + \frac{7}{32} \text{ in.} + 3\frac{10}{32} \text{ in.} + \frac{7}{32} \text{ in.} + 3\frac{10}{32} \text{ in.} + \frac{7}{32} \text{ in.}$$

$$= 9\frac{58}{32} \text{ in.} = 9\frac{29}{16} \text{ in.} = 10\frac{13}{16} \text{ in.}$$

Width:

$$\frac{7}{32} \text{ in.} + 3\frac{5}{16} \text{ in.} + \frac{7}{32} \text{ in.} = \frac{7}{32} \text{ in.} + 3\frac{10}{32} \text{ in.} + \frac{7}{32} \text{ in.} = 3\frac{24}{32} \text{ in.} = 3\frac{3}{4} \text{ in.}$$

83.

$$15\frac{3}{8} \text{ in.} + 7\frac{3}{4} \text{ in.} + 11\frac{1}{2} \text{ in.} + 7\frac{7}{32} \text{ in.} + 10\frac{5}{16} \text{ in.}$$

$$= 15\frac{12}{32} \text{ in.} + 7\frac{24}{32} \text{ in.} + 11\frac{16}{32} \text{ in.} + 7\frac{7}{32} \text{ in.} + 10\frac{10}{32} \text{ in.}$$

$$= 50\frac{69}{32} \text{ in.} = 52\frac{5}{32} \text{ in.}$$

85. a.

$$1\frac{3}{32} \text{ in.} + 1\frac{10}{32} \text{ in.} + 2\frac{12}{32} \text{ in.} + 1\frac{10}{32} \text{ in.} + 1\frac{3}{32} \text{ in.}$$

$$= 6\frac{38}{32} \text{ in.} = 7\frac{6}{32} \text{ in.} = 7\frac{3}{16} \text{ in.}$$

b.

$$10\frac{1}{2} \text{ in.} - 6\frac{5}{8} \text{ in.} - 2\frac{3}{16} \text{ in.}$$

$$= 10\frac{8}{16} \text{ in.} - 6\frac{10}{16} \text{ in.} - 2\frac{3}{16} \text{ in.}$$

$$= 9\frac{24}{16} \text{ in.} - 6\frac{10}{16} \text{ in.} - 2\frac{3}{16} \text{ in.}$$

$$= 1\frac{11}{16} \text{ in.}$$

87.

$$1\frac{1}{2} \text{ acres} - \frac{1}{2} \text{ acre} - \frac{1}{6} \text{ acre} - \frac{1}{3} \text{ acre}$$

$$= \frac{3}{2} \text{ acres} - \frac{1}{2} \text{ acre} - \frac{1}{6} \text{ acre} - \frac{1}{3} \text{ acre}$$

$$= \frac{9}{6} \text{ acres} - \frac{3}{6} \text{ acre} - \frac{1}{6} \text{ acre} - \frac{2}{6} \text{ acre}$$

$$= \frac{3}{6} \text{ acre} = \frac{1}{2} \text{ acre}$$

89.

$$\frac{3}{4} + \frac{1}{2} = \frac{3}{4} + \frac{4}{4}$$

$$= \frac{7}{4} = 1\frac{3}{4} \text{ sticks}$$

91.

$$3\frac{3}{8} - 2\frac{1}{4} = 3\frac{3}{8} - 2\frac{2}{8}$$

$$= 1\frac{1}{8} \text{ cups}$$

93.

$$1\frac{1}{2} + 3 - 1\frac{3}{4} - 2\frac{1}{2} - \frac{1}{8}$$

$$= 1\frac{4}{8} + 3 - 1\frac{6}{8} - 2\frac{4}{8} - \frac{1}{8}$$

$$= \frac{12}{8} + 3 - 1\frac{6}{8} - 2\frac{4}{8} - \frac{1}{8}$$

$$= \frac{1}{8} \text{ bag}$$

Section 1.8: Multiplication and Division of Fractions

1. 12

3. 9

5.

$$1\frac{3}{4} \times \frac{5}{16} = \frac{7}{4} \times \frac{5}{16}$$

$$= \frac{35}{64}$$

7. $\frac{2}{3}$

9. 10

11. $\frac{1}{8}$

13.

$$2\frac{1}{3} \times \frac{5}{8} \times \frac{6}{7}$$

$$= \frac{7}{3} \times \frac{5}{8} \times \frac{6}{7}$$

$$= \frac{5}{4} = 1\frac{1}{4}$$

15.

$$\frac{6}{11} \times \frac{26}{35} \times 1\frac{9}{13} \times \frac{7}{12}$$

$$= \frac{6}{11} \times \frac{26}{35} \times \frac{22}{13} \times \frac{7}{12}$$

$$= \frac{2}{5}$$

17.

$$\frac{3}{5} \div \frac{10}{12} = \frac{3}{5} \times \frac{12}{10}$$

$$= \frac{18}{25}$$

19.

$$4\frac{1}{2} \div \frac{1}{4} = \frac{9}{2} \div \frac{1}{4}$$

$$= \frac{9}{2} \times \frac{4}{1}$$

$$= 18$$

21.

$$15 \div \frac{3}{8}$$

$$= 15 \times \frac{8}{3}$$

$$= 40$$

23.

$$\frac{7}{11} \div \frac{3}{5} = \frac{7}{11} \times \frac{5}{3}$$

$$= \frac{35}{33} = 1\frac{2}{33}$$

25.

$$\frac{2}{5} \times 3\frac{2}{3} \div \frac{3}{4} = \frac{2}{5} \times \frac{11}{3} \times \frac{4}{3}$$

$$= \frac{88}{45} = 1\frac{43}{45}$$

27.

$$\frac{16}{5} \times \frac{3}{2} \times \frac{10}{4} \div 5\frac{1}{3}$$

$$= \frac{16}{5} \times \frac{3}{2} \times \frac{10}{4} \div \frac{16}{3}$$

$$= \frac{16}{5} \times \frac{3}{2} \times \frac{10}{4} \times \frac{3}{16}$$

$$= \frac{9}{4} = 2\frac{1}{4}$$

29.

$$\frac{7}{9} \times \frac{3}{8} \div \frac{28}{81}$$

$$= \frac{7}{9} \times \frac{3}{8} \times \frac{81}{28}$$

$$= \frac{27}{32}$$

31.

$$\frac{2}{7} \times \frac{5}{9} \times \frac{3}{10} \div 6$$

$$= \frac{2}{7} \times \frac{5}{9} \times \frac{3}{10} \times \frac{1}{6}$$

$$= \frac{1}{126}$$

33.

$$\frac{7}{16} \div \frac{3}{8} \times \frac{1}{2}$$

$$= \frac{7}{16} \times \frac{8}{3} \times \frac{1}{2}$$

$$= \frac{7}{12}$$

35. $\frac{3}{4} \times 42 \text{ gal} = \frac{126}{4} \text{ gal} = \frac{63}{2} \text{ gal} = 31\frac{1}{2} \text{ gal}$

37. $7 \times 6\frac{1}{2} \text{ in.} = 7 \times \frac{13}{2} \text{ in.} = \frac{91}{2} \text{ in.} = 45\frac{1}{2} \text{ in.}$

39.

$$\frac{684\frac{1}{4} \text{ mi}}{5\frac{2}{3} \text{ h}} = \frac{\frac{2737}{4} \text{ mi}}{\frac{17}{3} \text{ h}} = \frac{2737}{4} \times \frac{3}{17} \text{ mi/h}$$

$$= \frac{483}{4} \text{ mi/h} = 120\frac{3}{4} \text{ mi/h}$$

41. $9 \times 3\frac{2}{3} \text{ ft} = 9 \times \frac{11}{3} \text{ ft} = 33 \text{ ft}$

43.

$$\text{bd ft} = \frac{\overset{\text{number}}{\underset{\text{of boards}}{}} \times \overset{\text{thickness}}{\underset{(\text{in in.})}{}} \times \overset{\text{width}}{\underset{(\text{in in.})}{}} \times \overset{\text{length}}{\underset{(\text{in ft})}{}}}{12}$$

$$\text{bd ft} = \frac{10 \times 2 \text{ in.} \times 4 \text{ in.} \times 12 \text{ ft}}{12} = 80 \text{ bd ft}$$

45.

$$\text{bd ft} = \frac{\overset{\text{number}}{\underset{\text{of boards}}{}} \times \overset{\text{thickness}}{\underset{(\text{in in.})}{}} \times \overset{\text{width}}{\underset{(\text{in in.})}{}} \times \overset{\text{length}}{\underset{(\text{in ft})}{}}}{12}$$

$$\text{bd ft} = \frac{175 \times 1 \text{ in.} \times 8 \text{ in.} \times 14 \text{ ft}}{12} = 1633\frac{1}{3} \text{ bd ft}$$

47.

$$4\frac{9}{32} \text{ in.} - 2 \times \frac{7}{32} \text{ in.} = 4\frac{9}{32} \text{ in.} - \frac{14}{32} \text{ in.}$$

$$= 3\frac{41}{32} \text{ in.} - \frac{14}{32} \text{ in.}$$

$$= 3\frac{27}{32} \text{ in.}$$

49. There will be 15 spaces between the rivets.

$$\frac{28\frac{1}{8} \text{ in.}}{15} = 28\frac{1}{8} \text{ in.} \times \frac{1}{15}$$

$$= \frac{15}{8} \text{ in.} = 1\frac{7}{8} \text{ in.}$$

51. There will be $3+2+6+1=12$ cuts.

Total lengths of the pieces:

$$3 \times 2\frac{1}{8} \text{ in.} = 6\frac{3}{8} \text{ in.}$$

$$2 \times 5\frac{3}{4} \text{ in.} = 11\frac{1}{2} \text{ in.}$$

$$6 \times \frac{7}{8} \text{ in.} = 5\frac{1}{4} \text{ in.}$$

$$1 \times 3\frac{1}{2} \text{ in.} = 3\frac{1}{2} \text{ in.}$$

$$12 \times \frac{1}{16} \text{ in.} = \frac{3}{4} \text{ in.}$$

Remaining length:

$$36 \text{ in.} = 36 \text{ in.}$$

$$-6\frac{3}{8} \text{ in.} = -6\frac{3}{8} \text{ in.}$$

$$-11\frac{1}{2} \text{ in.} = -11\frac{4}{8} \text{ in.}$$

$$-5\frac{1}{4} \text{ in.} = -5\frac{2}{8} \text{ in.}$$

$$-3\frac{1}{2} \text{ in.} = -3\frac{4}{8} \text{ in.}$$

$$-\frac{3}{4} \text{ in.} = -\frac{6}{8} \text{ in.}$$

$$= \frac{69}{8} \text{ in.} = 8\frac{5}{8} \text{ in.}$$

53.

$$\text{Number of revolutions} = \frac{9\frac{9}{64} \text{ in.}}{\frac{3}{128} \text{ in.}}$$

$$= \frac{\frac{585}{64} \text{ in.}}{\frac{3}{128} \text{ in.}}$$

$$= \frac{585}{64} \times \frac{128}{3}$$

$$= 390 \text{ revolutions}$$

$$\text{Time} = 390 \text{ revolutions} \times \frac{1 \text{ min}}{45 \text{ revolutions}}$$

$$= \frac{26}{3} \text{ min} = 8\frac{2}{3} \text{ min}$$

55.

$$V = lwh$$

$$V = (4 \text{ ft})\left(2\frac{2}{3} \text{ ft}\right)\left(\frac{1}{4} \text{ ft}\right)$$

$$= (4 \text{ ft})\left(\frac{8}{3} \text{ ft}\right)\left(\frac{1}{4} \text{ ft}\right)$$

$$= \frac{8}{3} \text{ ft}^3 = 2\frac{2}{3} \text{ ft}^3$$

57.

$$\frac{7\frac{1}{2} \text{ h}}{6} = \frac{\frac{15}{2} \text{ h}}{6}$$

$$= \frac{15}{2} \text{ h} \times \frac{1}{6}$$

$$= \frac{5}{4} \text{ h} = 1\frac{1}{4} \text{ h}$$

59.

$$\text{Power} = (\text{voltage}) \times (\text{current})$$

$$\text{Power} = 12\frac{1}{2} \times 220$$

$$= \frac{25}{2} \times 220$$

$$= 2750 \text{ W}$$

61.

$$12 \times 8\frac{1}{2} \text{ ft} = 102 \text{ ft}$$

$$7 \times 18\frac{1}{2} \text{ ft} = 129\frac{1}{2} \text{ ft}$$

$$24 \times 1\frac{3}{4} \text{ ft} = 42 \text{ ft}$$

$$12 \times 6\frac{1}{2} \text{ ft} = 78 \text{ ft}$$

$$2 \times 34\frac{1}{4} \text{ ft} = 68\frac{1}{2} \text{ ft}$$

$$= 420 \text{ ft}$$

63.

$$\text{Current} = (\text{voltage}) \div (\text{resistance})$$

$$\text{Current} = 24 \div 10\frac{1}{2}$$

$$= 24 \div \frac{21}{2}$$

$$= 24 \times \frac{2}{21}$$

$$= \frac{16}{7} \text{ A} = 1\frac{2}{7} \text{ A}$$

65. There will be 18 spaces between the outlets.

$$\frac{130\frac{1}{2} \text{ ft}}{18} = \frac{\frac{261}{2} \text{ ft}}{18}$$

$$= \frac{261}{2} \text{ ft} \times \frac{1}{18}$$

$$= 7\frac{1}{4} \text{ ft or } 7\frac{1}{4} \text{ ft 3 in.}$$

67.

$$\frac{60 \text{ gal}}{\frac{3}{4} \text{ gal}} = 60 \times \frac{4}{3} = 80$$

$$80 \times \frac{1}{2} \text{ lb} = 40 \text{ lb}$$

69.

$$\frac{448 \text{ lb} \times \frac{1 \text{ bu}}{56 \text{ lb}}}{\frac{1}{20} \text{ acre}} = \frac{8 \text{ bu}}{\frac{1}{20} \text{ acre}}$$

$$= \frac{8}{\frac{1}{20}} \text{ bu/acre}$$

$$= 8 \times 20 \text{ bu/acre}$$

$$= 160 \text{ bu/acre}$$

71.

$$\frac{1}{5} \times 2\frac{1}{2} \text{ lb} = \frac{1}{5} \times \frac{5}{2} \text{ lb}$$

$$= \frac{1}{2} \text{ oz}$$

73. $\dfrac{15 \text{ mg}}{30 \text{ mg}} = \dfrac{1}{2}$ tablet

75.

$$2 \times 7\frac{1}{4} \text{ lb} = 2 \times \frac{29}{4} \text{ lb}$$

$$= \frac{58}{4} \text{ lb}$$

$$= \frac{29}{2} \text{ lb} = 14\frac{1}{2} \text{ lb}$$

77. $\dfrac{12 \text{ oz}}{\frac{1}{2} \text{ oz}} = 12 \times \dfrac{2}{1} = 24$ doses

79.

$$5 \times \frac{1}{2} \text{ tsp} = \frac{5}{2} \text{ tsp}$$

$$= 2\frac{1}{2} \text{ tsp}$$

81. a.

$$\frac{3 \text{ in.} - 1\frac{1}{2} \text{ in.}}{2} = \frac{1\frac{1}{2} \text{ in.}}{2}$$

$$= \frac{\frac{3}{2} \text{ in.}}{2}$$

$$= \frac{3}{2} \text{ in.} \times \frac{1}{2}$$

$$= \frac{3}{4} \text{ in.}$$

b.

$$\frac{3 \text{ in.} - 1\frac{1}{2} \text{ in.}}{2} = \frac{1\frac{1}{2} \text{ in.}}{2}$$

$$= \frac{\frac{3}{2} \text{ in.}}{2}$$

$$= \frac{3}{2} \text{ in.} \times \frac{1}{2}$$

$$= \frac{3}{4} \text{ in.}$$

83.

$$R_T = \cfrac{1}{\cfrac{1}{R_1} + \cfrac{1}{R_2}}$$

$$R_T = \cfrac{1}{\cfrac{1}{12\,\Omega} + \cfrac{1}{6\,\Omega}}$$

$$= \cfrac{1}{\cfrac{1}{12\,\Omega} + \cfrac{2}{12\,\Omega}}$$

$$= \cfrac{1}{\cfrac{3}{12\,\Omega}} = \cfrac{12\,\Omega}{3} = 4\,\Omega$$

85.

$$R_T = \cfrac{1}{\cfrac{1}{R_1} + \cfrac{1}{R_2} + \cfrac{1}{R_3} + \cfrac{1}{R_4}}$$

$$R_T = \cfrac{1}{\cfrac{1}{6\,\Omega} + \cfrac{1}{12\,\Omega} + \cfrac{1}{24\,\Omega} + \cfrac{1}{48\,\Omega}}$$

$$= \cfrac{1}{\cfrac{8}{48\,\Omega} + \cfrac{4}{48\,\Omega} + \cfrac{2}{48\,\Omega} + \cfrac{1}{48\,\Omega}}$$

$$= \cfrac{1}{\cfrac{15}{48\,\Omega}} = \cfrac{48\,\Omega}{15} = 3\frac{3}{15}\,\Omega = 3\frac{1}{5}\,\Omega$$

87.

$$\text{Red flowers} = 300 \times \frac{1}{4} = 75 \text{ flowers}$$

$$\text{White flowers} = 300 \times \frac{3}{4} = 225 \text{ flowers}$$

89.

$$\cfrac{1\frac{1}{2} \text{ cups}}{\frac{1}{4} \text{ cup}} = \cfrac{\frac{3}{2}}{\frac{1}{4}} = \frac{3}{2} \times \frac{4}{1} = 6 \text{ scoops}$$

91.

$$14 \text{ oz} \times \frac{1 \text{ lb}}{16 \text{ oz}} = \frac{14}{16} \text{ lb} = \frac{7}{8} \text{ lb}$$

$$16\frac{1}{4} \text{ lb} - 5\frac{1}{2} \text{ lb} = 15\frac{5}{4} \text{ lb} - 5\frac{2}{4} \text{ lb}$$

$$= 10\frac{3}{4} \text{ lb}$$

$$\cfrac{10\frac{3}{4} \text{ lb}}{\frac{7}{8} \text{ lb}} = \cfrac{\frac{43}{4} \text{ lb}}{\frac{7}{8} \text{ lb}}$$

$$= \left(\frac{43}{4}\right)\left(\frac{8}{7}\right)$$

$$= \frac{86}{7} = 12\frac{2}{7}$$

Number of whole steaks $= 12$

93.

$$10\frac{1}{3} \text{ gal} - 3 \times 2\frac{1}{2} \text{ gal}$$

$$= \frac{31}{3} \text{ gal} - 3 \times \frac{5}{2} \text{ gal}$$

$$= \frac{31}{3} \text{ gal} - \frac{15}{2} \text{ gal}$$

$$= \frac{62}{6} \text{ gal} - \frac{45}{6} \text{ gal}$$

$$= \frac{17}{6} \text{ gal} = 2\frac{5}{6} \text{ gal}$$

Section 1.9: The U.S. System of Weights and Measures

1. $3 \text{ ft} \times \dfrac{12 \text{ in.}}{1 \text{ ft}} + 7 \text{ in.} = 43 \text{ in.}$

3. $5 \text{ lb} \times \dfrac{16 \text{ oz}}{1 \text{ lb}} + 3 \text{ oz} = 83 \text{ oz}$

5. $4 \text{ qt} \times \dfrac{2 \text{ pt}}{1 \text{ qt}} + 1 \text{ pt} = 9 \text{ pt}$

7. $3 \text{ tbs} \times \dfrac{3 \text{ tsp}}{1 \text{ tbs}} = 9 \text{ tsp}$

9. $8 \text{ ft} \times \dfrac{12 \text{ in.}}{1 \text{ ft}} = 96 \text{ in.}$

11. $3 \text{ qt} \times \dfrac{2 \text{ pt}}{1 \text{ qt}} = 6 \text{ pt}$

13. $96 \text{ in.} \times \dfrac{1 \text{ ft}}{12 \text{ in.}} = 8 \text{ ft}$

15. $10 \text{ pt} \times \dfrac{1 \text{ qt}}{2 \text{ pt}} = 5 \text{ qt}$

17. $88 \text{ oz} \times \dfrac{1 \text{ lb}}{16 \text{ oz}} = 5\dfrac{1}{2} \text{ lb}$

19. $14 \text{ qt} \times \dfrac{1 \text{ gal}}{4 \text{ qt}} = 3\dfrac{1}{2} \text{ gal}$

21. $56 \text{ fl oz} \times \dfrac{1 \text{ cup}}{8 \text{ fl oz}} \times \dfrac{1 \text{ pt}}{2 \text{ cups}} = 3\dfrac{1}{2} \text{ pt}$

23. $92 \text{ ft} \times \dfrac{1 \text{ yd}}{3 \text{ ft}} = 30\dfrac{2}{3} \text{ yd}$

25. $2 \text{ mi} \times \dfrac{5280 \text{ ft}}{1 \text{ mi}} \times \dfrac{1 \text{ yd}}{3 \text{ ft}} = 3520 \text{ yd}$

27. $500 \text{ fl oz} \times \dfrac{1 \text{ cup}}{8 \text{ fl oz}} \times \dfrac{1 \text{ pt}}{2 \text{ cups}} \times \dfrac{1 \text{ qt}}{2 \text{ pt}} = 15\dfrac{5}{8} \text{ qt}$

29. $\dfrac{80 \text{ in.}}{12 \text{ in.}} = 6 \text{ r } 8 = 6 \text{ ft } 8 \text{ in.}$

31. $12\dfrac{3}{4} \text{ ft} \times \dfrac{12 \text{ in.}}{1 \text{ ft}} = \dfrac{51}{4} \text{ ft} \times \dfrac{12 \text{ in.}}{1 \text{ ft}} = 153 \text{ in.}$

33.

$$144 \text{ fl oz} + 24 \text{ fl oz} + 56 \text{ fl oz} = 224 \text{ fl oz}$$

$$224 \text{ fl oz} \times \dfrac{1 \text{ cup}}{8 \text{ fl oz}} \times \dfrac{1 \text{ pt}}{2 \text{ cups}} \times \dfrac{1 \text{ qt}}{2 \text{ pt}} = 7 \text{ qt}$$

35. $1 \text{ mi} \times \dfrac{5280 \text{ ft}}{1 \text{ mi}} \times \dfrac{\frac{1}{10}\,\Omega}{1000 \text{ ft}} = \dfrac{66}{125}\,\Omega$

37.

$$3\dfrac{3}{4} \text{ ft} \times 4\dfrac{2}{3} \text{ ft} = \dfrac{15}{4} \text{ ft} \times \dfrac{14}{3} \text{ ft} = \dfrac{35}{2} \text{ ft}$$

$$\dfrac{35}{2} \text{ ft} \times \dfrac{12 \text{ in.}}{1 \text{ ft}} \times \dfrac{12 \text{ in.}}{1 \text{ ft}} = 2520 \text{ in}^2$$

39. a. $2 \text{ mi} \times \dfrac{5280 \text{ ft}}{1 \text{ mi}} = 10{,}560 \text{ ft}$

b. $10{,}560 \text{ ft} \times \dfrac{1 \text{ yd}}{3 \text{ ft}} = 3520 \text{ yd}$

41. $3 \text{ lb} \times \dfrac{16 \text{ oz}}{1 \text{ lb}} = 48 \text{ oz}$

43. $153 \text{ ft} \times \dfrac{1 \text{ yd}}{3 \text{ ft}} = 51 \text{ yd}$

45. $561 \text{ ft} \times \dfrac{1 \text{ chain}}{66 \text{ ft}} = 8\dfrac{1}{2} \text{ chains}$

47. $15 \text{ drams} \times \dfrac{27\frac{17}{50} \text{ grains}}{1 \text{ dram}} = 410\dfrac{1}{10} \text{ grains}$

49. $4500 \dfrac{\text{ft}}{\text{h}} \times \dfrac{1 \text{ h}}{60 \text{ min}} = 75 \dfrac{\text{ft}}{\text{min}}$

51. $1\dfrac{1}{5} \dfrac{\text{mi}}{\text{s}} \times \dfrac{60 \text{ s}}{1 \text{ min}} = 72 \dfrac{\text{mi}}{\text{min}}$

53.

$$40 \dfrac{\text{mi}}{\text{h}} \times \dfrac{5280 \text{ ft}}{1 \text{ mi}} \times \dfrac{1 \text{ h}}{60 \text{ min}} \times \dfrac{1 \text{ min}}{60 \text{ s}}$$

$$= 58\dfrac{2}{3} \dfrac{\text{ft}}{\text{s}}$$

55. $24 \dfrac{\text{in.}}{\text{s}} \times \dfrac{1 \text{ ft}}{12 \text{ in.}} \times \dfrac{60 \text{ s}}{1 \text{ min}} = 120 \dfrac{\text{ft}}{\text{min}}$

57.

14 yd 5 ft 34 in.

= 14 yd 7 ft 10 in.

= 16 yd 1 ft 10 in.

59. $3 \times 1.5 \text{ tons} \times \dfrac{2000 \text{ lb}}{1 \text{ ton}} = 9000 \text{ lb}$

61.

$$4 \text{ rods} \times \dfrac{16.5 \text{ ft}}{1 \text{ rod}} = 66 \text{ ft}$$

$$\dfrac{66 \text{ ft}}{3 \text{ ft}} = 22 \text{ paces}$$

63. $7 \text{ gal} \times \dfrac{4 \text{ qt}}{1 \text{ gal}} = 28 \text{ qt}$

65.

$$2 \text{ gal} = 2 \text{ gal}$$

$$2 \text{ qt} \times \dfrac{1 \text{ gal}}{4 \text{ qt}} = \dfrac{1}{2} \text{ gal}$$

$$3 \text{ pt} \times \dfrac{1 \text{ qt}}{2 \text{ pt}} \times \dfrac{1 \text{ gal}}{4 \text{ qt}} = \dfrac{3}{8} \text{ gal}$$

$$\dfrac{\dfrac{1}{2} \text{ gal} = \dfrac{1}{2} \text{ gal}}{= 3\dfrac{3}{8} \text{ gal}}$$

Unit 1B Review

1. $\dfrac{9}{15} = \dfrac{3 \cdot 3}{3 \cdot 5} = \dfrac{3}{5}$

2. $\dfrac{48}{54} = \dfrac{2 \cdot 3 \cdot 8}{2 \cdot 3 \cdot 9} = \dfrac{8}{9}$

3. $\dfrac{27}{6} = 4 \text{ r } 3 = 4\dfrac{3}{6} = 4\dfrac{1}{2}$

4. $\dfrac{(3 \times 5) + 2}{5} = \dfrac{17}{5}$

5. $\dfrac{5}{6} + \dfrac{2}{3} = \dfrac{5}{6} + \dfrac{4}{6} = \dfrac{9}{6} = \dfrac{3}{2} = 1\dfrac{1}{2}$

6.

$5\dfrac{3}{8} - 2\dfrac{5}{12}$

$= 5\dfrac{9}{24} - 2\dfrac{10}{24}$

$= 4\dfrac{33}{24} - 2\dfrac{10}{24}$

$4\dfrac{23}{24}$

7. $\dfrac{4}{15}$

8.

$= \dfrac{3}{4} \div 1\dfrac{5}{8}$

$= \dfrac{3}{4} \div \dfrac{13}{8}$

$= \dfrac{3}{4} \times \dfrac{8}{13}$

$= \dfrac{6}{13}$

9.

$1\dfrac{2}{3} + 3\dfrac{5}{6} - 2\dfrac{1}{4}$

$= 1\dfrac{8}{12} + 3\dfrac{10}{12} - 2\dfrac{3}{12}$

$= 4\dfrac{18}{12} - 2\dfrac{3}{12}$

$= 2\dfrac{15}{12} = 3\dfrac{3}{12} = 3\dfrac{1}{4}$

10.

$4\dfrac{2}{3} \div 3\dfrac{1}{2} \times 1\dfrac{1}{2}$

$= \dfrac{14}{3} \div \dfrac{7}{2} \times \dfrac{3}{2}$

$= \dfrac{14}{3} \times \dfrac{2}{7} \times \dfrac{3}{2}$

$= \dfrac{4}{3} \times \dfrac{3}{2}$

$= 2$

11.

$7 \text{ in.} - 1\dfrac{7}{8} \text{ in.} - 1\dfrac{1}{2} \text{ in.} - 1\dfrac{1}{3} \text{ in.} - 1\dfrac{5}{12} \text{ in.}$

$= 7 \text{ in.} - 1\dfrac{21}{24} \text{ in.} - 1\dfrac{12}{24} \text{ in.} - 1\dfrac{8}{24} \text{ in.} - 1\dfrac{10}{24} \text{ in.}$

$= 7 \text{ in.} - 4\dfrac{51}{24} \text{ in.}$

$= 6\dfrac{24}{24} \text{ in.} - 6\dfrac{4}{24} \text{ in.}$

$= \dfrac{21}{24} \text{ in.} = \dfrac{7}{8} \text{ in.}$

12.

$72 \text{ in.} - 16\dfrac{3}{4} \text{ in.} - 24\dfrac{7}{8} \text{ in.} - 12\dfrac{5}{16} \text{ in.} - 3 \times \dfrac{1}{16} \text{ in.}$

$= 72 \text{ in.} - 16\dfrac{12}{16} \text{ in.} - 24\dfrac{14}{16} \text{ in.} - 12\dfrac{5}{16} \text{ in.} - \dfrac{3}{16} \text{ in.}$

$= 72 \text{ in.} - 16\dfrac{12}{16} \text{ in.} - 24\dfrac{14}{16} \text{ in.} - 12\dfrac{5}{16} \text{ in.} - \dfrac{3}{16} \text{ in.}$

$= 72 \text{ in.} - 53\dfrac{34}{24} \text{ in.}$

$= 71\dfrac{24}{24} \text{ in.} - 54\dfrac{10}{24} \text{ in.}$

$= 16\dfrac{14}{24} \text{ in.} = 17\dfrac{7}{8} \text{ in.}$

13.

$$P = 2l + 2w$$

$$P = 2\left(6\frac{1}{4}\text{ in.}\right) + 2\left(2\frac{2}{3}\text{ in.}\right)$$

$$= 2\left(\frac{25}{4}\text{ in.}\right) + 2\left(\frac{8}{3}\text{ in.}\right)$$

$$= \frac{25}{2}\text{ in.} + \frac{16}{3}\text{ in.}$$

$$= \frac{75}{6}\text{ in.} + \frac{32}{6}\text{ in.}$$

$$= \frac{107}{6}\text{ in.} = 17\frac{5}{6}\text{ in.}$$

14.

$$A = lw$$

$$A = \left(6\frac{1}{4}\text{ in.}\right)\left(2\frac{2}{3}\text{ in.}\right)$$

$$= \left(\frac{25}{4}\text{ in.}\right)\left(\frac{8}{3}\text{ in.}\right)$$

$$= \frac{50}{3}\text{ in}^2 = 16\frac{2}{3}\text{ in}^2$$

15. $4\text{ ft} \times \dfrac{12\text{ in.}}{1\text{ ft}} = 48\text{ in.}$

16. $24\text{ ft} \times \dfrac{1\text{ yd}}{3\text{ ft}} = 8\text{ yd}$

17. $3\text{ lb} \times \dfrac{16\text{ oz}}{1\text{ lb}} = 48\text{ oz}$

18. $20\text{ qt} \times \dfrac{1\text{ gal}}{4\text{ qt}} = 5\text{ gal}$

19. $\dfrac{60\text{ mi}}{1\text{ hr}} \times \dfrac{1\text{ hr}}{60\text{ min}} \times \dfrac{1\text{ min}}{60\text{ s}} \times \dfrac{5280\text{ ft}}{1\text{ mi}} = 88\text{ ft/s}$

20.

$$14\text{ ft } 4\text{ in.} = 13\text{ ft } 16\text{ in.}$$
$$\underline{8\text{ ft } 8\text{ in.} = 8\text{ ft } 8\text{ in.}}$$
$$= 5\text{ ft } 8\text{ in.}$$

Section 1.10: Addition and Subtraction of Decimal Fractions

1. four thousandths

3. five ten-thousandths

5. one and four hundred twenty-one hundred-thousandths

7. six and ninety-two thousandths

9. 5.02 ; $5\dfrac{2}{100} = 5\dfrac{1}{50}$

11. 71.0021 ; $71\dfrac{21}{10,000}$

13. 43.0101 ; $43\dfrac{101}{10,000}$

15. 0.375

17. $0.7\overline{3}$

19. 0.34

21. $1.\overline{27}$

23. $18.\overline{285714}$

25. $34.\overline{2}$

27. $\dfrac{7}{10}$

29. $\dfrac{11}{100}$

31. $\dfrac{8425}{10,000} = \dfrac{337}{400}$

33. $10\dfrac{76}{100} = 10\dfrac{19}{25}$

35. 150.000

37. 163.204

39. 86.6

41. 15.308

43. 8.68

45. 4.862

47. 10.0507

49. $6.25\text{ ft} - 2.4\text{ ft} - 2.4\text{ ft} = 1.45\text{ ft}$, so the remaining piece will be $1.45\text{ ft} \times 2.4\text{ ft}$.

51. $2.3\text{ h} + 3.1\text{ h} + 5.4\text{ h} = 10.8\text{ h}$

53.

$$\frac{3}{8}\text{ in.} - \frac{1}{16}\text{ in.} = \frac{6}{16}\text{ in.} - \frac{1}{16}\text{ in.}$$
$$= \frac{5}{16}\text{ in.} = 0.3125\text{ in.}$$

55.

$a = 2.69 \text{ cm} + 1.87 \text{ cm} = 4.56 \text{ cm}$

$b = 8.32 \text{ cm} - 3.45 \text{ cm} = 4.87 \text{ cm}$

57.

4.17 in.

1.30 in.

1.00 in.

<u>1.47 in.</u>

7.94 in.

61.

0.3 A

0.105 A

0.45 A

0.93 A

0.27 A

<u>0.55 A</u>

2.605 A

65. $1.625 \text{ in.} - 1.093 \text{ in.} = 0.532 \text{ in.}$

67. $\left(1.94 \text{ in.} - 1.50 \text{ in.}\right) \div 2 = 0.22 \text{ in.}$

69. $4.125 \text{ in.} - 0.007 \text{ in.} = 4.118 \text{ in.}$

71. $11.20 \text{ billion} - 6.11 \text{ billion} = 5.09 \text{ billion}$

73. 1317.5 bbl

75.

$1\dfrac{3}{4} \text{ gal} + 0.4 \text{ gal} + 0.75 \text{ gal} + 0.5 \text{ gal}$

$= 1.75 \text{ gal} + 0.4 \text{ gal} + 0.75 \text{ gal} + 0.5 \text{ gal}$

$= 3.4 \text{ gal}$

59.

$9.625 \text{ in.} = 9\dfrac{5}{8} \text{ in.}$

$9\dfrac{5}{8} \text{ in.} \div 2 = 4\dfrac{5}{32} \text{ in.} = 4.8125 \text{ in.}$

63.

15.7 Ω

40 Ω

25.5 Ω

0.6 Ω

1200 Ω

<u>115 Ω</u>

1396.8 Ω

77.

$2.5 \text{ lb} = 2.5 \text{ lb}$

$12 \text{ oz} \div 16 \text{ oz/lb} = 0.75 \text{ lb}$

$1.5 \text{ oz} \div 16 \text{ oz/lb} = 0.9375 \text{ lb}$

$0.7 \text{ lb} = 0.7 \text{ lb}$

$14 \text{ oz} \div 16 \text{ oz/lb} = 0.875 \text{ lb}$

$18 \text{ oz} \div 16 \text{ oz/lb} = \underline{1.125 \text{ lb}}$

$= 6.0125 \text{ lb}$

$= 6 \text{ lb}$

Section 1.11: Rounding Numbers

1. a. 1700

b. 1650

3. a. 3100

b. 3130

5. a. 18,700

b. 18,680

7. a. 3.1

b. 3.142

9. a. 0.1

b. 0.57

11. a. 0.1

b. 0.070

13. $600 \, ; \, 640 \, ; \, 636 \, ; \, 636.2 \, ; \, 636.18 \, ; \, 636.183$

15. $17,200 \, ; \, 17,160 \, ; \, 17,159 \, ; \, 17,159.2 \, ;$ $17,159.17 \, ; \, 17,159.167$

17. $1,543,700 \, ; \, 1,543,680 \, ; \, 1,543,679 \, ; \, \text{N/A};$ N/A; N/A

19. $10,600 \, ; \, 10,650 \, ; \, 10,650 \, ; \, 10,649.8 \, ;$ $10,649.83 \, ; \, \text{N/A}$

21. $600 \, ; \, 650 \, ; \, 650 \, ; \, 649.9 \, ; \, 649.90 \, ; \, 649.900$

23. $237,000$

25. 0.0328

27. 72

29. $1,462,000$

31. 0.0003376

33. 1.01

Section 1.12: Multiplication and Division of Decimal Fractions

1. 0.555

3. 10.5126

5. 9,280,000

7. 30

9. 15

11. 248.23

13. 3676.47

15. 7.80

17. 6.59

19.

$$\frac{8^2 - 6^2}{4 \cdot 8 + (7 + 9)}$$

$$= \frac{64 - 36}{32 + 16}$$

$$= \frac{28}{48} = \frac{7}{12}$$

21.

$$\frac{4 \cdot 5 \cdot 6 - 5 \cdot 2^3}{4^2 \cdot 5 + 5 \cdot 2^2}$$

$$= \frac{20 \cdot 6 - 5 \cdot 8}{16 \cdot 5 + 5 \cdot 4}$$

$$= \frac{120 - 40}{80 + 20} = \frac{80}{100} = \frac{4}{5}$$

23.

$$\frac{3.6 \text{ ft}}{3} = 1.2 \text{ ft}$$

25. $\dfrac{321.3 \text{ mi}}{2.7 \text{ h}} = 119 \text{ mi/h}$

27. $\dfrac{475 \text{ mi}}{17.12 \text{ gal}} = 27.7 \text{ mi/gal}$

29.

$$12 \times 8\frac{7}{8} \text{ in.} = 12 \times 8.875 \text{ in.}$$

$$= 106.5 \text{ in.}$$

$$\frac{106.5 \text{ in.}}{11} = 9.682 \text{ in.}$$

31. a. $8 \times 4.72 \text{ m} = 37.76 \text{ m}$

   b. $2 \times 4.72 \text{ m} = 9.44 \text{ m}$

41. $4.62 \text{ in.} + 7 \times 0.47 \text{ in.} + 6 \times 6.44 \text{ in.} + 4.65 \text{ in.} = 51.20 \text{ in.}$

43. $6 \times 56.25 \text{ in}^3 = 337.5 \text{ in}^3$

47. a. $45,000 \text{ mi} \times \dfrac{0.062 \text{ in.}}{15,000 \text{ mi}} = 0.186 \text{ in.}$

33.

$$n = \frac{1}{p}$$

$$n = \frac{1}{0.0125}$$

$$= 80 \text{ threads/in.}$$

35.

$$32.63 \text{ in.} - 8 \times 3.56 \text{ in.} - 8 \times 0.15 \text{ in.}$$

$$= 2.95 \text{ in.}$$

37. $\dfrac{18 \text{ in.}}{0.0060 \text{ in.}} = 3000 \text{ sheets}$

38.

$$(45 \text{ ft } 3 \text{ in.})(64 \text{ ft } 6 \text{ in.})$$

$$= (45.25 \text{ ft})(64.5 \text{ ft})$$

$$= 2918.625 \text{ ft}^2$$

39.

$$V = lwh$$

$$V = (87 \text{ ft})(42 \text{ ft})(8 \text{ ft})$$

$$= 29,232 \text{ ft}^3$$

$$\text{Cost} = 29,232 \text{ ft}^3 \times \left(\frac{1 \text{ yd}}{3 \text{ ft}}\right)^3 \times \frac{\$4.50}{1 \text{ yd}^3}$$

$$= \$4872.00$$

40.

$$\frac{2.640 \text{ in.} - 2.640 \text{ in.}}{0.018 \text{ in.}}$$

$$= \frac{0.252 \text{ in.}}{0.018 \text{ in.}}$$

$$= 14 \text{ cuts}$$

45. $\dfrac{2.0 \text{ L}}{4} = 0.5 \text{ L}$

47. (continued)

b.

$$60{,}000 \text{ mi} \times \frac{0.062 \text{ in.}}{15{,}000 \text{ mi}} = 0.248 \text{ in.}$$

$$\text{Thickness} = 0.375 \text{ in.} - 0.248 \text{ in.}$$
$$= 0.127 \text{ in.}$$

51. The cost of one head of cattle is
$550 \text{ lb} \times \$1.45/\text{lb} = \$797.50$.

The revenue of one head of cattle is
$(550 \text{ lb} + 500 \text{ lb}) \times \$1.20/\text{lb} = \$1260.00$.

The expected profit is \$150, so the cost of the weight gain is $\$1260.00 - \$797.50 - \$150.00 = \$312.00$.

The cost of weight gain per pound is
$\dfrac{\$312.00}{500 \text{ lb}} = \$0.625/\text{lb}$.

53. $2 \times \pi \times 60 \text{ Hz} \times 0.25 \text{ H} = 94.2 \ \Omega$

55. $(6.4 \text{ V})(0.045 \text{ A}) = 0.288 \text{ W}$

49. $150 \text{ acres} \times 1.6 \dfrac{\text{gal}}{\text{acre}} = 240 \text{ gal}$

57. $\dfrac{220 \text{ V}}{35.5 \ \Omega} = 6.20 \text{ A}$

59. $\dfrac{115 \text{ V}}{0.84 \text{ A}} = 136.9 \ \Omega$

61. $3 \times 0.1 \text{ mg} = 0.3 \text{ mg}$

63. $\dfrac{0.5 \text{ mg}}{0.1 \text{ mg}} = 5 \text{ tablets}$

65. $350 \text{ mi} \times \dfrac{0.868 \text{ naut. mi}}{1 \text{ mi}} = 303.8 \text{ naut. mi}$

67. $4.00 \text{ ft} \times 8.00 \text{ ft} \times 40.32 \dfrac{\text{lb}}{\text{ft}^2} = 1290 \text{ lb}$

69. $312{,}780{,}968 \text{ people} \times 4.4 \text{ lb/person} \times \dfrac{1 \text{ ton}}{2000 \text{ lb}} = 688{,}000 \text{ tons}$

71.

$$V = lwh$$

$$V = (4 \text{ ft})(8 \text{ ft})(16 \text{ in.}) \times \frac{1 \text{ ft}}{12 \text{ in.}}$$

$$= \frac{512}{12} \text{ ft}^3 = 42.7 \text{ ft}^3$$

73.

$$200 \times 1.5 \text{ oz} = 300 \text{ oz}$$

$$5 \text{ lb} \times \frac{16 \text{ oz}}{1 \text{ lb}} = 80 \text{ oz}$$

$$\frac{300 \text{ oz}}{80 \text{ oz}} = 3.75 \text{ bags}$$

Section 1.13: Percent

1. 0.27

3. 0.06

5. 1.56

7. 0.292

9. 0.087

11. 9.478

13. 0.0028

15. 0.00068

17. $4\dfrac{1}{4}\% = 4.25\% = 0.0425$

19. $\dfrac{3}{8}\% = 0.375\% = 0.00375$

21. 54%

23. 8%

25. 62%

27. 217%

29. 435%

31. 18.5%

33. 29.7%

35. 519%

37. 1.87%

39. 0.29%

41. $\dfrac{4}{5} = 0.8 = 80\%$

43. $\dfrac{1}{8} = 0.125 = 12\dfrac{1}{2}\% \text{ or } 12.5\%$

45. $\dfrac{1}{6} = 0.16 \text{ r } 4 = 16\dfrac{4}{6}\% = 16\dfrac{2}{3}\%$

47. $\dfrac{4}{9} = 0.44 \text{ r } 4 = 44\dfrac{4}{9}\%$

49. $\dfrac{3}{5} = 0.60 = 60\%$

51. $\dfrac{13}{40} = 0.325 = 32.5\%$ or $32\dfrac{1}{2}\%$

53. $\dfrac{7}{16} = 0.4375 = 43.75\%$ or $43\dfrac{3}{4}\%$

55. $\dfrac{96}{40} = 2.40 = 240\%$

57. $1\dfrac{3}{4} = 1.75 = 175\%$

59. $2\dfrac{5}{12} = \dfrac{29}{12} = 2.41\text{ r }8 = 241\dfrac{8}{12}\% = 241\dfrac{2}{3}\%$

61. $75\% = \dfrac{75}{100} = \dfrac{3}{4}$

63. $16\% = \dfrac{16}{100} = \dfrac{4}{25}$

65. $60\% = \dfrac{60}{100} = \dfrac{3}{5}$

67. $93\% = \dfrac{93}{100}$

69. $275\% = \dfrac{275}{100} = \dfrac{11}{4} = 2\dfrac{3}{4}$

71. $125\% = \dfrac{125}{100} = \dfrac{5}{4} = 1\dfrac{1}{4}$

73. $10\dfrac{3}{4}\% = \dfrac{43}{4}\% = \dfrac{43}{4} \times \dfrac{1}{100} = \dfrac{43}{400}$

75. $10\dfrac{7}{10}\% = \dfrac{107}{10}\% = \dfrac{107}{10} \times \dfrac{1}{100} = \dfrac{107}{1000}$

77. $17\dfrac{1}{4}\% = \dfrac{69}{4}\% = \dfrac{69}{4} \times \dfrac{1}{100} = \dfrac{69}{400}$

79. $16\dfrac{1}{6}\% = \dfrac{97}{6}\% = \dfrac{97}{6} \times \dfrac{1}{100} = \dfrac{97}{600}$

81.

Fraction	Decimal	Percent
$\dfrac{3}{8}$	0.375	37.5%
$\dfrac{45}{100} = \dfrac{9}{20}$	0.45	45%
$\dfrac{18}{100} = \dfrac{9}{50}$	0.18	18%
$1\dfrac{2}{5}$	1.4	140%
$1\dfrac{8}{100} = 1\dfrac{2}{25}$	1.08	108%
$\dfrac{1675}{1000} = \dfrac{67}{40}$	0.1675	$16\dfrac{3}{4}\%$

Section 1.14: Rate, Base, and Part

1. $P = 60$; $R = 25\%$; $B = 240$

3. $P = 108$; $R = 40\%$; $B = 270$

5. $P = $ unknown; $R = 4\%$; $B = 28,000$

7. $P = 21$; $R = 60\%$; $B = $ unknown

9. $P = 2050$; $R = 6\%$; $B = $ unknown

11.

$P = BR$

$P = (\$32,500)(0.08)$

$\quad = \$2600$

Her new salary is $\$32,500 + \$2600 = \$35,100.$

13. a.

10%; $\$5.49 + \$3.28 + \$7.22 + \$2.12 = \$18.11$

$\$18.11 - 0.10 \times \$18.11 = \$16.30$

20%; $\$12.57 + \$22.12 + \$17.88 = \$52.57$

$\$52.57 - 0.20 \times \$52.57 = \$42.06$

30%; $\$38.42 + \$40.12 + \$35.18 = \$113.72$

$\$113.72 - 0.30 \times \$113.72 = \$79.60$

Total; $\$16.30 + \$42.06 + \$79.61 = \$137.96$

b. $137.96 + 0.0625 \times 137.96 = \$146.58$

15.

$$880 \text{ yd} \times \frac{3 \text{ ft}}{1 \text{ yd}} = 2650 \text{ ft}$$

$$R = \frac{P}{B}$$

$$R = \frac{2650 \text{ ft}}{5280 \text{ ft}}$$

$$= 0.5 = 50\%$$

17.

$$B = \frac{P}{R}$$

$$B = \frac{\$72}{0.045}$$

$$= \$1600$$

19.

$$P = BR$$

$$P = (48)(2.35)$$

$$= 112.8$$

21.

$$P = BR$$

$$P = (32 \text{ V})(0.28)$$

$$= 8.96 \text{ V}$$

23.

$$R = \frac{P}{B}$$

$$R = \frac{97}{130}$$

$$= 0.746 = 74.6\%$$

25.

$$R = \frac{P}{B}$$

$$R = \frac{24 \text{ h}}{65 \text{ h}}$$

$$= 0.369 = 36.9\%$$

27.

$$R = \frac{P}{B}$$

$$R = \frac{0.3 \text{ qt}}{4.5 \text{ qt}}$$

$$= 0.067 = 6.7\%$$

29.

$$R = \frac{P}{B}$$

$$R = \frac{2400 \text{ ft}^3 - 1920 \text{ ft}^3}{2400 \text{ ft}^3}$$

$$= \frac{480 \text{ ft}^3}{2400 \text{ ft}^3}$$

$$= 0.20 = 20\%$$

31.

$$B = \frac{P}{R}$$

$$B = \frac{20 \text{ ft}}{0.03}$$

$$= 666.7 \text{ ft}$$

$$A = 666.7 \text{ ft} + 100 \text{ ft} = 766.7 \text{ ft}$$

33.

$$\text{Chemical: } 160 \text{ acre} \times \frac{2\frac{3}{4} \text{ lb}}{1 \text{ acre}}$$

$$= 440 \text{ lb}$$

Active ingredients: $440 \text{ lb} \times 0.80 = 352 \text{ lb}$

Inert ingredients: $440 \text{ lb} - 352 \text{ lb} = 88 \text{ lb}$

35.

$$7310 \text{ lb} \times \frac{1 \text{ gal}}{8.6 \text{ lb}} = 850 \text{ gal}$$

$$\text{Butterfat} = 850 \text{ gal} \times 0.42 = 35.7 \text{ gal}$$

37.

$$R = \frac{P}{B}$$

$$R = \frac{150 - 39}{150}$$

$$= 74\%$$

39.

$$P = BR$$

$$P = (250 \text{ ml})(0.03)$$

$$= 7.5 \text{ ml}$$

41.

$$R = \frac{P}{B}$$

$$R = \frac{25 \text{ ml}}{1000 \text{ ml}}$$

$$= 0.025 = 2.5\%$$

43.

$$\text{Percent increase} = \frac{\text{change}}{\text{original value}} \times 100\%$$

$$\text{Percent increase} = \frac{115 \text{ lb/in}^2 - 75 \text{ lb/in}^2}{75 \text{ lb/in}^2} \times 100\%$$

$$= 53.3\%$$

45.

$$\text{Percent decrease} = \frac{\text{change}}{\text{original value}} \times 100\%$$

$$\text{Percent decrease} = \frac{\$25.50 - \$21.88}{\$25.50} \times 100\%$$

$$= 14.2\%$$

47. First item: $\$100.00 - 0.55 \times \$100.00 = \$45.$

Second item:

$$\$100.00 - 0.40 \times \$100.00 = \$60.00$$
$$\$60.00 - 0.15 \times \$60.00 = \$51.00$$

49.

$$P = BR$$
$$P = (1640 \text{ lb})(0.95)$$
$$= 1558 \text{ lb}$$

51.

$$R = \frac{P}{B}$$
$$R = \frac{187}{250}$$
$$= 0.748 = 74.8\%$$

53. a.

$$P = BR$$
$$P = (25 \text{ deer/mi}^2)(0.40)$$
$$= 10 \text{ deer/mi}^2$$

$$\text{Population} = 25 \text{ deer/mi}^2 + 10 \text{ deer/mi}^2$$
$$= 35 \text{ deer/mi}^2$$

b.

$$P = BR$$
$$P = (35 \text{ deer/mi}^2)(0.40)$$
$$= 14 \text{ deer/mi}^2$$

$$\text{Population} = 35 \text{ deer/mi}^2 + 14 \text{ deer/mi}^2$$
$$= 45 \text{ deer/mi}^2$$

55.

$$\text{Total cost} = \$5.66$$

$$B = \frac{P}{R}$$

$$B = \frac{\$5.66}{0.34}$$

$$= \$16.65$$

57.

	Total Cost
	$22 \times \$1.33 = \$29.26$
	$14 \times \$3.89 = \$54.46$
	$12 \times \$6.49 = \$77.88$
	$6 \times \$7.43 = \$44.58$
	$6 \times \$8.76 = \$52.56$
	$6 \times \$5.54 = \$33.24$
	$5 \times \$6.45 = \$32.25$
	$4 \times \$2.09 = \$8.36$
	$120 \times \$1.69 = \$202.80$
	$32 \times \$48.00 = \$1536$
Total	$2,071.39
Less 5% Cash Discount Net 30 Days	$103.57
Net Total	$1,967.82

59.

	$66 \times \$7.97 = \$526.02$
	$30 \times \$3.95 = \$118.50$
	$14 \times \$3.39 = \$47.46$
	$17 \times \$6.59 = \$112.03$
	$4 \times \$12.10 = \$48.40$
	$9 \times \$5.39 = \$48.51$
	$7 \times \$4.97 = \$34.79$
	$10 \times \$11.97 = \$119.70$
	$6 \times \$16.89 = \$101.34$
	$11 \times \$18.55 = \$204.05$
	$15 \times \$24.25 = \$363.75$
	$27 \times \$16.95 = \$457.65$
	$7 \times \$14.39 = \$100.73$
	$1 \times \$24.96 = \$24.96$
	$10 \times \$10.37 = \$103.7$
	$27 \times \$19.85 = \$535.95$
	$7 \times \$12.25 = \$85.75$
	$1 \times \$17.85 = \$17.85$
	$7 \times \$12.19 = \$85.33$
	$8 \times \$3.49 = \$27.92$
	$3 \times \$17.65 = \$52.95$
	$80 \times \$17.29 = \$1383.2$
	$7 \times \$20.65 = \$144.55$
	$\$5428.59 - \$108.57$
	$1 \times \$33.59 = \$33.59$
	$3 \times \$34.97 = \$104.91$
	$250 \times \$2.18 = \$545$
Subtotal	$\$5428.59$
Less 2% Discount	$\$5428.59 \times 0.02$ $= \$108.57$
Subtotal	$\$5428.59 - \$108.57$ $= \$5320.02$
5 ¾% Sales Tax	$\$5320.02 \times 0.0575 = \$305.90$
NET TOTAL	$\$5320.02 + \$305.90 = \$5625.92$

Section 1.15: Powers and Roots

1. 225

3. 222

5. 0.00000661

7. 729

9. 562

11. 0.00483

13. 157

15. 2.96

17. 68.9

19. 42.4

21. 0.198

Section 1.16: Applications Involving Percent: Business and Personal Finance

1. a.

$$i = prt$$
$$i = (\$2000)(0.05)(3)$$
$$= \$300$$

b.

$$\text{payment} = \frac{\text{principle} + \text{interest}}{\text{loan period}}$$

$$\text{payment} = \frac{\$2000 + \$300}{36}$$

$$= \$63.89$$

3.

$$A = P\left(1 + \frac{r}{n}\right)^{nt}$$

$$A = \$7500\left(1 + \frac{0.065}{4}\right)^{(4)(4)}$$

$$= \$7500(1.01625)^{16}$$

$$= \$9706.67$$

5.

$$A = P\left(1 + \frac{r}{n}\right)^{nt}$$

$$A = \$15,000\left(1 + \frac{0.055}{2}\right)^{(2)(8)}$$

$$= \$15,000(1.0275)^{16}$$

$$= \$23,152.64$$

7.

$$P = \$150,000$$
$$i = 0.065/12$$
$$n = 30 \times 12 = 360$$

$$A = P\left(\frac{i(1+i)^n}{(1+i)^n - 1}\right)$$

$$A = \$150,000\left(\frac{\left(\frac{0.065}{12}\right)\left(1 + \frac{0.065}{12}\right)^{360}}{\left(1 + \frac{0.065}{12}\right)^{360} - 1}\right)$$

$$= \$948.10$$

9.

$$\text{Price} = 275 \text{ acres} \times \$4100/\text{acre}$$
$$= \$1,127,500$$
$$P = \$1,127,500 \times 0.75 = \$845,625$$
$$i = 0.0675$$
$$n = 20$$

$$A = P\left(\frac{i(1+i)^n}{(1+i)^n - 1}\right)$$

$$A = \$845,625\left(\frac{0.0675(1+0.0675)^{20}}{(1+0.0675)^{20} - 1}\right)$$

$$= \$78,276.71$$

The annual payment is $\$6429.83 \times 12$
$= \$77,157.96$.

11. a.

$$P = \$24,000$$
$$i = 0.0075/12$$
$$n = 3 \times 12 = 36$$

$$A = P\left(\frac{i(1+i)^n}{(1+i)^n - 1}\right)$$

$$A = \$24,000\left(\frac{\left(\frac{0.0075}{12}\right)\left(1 + \frac{0.0075}{12}\right)^{36}}{\left(1 + \frac{0.0075}{12}\right)^{36} - 1}\right)$$

$$= \$674.40$$

Total payment = $\$674.40 \times 36 = \$24,278.40$

11. (continued)

b.

$$P = \$24,000 - \$1500 = \$22,500$$

$$i = 0.085/12$$

$$n = 3 \times 12 = 36$$

$$A = P\left(\frac{i(1+i)^n}{(1+i)^n - 1}\right)$$

$$A = \$22,500\left(\frac{\left(\frac{0.0.085}{12}\right)\left(1+\frac{0.0.085}{12}\right)^{36}}{\left(1+\frac{0.0.085}{12}\right)^{36} - 1}\right)$$

$$= \$710.27$$

Total payment $= \$710.27 \times 36 = \$25,569.71$

Choice a costs $\$25,569.71 - \$24,278.51 = \$1291.20$ less.

13.

$$P = \$220,500 - \$4500 - \$9500 - \$8000$$

$$= \$198,500$$

$$i = 0.08$$

$$n = 4$$

$$A = P\left(\frac{i(1+i)^n}{(1+i)^n - 1}\right)$$

$$A = \$198,500\left(\frac{(0.08)(1+0.08)^4}{(1+0.08)^4 - 1}\right)$$

$$= \$59,931.28$$

15.

$$A = P\left(1+\frac{r}{n}\right)^{nt}$$

$$A = \$30,000\left(1+\frac{0.05}{1}\right)^{(1)(3)}$$

$$= \$30,000(1.05)^3$$

$$= \$34,728.75$$

17.

$$A = P\left(1+\frac{r}{n}\right)^{nt}$$

$$A = \$30,000\left(1+\frac{0.05}{365}\right)^{(365)(3)}$$

$$= \$34,854.67$$

19.

$$A = P\left(1+\frac{r}{n}\right)^{nt}$$

$$A = \$8400\left(1+\frac{0.035}{12}\right)^{(12)(5)}$$

$$= \$10,003.92$$

21.

$$P = \$37,500 - \$37,500 \times 0.10 + \$37,500 \times 0.06$$

$$= \$36,000$$

$$i = 0.042/12 = 0.0035$$

$$n = 3 \times 12 = 36$$

$$A = P\left(\frac{i(1+i)^n}{(1+i)^n - 1}\right)$$

$$A = \$36,000\left(\frac{0.0035(1+0.0035)^{36}}{(1+0.0035)^{36} - 1}\right)$$

$$= \$1066.07$$

23.

Discount amount $= (0.03)(\$15{,}870) = \$476.10$

$$\text{Interest} = \frac{\text{Discount amount}}{\text{Invoice amount} - \text{Discount amount}} \times \frac{\text{Number of days per year}}{\text{Number of days paid early}}$$

$$\text{Interest} = \frac{\$476.10}{\$15{,}870 - \$476.10} \times \frac{365}{20} = 56.4\%$$

25.

Discount amount $= (0.025)(\$129{,}115.23) = \$3227.88$

$$\text{Interest} = \frac{\text{Discount amount}}{\text{Invoice amount} - \text{Discount amount}} \times \frac{\text{Number of days per year}}{\text{Number of days paid early}}$$

$$\text{Interest} = \frac{\$3227.88}{\$129{,}115.23 - \$3227.88} \times \frac{365}{20} = 46.8\%$$

27.

Discount amount $= (0.01)(\$21{,}500) = \$215$

$$\text{Interest} = \frac{\text{Discount amount}}{\text{Invoice amount} - \text{Discount amount}} \times \frac{\text{Number of days per year}}{\text{Number of days paid early}}$$

$$\text{Interest} = \frac{\$215}{\$21{,}500 - \$215} \times \frac{365}{10} = 36.9\%$$

Unit 1C Review

1. 1.625

2. $\dfrac{45}{100} = \dfrac{9}{20}$

3. 10.129

4. 116.935

5. 5.854

6. 55.6 ft $-$ 15.0 ft $-$ 15.0 ft $=$ 25.6 ft

7.

   55.6 ft
   15.0 ft
   15.0 ft
   9.5 ft
   25.6 ft
   9.5 ft
   15.0 ft
   15.0 ft
   ‾‾‾‾‾‾
   160.2 ft

8. a. 45.1

   b. 45.06

9. a. 45.1

   b. 45.06

10. 0.11515

11. 18.85

12. 18.5 in. $\div$ 2.75 in. $=$ 6 r 2 . Six cables could be cut and there would be 2 in. remaining.

13. 0.25

14. 72.4

15.

$P = BR$

$P = (420)(0.165)$

$\quad = 69.3$

16.

$B = \dfrac{P}{R}$

$B = \dfrac{240}{0.12}$

$\quad = 2000$

17.

$R = \dfrac{P}{B}$

$R = \dfrac{96 \text{ yd}}{240 \text{ yd}}$

$\quad = 40.0\%$

18.

$$P = BR$$
$$P = (\$16.50)(0.06)$$
$$= \$0.99$$

Her new salary is $\$16.50 + \$0.99 = \$17.49/h$.

19. 2110

20. 9.40

Chapter 1 Review

1. 8243

2. 55,197

3. 9,178,000

4. 226 r 240

5.

$$12 - 3(5 - 2)$$
$$= 12 - 3(3)$$
$$= 12 - 9$$
$$= 3$$

6.

$$(6 + 4)8 \div 2 + 3$$
$$= (10)8 \div 2 + 3$$
$$= 80 \div 2 + 3$$
$$= 40 + 3$$
$$= 43$$

7.

$$18 \div 2 \times 5 \div 3 - 6 + 4 \times 7$$
$$= 9 \times 5 \div 3 - 6 + 28$$
$$= 45 \div 3 - 6 + 28$$
$$= 15 - 6 + 28$$
$$= 37$$

8.

$$18 / (5 - 3) + (6 - 2) \times 8 - 10$$
$$= 18 / 2 + 4 \times 8 - 10$$
$$= 9 + 32 - 10$$
$$= 31$$

9.

Area of upper rectangle: $12 \text{ cm} \times 5 \text{ cm} = 60 \text{ cm}^2$

Area of lower rectangle: $10 \text{ cm} \times 28 \text{ cm} = \underline{280 \text{ cm}^2}$

Total area: $= 340 \text{ cm}^2$

10.

Volume of left box: $10 \text{ cm} \times 1 \text{ cm} \times 1 \text{ cm} = 10 \text{ cm}^3$

Volume of middle box: $10 \text{ cm} \times 1 \text{ cm} \times 1 \text{ cm} = 10 \text{ cm}^2$

Volume of right box: $10 \text{ cm} \times 1 \text{ cm} \times 1 \text{ cm} = \underline{10 \text{ cm}^2}$

Total Volume: $= 30 \text{ cm}^2$

11.

$$C = \frac{5}{9}(F - 32)$$
$$C = \frac{5}{9}(50 - 32)$$
$$= \frac{5}{9}(18)$$
$$= 10$$

12.

$$P = \frac{Fs}{t}$$
$$P = \frac{(600)(50)}{10}$$
$$= \frac{30,000}{10}$$
$$= 3000$$

13. $4 + 6 + 0 = 10$ is not divisible by 3, so 28 is not divisible by 3.

14. $54 = 2 \cdot 3 \cdot 3 \cdot 3$

15. $330 = 2 \cdot 3 \cdot 5 \cdot 11$

16. $\dfrac{36}{56} = \dfrac{9 \cdot 4}{14 \cdot 4} = \dfrac{9}{14}$

17. $\dfrac{180}{216} = \dfrac{5 \cdot 36}{6 \cdot 36} = \dfrac{5}{6}$

18. $4\dfrac{1}{6}$

19. $3\dfrac{18}{5} = 3 + \dfrac{18}{5} = 3 + 3\dfrac{3}{5} = 6\dfrac{3}{5}$

20. $2\dfrac{5}{8} = \dfrac{(2 \times 8) + 5}{8} = \dfrac{21}{8}$

21. $3\dfrac{7}{16} = \dfrac{(3 \times 16) + 7}{16} = \dfrac{55}{16}$

22. $\dfrac{16}{8} = 2$

23.

$\dfrac{1}{4} + \dfrac{5}{12} + \dfrac{5}{6}$

$= \dfrac{3}{12} + \dfrac{5}{12} + \dfrac{10}{12}$

$= \dfrac{18}{12} = \dfrac{3}{2} = 1\dfrac{1}{2}$

24.

$\dfrac{29}{36} - \dfrac{7}{30}$

$= \dfrac{145}{180} - \dfrac{42}{180}$

$= \dfrac{103}{180}$

25.

$5\dfrac{3}{16} + 9\dfrac{5}{12}$

$= 5\dfrac{9}{48} + 9\dfrac{20}{48}$

$= 14\dfrac{29}{48}$

26.

$6\dfrac{3}{8} - 4\dfrac{7}{12}$

$= 6\dfrac{9}{24} - 4\dfrac{14}{24}$

$= 5\dfrac{33}{24} - 4\dfrac{14}{24}$

$= 1\dfrac{19}{24}$

27.

$18 - 6\dfrac{2}{5}$

$= 17\dfrac{5}{5} - 6\dfrac{2}{5}$

$= 11\dfrac{3}{5}$

28.

$16\dfrac{2}{3} + 1\dfrac{1}{4} - 12\dfrac{11}{12}$

$= 16\dfrac{8}{12} + 1\dfrac{3}{12} - 12\dfrac{11}{12}$

$= 17\dfrac{11}{12} - 12\dfrac{11}{12}$

$= 5$

29. $\dfrac{1}{4}$

30.

$3\dfrac{6}{7} \times 4\dfrac{2}{3}$

$= \dfrac{27}{7} \times \dfrac{14}{3}$

$= 18$

31.

$\dfrac{3}{8} \div 6$

$= \dfrac{3}{8} \times \dfrac{1}{6}$

$= \dfrac{1}{16}$

32.

$$\frac{2}{3} \div 1\frac{7}{9}$$

$$= \frac{2}{3} \div \frac{16}{9}$$

$$= \frac{2}{3} \times \frac{9}{16}$$

$$= \frac{3}{8}$$

33.

$$1\frac{4}{5} \div 1\frac{9}{16} \times 11\frac{2}{3}$$

$$= \frac{9}{5} \div \frac{25}{16} \times \frac{35}{3}$$

$$= \frac{9}{5} \times \frac{16}{25} \times \frac{35}{3}$$

$$= \frac{144}{125} \times \frac{35}{3}$$

$$= \frac{336}{25} = 13\frac{11}{25}$$

34.

$$A = 12\frac{5}{16} \text{ in.} - 4\frac{3}{8} \text{ in.} - 4\frac{9}{16} \text{ in.}$$

$$= 12\frac{5}{16} \text{ in.} - 4\frac{6}{16} \text{ in.} - 4\frac{9}{16} \text{ in.}$$

$$= 12\frac{5}{16} \text{ in.} - 8\frac{15}{16} \text{ in.}$$

$$= 11\frac{21}{16} \text{ in.} - 8\frac{15}{16} \text{ in.}$$

$$= 3\frac{6}{16} \text{ in.} = 3\frac{3}{8} \text{ in.}$$

$$B = 9\frac{3}{32} \text{ in.} - 6\frac{5}{32} \text{ in.} + 2\frac{1}{2} \text{ in.}$$

$$= 9\frac{3}{32} \text{ in.} + 2\frac{16}{32} \text{ in.} - 6\frac{5}{32} \text{ in.}$$

$$= 11\frac{19}{32} \text{ in.} - 6\frac{5}{32} \text{ in.}$$

$$= 5\frac{14}{32} \text{ in.} = 5\frac{7}{16} \text{ in.}$$

35. $6 \text{ lb } 9 \text{ oz} = \left(6 \text{ lb} \times \frac{16 \text{ oz}}{1 \text{ lb}}\right) + 9 \text{ oz} = 105 \text{ oz}$

36. $168 \text{ ft} \times \frac{12 \text{ in.}}{1 \text{ ft}} = 2016 \text{ in.}$

37. $72 \text{ ft} \times \frac{1 \text{ yd}}{3 \text{ ft}} = 24 \text{ yd}$

38. $36 \text{ mi} \times \frac{1760 \text{ yd}}{3 \text{ mi}} = 63,360 \text{ yd}$

39. 0.5625

40. 0.416

41. $\frac{45}{100} = \frac{9}{20}$

42. $19\frac{625}{1000} = 19\frac{5}{8}$

43. 168.278

44. 17.25

45. 68.665

46. 33.72

47. 3206.5

48. 1.9133

49. 3.18

50. 20.6

51. a. 200

　　b. 248.2

　　c. 250

52. a. 5.6

　　b. 5.65

　　c. 5.6491

53. $15\% = \frac{15}{100} = 0.15$

54. $8\frac{1}{4}\% = 8.25\% = 0.0825$

55. 6.5%

56. 120%

57.

$$P = BR$$

$$P = (\$12,000)(0.0875)$$

$$= \$1050$$

58.

Fraction	Decimal	Percent
$\dfrac{1}{4}$	0.25	25%
$\dfrac{3}{8}$	0.375	$37\dfrac{1}{2}\%$
$\dfrac{5}{6}$	$0.83\dfrac{1}{3}$	$83\dfrac{1}{3}\%$
$8\dfrac{3}{4}$	8.75	875%
$2\dfrac{2}{5}$	2.4	240%
$\dfrac{3}{2000}$	0.0015	0.15%

59.

$$R = \frac{P}{B}$$
$$R = \frac{\$32{,}000}{\$84{,}000}$$
$$= 38.1\%$$

60.

$$R = \frac{P}{B}$$
$$R = \frac{\dfrac{11}{64}}{\dfrac{13}{32}} = \frac{11}{64} \times \frac{32}{13}$$
$$= 42.3\%$$

Chapter 1 Test

1. 5729

3. 2,584.450

7.

Area of upper rectangle: $10 \text{ m} \times 40 \text{ m} = 400 \text{ m}^2$

Area of middle rectangle: $10 \text{ m} \times 15 \text{ m} = 150 \text{ m}^2$

Area of lower rectangle: $10 \text{ m} \times 20 \text{ m} = \underline{200 \text{ m}^2}$

Total area: $= 750 \text{ m}^2$

9. $\dfrac{120 \text{ V}}{40 \ \Omega} = 3 \text{ A}$

61. 60 tons $\times 0.80 = 48$ tons

62.

$$6 \times \left(3\frac{1}{16} \text{ in.}\right) + 5 \times \left(\frac{1}{4} \text{ in.}\right) + 2 \times \left(1\frac{1}{8} \text{ in.}\right)$$
$$= 6 \times \left(\frac{49}{16} \text{ in.}\right) + 5 \times \left(\frac{1}{4} \text{ in.}\right) + 2 \times \left(\frac{9}{8} \text{ in.}\right)$$
$$= \frac{147}{8} \text{ in.} + \frac{5}{4} \text{ in.} + \frac{9}{4} \text{ in.}$$
$$= \frac{147}{8} \text{ in.} + \frac{10}{8} \text{ in.} + \frac{18}{8} \text{ in.}$$
$$= 21\frac{7}{8} \text{ in.}$$

63. $\dfrac{7}{8} \text{ in.} - \dfrac{9}{16} \text{ in.} = \dfrac{14}{16} \text{ in.} - \dfrac{9}{16} \text{ in.} = \dfrac{5}{16} \text{ in.}$

64.

Height = 20 in. + 2 × 5 in. = 30 in.

Length = 4 × 10 in. + 1 in. = 41 in.

The sheet of cardboard would have to be 30 in. × 41 in.

65. 4020

66. 139

5.

$$8 + 2(5 \times 6 + 8)$$
$$= 8 + 2(30 + 8)$$
$$= 8 + 2(38)$$
$$= 8 + 76 = 84$$

11.

$$t = \frac{d}{r}$$
$$t = \frac{1050}{21}$$
$$= 50$$

13. $90 = 2 \cdot 3 \cdot 3 \cdot 5$

15. $\dfrac{30}{64} = \dfrac{15 \cdot 2}{32 \cdot 2} = \dfrac{15}{32}$

17. $\dfrac{23}{6} = 3 \text{ r } 5 = 3\dfrac{5}{6}$

19. $\dfrac{3}{8} + \dfrac{1}{4} = \dfrac{3}{8} + \dfrac{2}{8} = \dfrac{5}{8}$

21.

$$3\frac{1}{8} = 3\frac{1}{8}$$
$$2\frac{1}{2} = 2\frac{4}{8}$$
$$\underline{4\frac{3}{4} = 4\frac{6}{8}}$$
$$9\frac{11}{8} = 10\frac{3}{8}$$

23.

$$3\frac{5}{8} + 2\frac{3}{16} - 1\frac{1}{4}$$
$$= 3\frac{10}{16} + 2\frac{3}{16} - 1\frac{4}{16}$$
$$= 5\frac{13}{16} - 1\frac{4}{16}$$
$$= 4\frac{9}{16}$$

25.

$$\frac{3}{8} \div 3\frac{5}{16} = \frac{3}{8} \div \frac{53}{16}$$
$$= \frac{3}{8} \times \frac{16}{53}$$
$$= \frac{6}{53}$$

27.

$$3\frac{5}{8} + 1\frac{3}{4} \times 6\frac{1}{5} = \frac{29}{8} + \frac{7}{4} \times \frac{31}{5}$$
$$= \frac{29}{8} + \frac{217}{20}$$
$$= \frac{145}{40} + \frac{434}{40}$$
$$= \frac{579}{40} = 14\frac{19}{40}$$

29.

$$3\frac{5}{8}\,A + 2\frac{3}{4}\,A + 4\frac{5}{16}\,A$$
$$= 3\frac{10}{16}\,A + 2\frac{12}{16}\,A + 4\frac{5}{16}\,A$$
$$= 9\frac{27}{16}\,A = 10\frac{11}{16}\,A$$

31. $3 \text{ lb }\ 5 \text{ oz} = \left(3 \text{ lb} \times \dfrac{16 \text{ oz}}{1 \text{ lb}} \right) + 5 \text{ oz} = 53 \text{ oz}$

33. $2.12 = 2\dfrac{12}{100} = 2\dfrac{3}{25}$

35.   397.19

37.   8.0784

39.

$$B = \frac{P}{R}$$
$$B = \frac{59.45}{0.41}$$
$$= 145$$

41.

$$P = BR$$
$$P = (\$612)(0.067)$$
$$= \$41$$

Her new salary is $\$612 + \$41 = \$653$.

43.   6.73

Chapter 2: Signed Numbers and Powers of 10

Section 2.1: Addition of Signed Numbers

1.	3	**25.**	−3	**49.**	−20
3.	6	**27.**	−1	**51.**	−2
5.	4	**29.**	−4	**53.**	0
7.	17	**31.**	−6	**55.**	−5
9.	15	**33.**	−2	**57.**	−19
11.	10	**35.**	4	**59.**	−12
13.	7	**37.**	9	**61.**	7
15.	−2	**39.**	3	**63.**	11
17.	−12	**41.**	4	**65.**	4
19.	6	**43.**	7	**67.**	4
21.	−9	**45.**	−7	**69.**	−14
23.	4	**47.**	5		

Section 2.2: Subtraction of Signed Numbers

1.	−2	**21.**	−7	**41.**	8
3.	11	**23.**	15	**43.**	2
5.	12	**25.**	−16	**45.**	−15
7.	6	**27.**	−10	**47.**	−3
9.	18	**29.**	−2	**49.**	−23
11.	−6	**31.**	14	**51.**	−4
13.	−4	**33.**	23	**53.**	−1
15.	0	**35.**	−15	**55.**	2
17.	1	**37.**	8	**57.**	−8
19.	−10	**39.**	−4	**59.**	−10

Section 2.3: Multiplication and Division of Signed Numbers

1.	24	**29.**	54	**57.**	2
3.	−18	**31.**	−6	**59.**	9
5.	−35	**33.**	24	**61.**	−4
7.	27	**35.**	27	**63.**	3
9.	−72	**37.**	−63	**65.**	17
11.	27	**39.**	−9	**67.**	40
13.	0	**41.**	−6	**69.**	15
15.	49	**43.**	36	**71.**	7
17.	−300	**45.**	30	**73.**	4
19.	−13	**47.**	−168	**75.**	−3
21.	−6	**49.**	−162	**77.**	10
23.	48	**51.**	5	**79.**	8
25.	21	**53.**	−9		
27.	−16	**55.**	8		

Section 2.4: Signed Fractions

1.

$$\frac{1}{8} + \left(-\frac{5}{16}\right) = \frac{2}{16} + \left(-\frac{5}{16}\right)$$

$$= -\frac{3}{16}$$

3.

$$\frac{1}{2} + \left(-\frac{7}{16}\right) = \frac{8}{16} + \left(-\frac{7}{16}\right)$$

$$= \frac{1}{16}$$

7.

$$\left(-3\frac{2}{3}\right) + \left(-\frac{4}{9}\right) + \left(4\frac{5}{6}\right) = \left(-3\frac{12}{18}\right) + \left(-\frac{8}{18}\right) + \left(4\frac{15}{18}\right)$$

$$= \left(-3\frac{20}{18}\right) + \left(4\frac{15}{18}\right)$$

$$= \left(-4\frac{2}{18}\right) + \left(4\frac{15}{18}\right)$$

$$= \frac{13}{18}$$

9.

$$\left(-\frac{1}{4}\right) - \left(-\frac{1}{5}\right) = \left(-\frac{5}{20}\right) + \frac{4}{20}$$

$$= -\frac{1}{20}$$

11.

$$\left(1\frac{3}{8}\right) - \left(+\frac{5}{16}\right) = \left(1\frac{6}{16}\right) + \left(-\frac{5}{16}\right)$$

$$= 1\frac{1}{16}$$

13.

$$\left(+1\frac{3}{4}\right) - (-4) = \left(+1\frac{3}{4}\right) + \frac{16}{4}$$

$$= 1\frac{19}{4}$$

$$= 1 + 4\frac{3}{4}$$

$$= 5\frac{3}{4}$$

5.

$$\left(-5\frac{3}{4}\right) + \left(-6\frac{2}{5}\right) = \left(-5\frac{15}{20}\right) + \left(-6\frac{8}{20}\right)$$

$$= -11\frac{23}{20}$$

$$= -\left(11 + 1\frac{3}{20}\right)$$

$$= -11\frac{3}{20}$$

15.

$$\left(-\frac{2}{3}\right) + \left(-\frac{5}{6}\right) - \frac{1}{4} = \left(-\frac{8}{12}\right) + \left(-\frac{10}{12}\right) - \frac{3}{12}$$

$$= -\frac{21}{12}$$

$$= -1\frac{8}{12}$$

$$= -1\frac{3}{4}$$

17. $-\frac{1}{63}$

19.

$$\left(-3\frac{1}{3}\right)\left(-1\frac{4}{5}\right) = \left(-\frac{10}{3}\right)\left(-\frac{9}{5}\right)$$

$$= 6$$

21.

$$\frac{4}{5} \div \left(-\frac{8}{9}\right) = \frac{4}{5} \times \left(-\frac{9}{8}\right)$$

$$= -\frac{9}{10}$$

23.

$$\left(-\frac{7}{9}\right) \div \left(-\frac{8}{3}\right) = \left(-\frac{7}{9}\right)\left(-\frac{3}{8}\right)$$

$$= \frac{7}{24}$$

25.

$$\left(\frac{-1}{4}\right) + \left(\frac{1}{-5}\right) = \left(-\frac{1}{4}\right) + \left(-\frac{1}{5}\right)$$

$$= \left(-\frac{5}{20}\right) + \left(-\frac{4}{20}\right)$$

$$= \frac{-9}{20}$$

27.

$$\frac{3}{4} + \left(\frac{-3}{8}\right) = \frac{6}{8} + \left(-\frac{3}{8}\right)$$

$$= \frac{3}{8}$$

29.

$$\frac{5}{8} - \left(\frac{-5}{8}\right) = \frac{5}{8} + \left(\frac{5}{8}\right)$$

$$= \frac{10}{8}$$

$$= \frac{5}{4}$$

$$= 1\frac{1}{4}$$

31.

$$\left(\frac{-6}{8}\right) - (-4) = \left(-\frac{6}{8}\right) + \frac{32}{8}$$

$$= \frac{26}{8}$$

$$= 3\frac{2}{8}$$

$$= 3\frac{1}{4}$$

33.

$$\left(\frac{-1}{4}\right)\left(\frac{1}{-5}\right) = \left(-\frac{1}{4}\right)\left(-\frac{1}{5}\right)$$

$$= \frac{1}{20}$$

35.

$$\left(\frac{-5}{-8}\right)\left(-5\frac{1}{3}\right) = \left(\frac{5}{8}\right)\left(-\frac{16}{3}\right)$$

$$= -\frac{10}{3}$$

$$= -3\frac{1}{3}$$

37.

$$32 \div \left(\frac{-2}{3}\right) = \left(\frac{32}{1}\right)\left(-\frac{3}{2}\right)$$

$$= -48$$

39.

$$\left(\frac{-2}{-3}\right) \div \left(\frac{2}{-3}\right) = \left(\frac{2}{3}\right)\left(-\frac{3}{2}\right)$$

$$= -1$$

41.

$$\left(\frac{-2}{3}\right) + \left(-\frac{5}{6}\right) + \frac{1}{4} + \frac{1}{8} = \left(-\frac{16}{24}\right) + \left(-\frac{20}{24}\right) + \frac{6}{24} + \frac{3}{24}$$

$$= \left(-\frac{36}{24}\right) + \frac{9}{24}$$

$$= -\frac{27}{24}$$

$$= -1\frac{3}{24}$$

$$= -1\frac{1}{8}$$

43. $\frac{1}{4}$

45.

$$\left(\frac{-2}{3}\right)+\left(-\frac{1}{2}\right)\left(\frac{5}{-6}\right)=\left(-\frac{2}{3}\right)+\frac{5}{12}$$

$$=\left(-\frac{8}{12}\right)+\frac{5}{12}$$

$$=-\frac{3}{12}$$

$$=-\frac{1}{4}$$

Section 2.5: Powers of 10

1.
$$10^4 \cdot 10^9 = 10^{4+9}$$
$$= 10^{13}$$

3.
$$\frac{10^4}{10^8} = 10^{4-8}$$
$$= 10^{-4}$$
$$= \frac{1}{10^4}$$

5. 10^3

7.
$$\left(10^4\right)^3 = 10^{4\cdot3}$$
$$= 10^{12}$$

9.
$$10^{-6} \cdot 10^{-4} = 10^{(-6)+(-4)}$$
$$= 10^{-10}$$
$$= \frac{1}{10^{10}}$$

11.
$$\frac{10^{-3}}{10^{-6}} = 10^{(-3)-(8)-6}$$
$$= 10^3$$

13.
$$\left(10^{-3}\right)^4 = 10^{(-3)\cdot4}$$
$$= 10^{-12} = \frac{1}{10^{12}}$$

15.
$$\left(\frac{10^6}{10^8}\right)^2 = \left(10^{-2}\right)^2$$
$$= 10^{-4}$$
$$= \frac{1}{10^4}$$

17.
$$\frac{\left(10^0\right)^3}{10^{-2}} = \frac{(1)^3}{10^{-2}}$$
$$= \frac{1}{10^{-2}}$$
$$= 10^2$$

19.
$$10^2 \cdot 10^{-5} \cdot 10^{-3} = 10^{-6}$$
$$= \frac{1}{10^6}$$

21. $10^3 \cdot 10^4 \cdot 10^{-5} \cdot 10^3 = 10^5$

23.
$$\frac{10^3 \cdot 10^2 \cdot 10^{-7}}{10^5 \cdot 10^{-3}} = \frac{10^{-2}}{10^2}$$
$$= 10^{-4}$$
$$= \frac{1}{10^4}$$

25.
$$\frac{10^8 \cdot 10^{-6} \cdot 10^{10} \cdot 10^0}{10^4 \cdot 10^{-17} \cdot 10^8} = \frac{10^{12}}{10^{-5}}$$
$$= 10^{17}$$

27.

$$\frac{\left(10^{-9}\right)^{-2}}{10^{16}\cdot10^{-4}}=\frac{10^{18}}{10^{12}}$$
$$=10^{6}$$

29.

$$\left(\frac{10^{5}\cdot10^{-2}}{10^{-4}}\right)^{2}=\left(\frac{10^{3}}{10^{-4}}\right)^{2}$$
$$=\left(10^{7}\right)^{2}$$
$$=10^{14}$$

Section 2.6: Scientific Notation

1. 3.56×10^{2}

3. 6.348×10^{2}

5. 8.25×10^{-3}

7. 7.4×10^{0}

9. 7.2×10^{-5}

11. 7.1×10^{5}

13. 4.5×10^{-6}

15. 3.4×10^{-8}

17. 6.4×10^{5}

19. 75,500

21. 5310

23. 0.078

25. 0.000555

27. 64

29. 960

31. 5.76

33. 0.0000064

35. 50,000,000,000

37. 0.00000062

39. 2,500,000,000,000

41. 0.000000000033

43. 0.0048

45. 0.00091

47. 0.00037

49. 0.0613

51. 0.0009

53. 1.0009

55. 0.00000000998

57. 0.000271

59.

$$\left(4\times10^{-6}\right)\left(6\times10^{-10}\right)=(4)(6)\times\left(10^{-6}\right)\left(10^{-10}\right)$$
$$=24\times10^{-16}$$
$$=2.4\times10^{-15}$$

63.

$$\frac{\left(4\times10^{-5}\right)\left(6\times10^{-3}\right)}{\left(3\times10^{-10}\right)\left(8\times10^{8}\right)}=\frac{(4)(6)}{(3)(8)}\times\frac{\left(10^{-5}\right)\left(10^{-3}\right)}{\left(10^{-10}\right)\left(10^{8}\right)}$$
$$=1\times10^{-6}$$

61.

$$\frac{4.5\times10^{16}}{1.5\times10^{-8}}=\frac{4.5}{1.5}\times\frac{10^{16}}{10^{-8}}$$
$$=3\times10^{24}$$

65.

$$\left(1.2\times10^{6}\right)^{3}=(1.2)^{3}\times\left(10^{6}\right)^{3}$$
$$=1.73\times10^{18}$$

67.

$$\left(6.2\times10^{-5}\right)\left(5.2\times10^{-6}\right)\left(3.5\times10^{8}\right)=(6.2)(5.2)(3.5)\times\left(10^{-5}\right)\left(10^{-6}\right)\left(10^{8}\right)$$
$$=113\times10^{-3}$$
$$=1.13\times10^{-1}$$

69.

$$\left(\frac{2.5\times10^{-4}}{7.5\times10^{8}}\right)^{2} = \left(\frac{2.5}{7.5}\times\frac{10^{-4}}{10^{8}}\right)^{2}$$

$$= \left(\frac{1}{3}\times10^{-12}\right)^{2}$$

$$= \left(\frac{1}{3}\right)^{2}\times\left(10^{-12}\right)^{2}$$

$$= 0.111\times10^{-24}$$

$$= 1.11\times10^{-24}$$

71.

$$(18,000)(0.00005) = \left(1.8\times10^{4}\right)\left(5\times10^{-5}\right)$$

$$= (1.8)(5)\times\left(10^{4}\right)\left(10^{-5}\right)$$

$$= 9\times10^{-1}$$

73.

$$\frac{2,400,000}{36,000} = \frac{2.4\times10^{6}}{3.6\times10^{4}}$$

$$= \frac{2.4}{3.6}\times\frac{10^{6}}{10^{4}}$$

$$= 0.667\times10^{2}$$

$$= 6.67\times10^{1}$$

75.

$$\frac{84,000\times0.0004\times142,000}{0.002\times3200} = \frac{\left(8.4\times10^{4}\right)\left(4\times10^{-4}\right)\left(1.42\times10^{5}\right)}{\left(2\times10^{-3}\right)\left(3.2\times10^{3}\right)}$$

$$= \frac{(8.4)(4)(1.42)}{(2)(3.2)}\times\frac{\left(10^{4}\right)\left(10^{-4}\right)\left(10^{5}\right)}{\left(10^{-3}\right)\left(10^{3}\right)}$$

$$= 7.46\times10^{5}$$

77.

$$\left(\frac{48,000\times0.0144}{0.0064}\right)^{2} = \left(\frac{4.8\times10^{4}\times1.44\times10^{-2}}{6.4\times10^{-3}}\right)^{2}$$

$$= \left(\frac{(4.8)(1.44)}{6.4}\times\frac{\left(10^{4}\right)\left(10^{-2}\right)}{10^{-3}}\right)^{2}$$

$$= \left(1.08\times10^{5}\right)^{2}$$

$$= (1.08)^{2}\times\left(10^{5}\right)^{2}$$

$$= 1.17\times10^{10}$$

79.

$$\left(\frac{1.3\times10^{4}}{\left(2.6\times10^{-3}\right)\left(5.1\times10^{8}\right)}\right)^{5} = \left(\frac{1.3}{(2.6)(5.1)}\times\frac{10^{4}}{\left(10^{-3}\right)\left(10^{8}\right)}\right)^{5}$$

$$= \left(9.8\times10^{-1}\right)^{5}$$

$$= (9.8)^{5}\times\left(10^{-1}\right)^{5}$$

$$= 904\times10^{-5}$$

$$= 9.04\times10^{-3}$$

81.

$$\left(\frac{18.4 \times 2100}{0.036 \times 950}\right)^8 = \left(\frac{\left(1.84 \times 10^1\right)\left(2.1 \times 10^3\right)}{\left(3.6 \times 10^{-2}\right)\left(9.5 \times 10^2\right)}\right)^8$$

$$= \left(\frac{(1.84)(2.1)}{(3.6)(9.5)} \times \frac{\left(10^1\right)\left(10^3\right)}{\left(10^{-2}\right)\left(10^2\right)}\right)^8$$

$$= \left(0.113 \times 10^4\right)^8$$

$$= (0.113)^8 \times \left(10^4\right)^8$$

$$= 2.66 \times 10^{32}$$

Section 2.7: Engineering Notation

1. 28×10^3

3. 3.45×10^6

5. 220×10^9

7. 6.6×10^{-3}

9. 76.5×10^{-9}

11. 975×10^{-3}

13. $57,700$

15. $4,940,000,000,000$

17. $567,000,000$

19. 0.000026

21. 0.000000005945

23. 0.00000000001064

25.

$$\left(35.5 \times 10^6\right)\left(420 \times 10^9\right)$$

$$= (35.5)(420) \times \left(10^6\right)\left(10^9\right)$$

$$= 14,900 \times 10^{15}$$

$$= 14.9 \times 10^{18}$$

27. 19.7×10^{-6}

29.

$$\frac{70.5 \times 10^6}{120 \times 10^{-9}} = \frac{70.5}{120} \times \frac{10^6}{10^{-9}}$$

$$= 0.588 \times 10^{15}$$

$$= 588 \times 10^{12}$$

31. 15.6×10^{-18}

33. 339×10^6

35. 123×10^{21}

37. 8.97×10^6

39. 1.31×10^{12}

Chapter 2 Review

1. 5

2. 16

3. 13

4. 3

5. −8

6. −3

7. −13

8. −3

9. −11

10. 19

11. −2

12. 0

13. −19

14. −24

15. 36

16. 72

17. -84

18. 6

19. -6

20. 5

21.

$$\left(-\frac{6}{7}\right)-\left(\frac{5}{-6}\right)=\left(-\frac{6}{7}\right)-\left(-\frac{5}{6}\right)$$
$$=\left(-\frac{36}{42}\right)+\left(\frac{35}{42}\right)$$
$$=-\frac{1}{42}$$

22.

$$\frac{-3}{16}\div\left(-2\frac{1}{4}\right)=-\frac{3}{16}\div\left(-\frac{9}{4}\right)$$
$$=-\frac{3}{16}\times\left(-\frac{4}{9}\right)$$
$$=\frac{1}{12}$$

23.

$$\frac{-5}{8}+\left(-\frac{5}{6}\right)-\left(+1\frac{2}{3}\right)$$
$$=\left(-\frac{5}{8}\right)+\left(-\frac{5}{6}\right)-\left(\frac{5}{3}\right)$$
$$=\left(-\frac{15}{24}\right)+\left(-\frac{20}{24}\right)-\left(\frac{40}{24}\right)$$
$$=-\frac{25}{8}$$
$$=-3\frac{1}{8}$$

24.

$$\left(-\frac{9}{16}\right)\left(2\frac{2}{3}\right)=\left(-\frac{9}{16}\right)\left(\frac{8}{3}\right)$$
$$=-\frac{3}{2}$$
$$=-1\frac{1}{2}$$

25.

$$10^9\cdot10^{-14}=10^{9+(-14)}$$
$$=10^{-5}$$
$$=\frac{1}{10^5}$$

26.

$$10^6\div10^{-3}=10^{6-(-3)}$$
$$=10^9$$

27.

$$\left(10^{-4}\right)^3=10^{(-4)(3)}$$
$$=10^{-12}$$
$$=\frac{1}{10^{12}}$$

28.

$$\frac{\left(10^{-3}\cdot10^5\right)^3}{10^6}=\frac{\left(10^2\right)^3}{10^6}$$
$$=\frac{10^6}{10^6}$$
$$=1$$

29. 4.76×10^5

30. 1.4×10^{-3}

31. 0.0000535

32. $61,000,000$

33. 0.00105

34. 0.06

35. 0.000075

36. 0.00183

37.

$$\left(9.5\times10^{10}\right)\left(4.6\times10^{-13}\right)$$
$$=(9.5)(4.6)\times\left(10^{10}\right)\left(10^{-13}\right)$$
$$=43.7\times10^{-3}$$
$$=4.37\times10^{-2}$$

38.

$$\frac{8.4\times10^8}{3\times10^{-6}}=\frac{8.4}{3}\times\frac{10^8}{10^{-6}}$$
$$=2.8\times10^{14}$$

39.

$$\frac{(50,000)(640,000,000)}{(0.0004)^2}$$

$$= \frac{(5 \times 10^4)(6.4 \times 10^8)}{(4 \times 10^{-4})^2}$$

$$= \frac{(5 \times 10^4)(6.4 \times 10^8)}{(4)^2 \times (10^{-4})^2}$$

$$= \frac{(5 \times 10^4)(6.4 \times 10^8)}{16 \times 10^{-8}}$$

$$= \frac{(5)(6.4)}{16} \times \frac{(10^4)(10^8)}{10^{-8}}$$

$$= 2 \times 10^{20}$$

40.

$$(4.5 \times 10^{-8})^2 = (4.5)^2 \times (10^{-8})^2$$

$$= 20.3 \times 10^{-16}$$

$$= 2.03 \times 10^{-15}$$

41.

$$(2 \times 10^9)^4 = (2)^4 \times (10^9)^4$$

$$= 16 \times 10^{36}$$

$$= 1.6 \times 10^{37}$$

42.

$$\left(\frac{1.2 \times 10^{-2}}{3 \times 10^{-5}}\right)^3 = \left(\frac{1.2}{3} \times \frac{10^{-2}}{10^{-5}}\right)^3$$

$$= (0.4 \times 10^3)^3$$

$$= (0.4)^3 \times (10^3)^3$$

$$= 0.064 \times 10^9$$

$$= 6.4 \times 10^7$$

43. 275×10^3

44. 32×10^6

45. $450 \times 10 - 6$

46. $31,600,000$

47. 0.746

48.

$$(39.4 \times 10^6)(120 \times 10^{-3})$$

$$= (39.4)(120) \times (10^6)(10^{-3})$$

$$= 4730 \times 10^3$$

$$= 4.73 \times 10^6$$

49.

$$\frac{84.5 \times 10^{-9}}{3.48 \times 10^6} = \frac{84.5}{3.48} \times \frac{10^{-9}}{10^6}$$

$$= 24.3 \times 10^{-15}$$

50. 46.1×10^3

Chapter 2 Test

1. 2

3. 12

5. 353

7. −2

9. −330

11. −5

13. −2

15. −5

17. −11

19.

$$2\frac{1}{5} - \left(-1\frac{3}{10}\right) + 2\frac{3}{5}$$

$$= \frac{11}{5} - \left(-\frac{13}{10}\right) + \frac{13}{5}$$

$$= \frac{22}{10} + \left(\frac{13}{10}\right) + \frac{26}{10}$$

$$= \frac{61}{10}$$

$$= 6\frac{1}{10}$$

21. 1.82×10^{-5}

23.
$$\left(10^{-3}\right)\left(10^{6}\right) = 10^{-3+6}$$
$$= 10^{3}$$

25.
$$\left(10^{2}\right)^{4} = 10^{2(4)}$$
$$= 10^{8}$$

27.
$$\frac{\left(10^{-4}\right)\left(10^{-8}\right)^{2}}{\left(10^{4}\right)^{-6}} = \frac{\left(10^{-4}\right)\left(10^{-16}\right)}{10^{-24}}$$
$$= \frac{10^{-20}}{10^{-24}}$$
$$= 10^{4}$$

29. 3.62×10^{5}

31. 825×10^{3}

33. 0.00088

35. 11.7×10^{24}

Cumulative Review Chapters 1-2

1.
$$16 \div 8 + 5 \times 2 - 3 + 7 \times 9 = 2 + 10 - 3 + 63$$
$$= 72$$

2. Splitting the figure into 3 vertical rectangles, the area is $2\ \text{cm} \times 7\ \text{cm} + 4\ \text{cm} \times 3\ \text{cm} + 3\ \text{cm} \times 7\ \text{cm} = 47\ \text{cm}^2$.

3. $520\,\Omega + 55\,\Omega + 60\,\Omega + 75\,\Omega + 3040\,\Omega = 3750\,\Omega$

4. No, since the sum of the digits, $2 + 3 + 0 + 6 = 11$, is not divisible by 3, 2306 is not divisible by 6.

5. $630 = 2 \cdot 3 \cdot 3 \cdot 5 \cdot 7$

6. $3\frac{5}{9}$

7.
$$A = \left(\frac{a+b}{2}\right)h$$
$$A = \left(\frac{40\ \text{ft} + 72\ \text{ft}}{2}\right)(80\ \text{ft})$$
$$= \left(\frac{112\ \text{ft}}{2}\right)(80\ \text{ft})$$
$$= (56\ \text{ft})(80\ \text{ft})$$
$$= 4480\ \text{ft}^2$$

8. $\frac{13}{16}$

9.
$$\frac{3}{8} + \frac{1}{4} + \frac{7}{16} = \frac{6}{16} + \frac{4}{16} + \frac{7}{16}$$
$$= \frac{17}{16} = 1\frac{1}{16}$$

10.
$$6\frac{1}{2} - 4\frac{5}{8} = \frac{13}{2} - \frac{37}{8}$$
$$= \frac{52}{8} - \frac{37}{8}$$
$$= \frac{15}{8} = 1\frac{7}{8}$$

11. $\frac{1}{20}$

12. $5 \times 16 + 3 = 83$ oz

13. a. 600

b. 615.3

c. 620

c. 615.288

14.
$$7\frac{2}{5}\% = \frac{37}{5}\%$$
$$= 7.4\%$$
$$= 0.074$$

15.
$$P = BR$$
$$P = (\$14{,}000)(0.285)$$
$$= \$3{,}990$$

16.

$$B = \frac{P}{R}$$

$$B = \frac{212}{0.32}$$

$$= 613$$

17.

$$R = \frac{P}{B}$$

$$R = \frac{58}{615}$$

$$= 0.0943$$

$$= 9.43\%$$

18.

$$R = \frac{P}{B}$$

$$R = \frac{\$6800 - \$6375}{\$6800}$$

$$= 0.0625$$

$$= 6.25\%$$

Joy received a 6.25% markdown.

19. 10

20. −432

21.

$$-\frac{3}{8} + \left(-\frac{1}{4}\right) - 2\frac{5}{16} = -\frac{3}{8} + \left(-\frac{1}{4}\right) - \frac{37}{16}$$

$$= -\frac{6}{16} + \left(-\frac{4}{16}\right) - \frac{37}{16}$$

$$= -\frac{47}{16}$$

$$= -2\frac{15}{16}$$

22. $-\dfrac{25}{64}$

23. 3.1818×10^5

24. 0.00213

25.

$$\frac{\left(10^{-4} \times 10^3\right)^{-2}}{10^6} = \frac{\left(10^{-1}\right)^{-2}}{10^6}$$

$$= \frac{10^2}{10^6}$$

$$= 10^{-4}$$

$$= \frac{1}{10^4}$$

26. 4.5×10^3

27. 270×10^{-6}

28. 0.000000281

29. 16,300,000

30.

$$\left(4.62 \times 10^4\right)\left(1.52 \times 10^6\right)$$

$$= (4.62)(1.52) \times \left(10^4\right)\left(10^6\right)$$

$$= 7.02 \times 10^{10}$$

31.

$$\frac{5.61 \times 10^7}{1.18 \times 10^{10}} = \frac{5.61}{1.18} \times \frac{10^7}{10^{10}}$$

$$= 4.75 \times 10^{-3}$$

32. 3.46×10^{-15}

33. 2.07×10^{-3}

Chapter 3: The Metric System

Section 3.1: Introduction to the Metric System

1. kilo

3. centi

5. milli

7. mega

9. h

11. d

13. c

15. μ

17. 65 mg

19. 82 cm

21. 36μ A

23. 19 hL

25. 18 meters

27. 36 kilograms

29. 24 picoseconds

31. 135 millilitres

33. 45 milliamperes

35. metre

37. ampere

39. litre and cubic metre

Section 3.2: Length

1. 1 metre

3. 1 kilometre

5. 1 centimetre

7. 1000

9. 0.01

11. 0.001

13. 0.001

15. 10

17. 100

19. cm

21. mm

23. m

25. km

27. mm

29. mm

31. km

33. m

35. cm

37. mm; mm

39. A: 52 mm; 5.2 cm

    B: 11 mm; 1.1 cm

    C: 137 mm; 13.7 cm

    D: 95 mm; 9.5 cm

    E: 38 mm; 3.8 cm

    F: 113 mm; 11.3 cm

41. 52 mm; 5.2 cm

43. 79 mm; 7.9 cm

45. 65 mm; 6.5 cm

47. 102 mm; 10.2 cm

49. $675 \text{ m} \times \dfrac{1 \text{ km}}{1000 \text{ m}} = 0.675 \text{ km}$

51. $1540 \text{ mm} \times \dfrac{1 \text{ m}}{1000 \text{ mm}} = 1.54 \text{ m}$

53. $65 \text{ cm} \times \dfrac{1 \text{ m}}{100 \text{ cm}} = 0.65 \text{ m}$

55. $7.3 \text{ m} \times \dfrac{100 \text{ cm}}{1 \text{ m}} = 730 \text{ cm}$

57. $1250 \text{ m} \times \dfrac{1 \text{ km}}{1000 \text{ m}} = 1.25 \text{ km}$

59. $275 \text{ mm} \times \dfrac{1 \text{ cm}}{10 \text{ mm}} = 27.5 \text{ cm}$

61. $125 \text{ mm} \times \dfrac{1 \text{ cm}}{10 \text{ mm}} = 12.5 \text{ cm}$

63. Answers will vary.

Section 3.3: Mass and Weight

1. 1 gram

3. 1 kilogram

5. 1 milligram

7. $1 \text{ g} \times \dfrac{1000 \text{ mg}}{1 \text{ g}} = 1000 \text{ mg}$

9. $1 \text{ cg} \times \dfrac{1 \text{ g}}{100 \text{ cg}} = 0.01 \text{ g}$

11. $1 \text{ metric ton} \times \dfrac{1000 \text{ kg}}{1 \text{ metric ton}} = 1000 \text{ kg}$

13. $1 \text{ mg} \times \dfrac{1000 \text{ } \mu\text{g}}{1 \text{ mg}} = 1000 \text{ } \mu\text{g}$

15. g

17. kg

19. g

21. metric ton

23. mg

25. g

27. mg

29. g

31. kg

33. metric ton

48

35. $875 \text{ g} \times \dfrac{1 \text{ kg}}{1000 \text{ g}} = 0.875 \text{ kg}$

37. $85 \text{ g} \times \dfrac{1000 \text{ mg}}{1 \text{ g}} = 85{,}000 \text{ mg}$

39. $3.6 \text{ kg} \times \dfrac{1000 \text{ g}}{1 \text{ kg}} = 3600 \text{ g}$

41. $270 \text{ mg} \times \dfrac{1 \text{ g}}{1000 \text{ mg}} = 0.270 \text{ g}$

43. $885 \text{ } \mu\text{g} \times \dfrac{1 \text{ mg}}{1000 \text{ } \mu\text{g}} = 0.885 \text{ mg}$

45. $375 \text{ } \mu\text{g} \times \dfrac{1 \text{ mg}}{1000 \text{ } \mu\text{g}} = 0.3755 \text{ mg}$

47. $2.5 \text{ metric ton} \times \dfrac{1000 \text{ kg}}{1 \text{ metric ton}} = 2500 \text{ kg}$

49. $225{,}000 \text{ kg} \times \dfrac{1 \text{ metric ton}}{1000 \text{ kg}} = 225 \text{ metric tons}$

51. Answers will vary.

Section 3.4: Volume and Area

1. 1 litre

3. 1 cubic centimetre

5. 1 square kilometre

7. $1 \text{ L} \times \dfrac{1000 \text{ mL}}{1 \text{ L}} = 1000 \text{ mL}$

9. $1 \text{ m}^3 \times \left(\dfrac{100 \text{ cm}}{1 \text{ m}}\right)^3 = 1{,}000{,}000 \text{ cm}^3$

11. $1 \text{ cm}^2 \times \left(\dfrac{10 \text{ mm}}{1 \text{ cm}}\right)^2 = 100 \text{ mm}^2$

13. $1 \text{ m}^3 \times \left(\dfrac{100 \text{ cm}}{1 \text{ m}}\right)^3 \times \dfrac{1 \text{ L}}{1000 \text{ cm}^3} = 1000 \text{ L}$

15. L

17. m^2

19. cm^2

21. m^3

23. mL

25. ha

27. m^3

29. L

31. m^2

33. ha

35. cm^2

37. cm^2

39. $1500 \text{ mL} \times \dfrac{1 \text{ L}}{1000 \text{ mL}} = 1.5 \text{ L}$

41. $1.5 \text{ m}^3 \times \left(\dfrac{100 \text{ cm}}{1 \text{ m}}\right)^3 = 1{,}500{,}000 \text{ cm}^3$

43. $85 \text{ cm}^3 \times \dfrac{1 \text{ mL}}{1 \text{ cm}^3} = 85 \text{ mL}$

45. $85{,}000 \text{ m}^2 \times \left(\dfrac{1 \text{ km}}{1000 \text{ m}}\right)^2 = 0.085 \text{ km}^2$

47. $85{,}000 \text{ m}^2 \times \dfrac{1 \text{ ha}}{10{,}000 \text{ m}^2} = 8.5 \text{ ha}$

49. $500 \text{ mL} \times \dfrac{1 \text{ L}}{1000 \text{ mL}} \times \dfrac{1 \text{ kg}}{1 \text{ L}} \times \dfrac{1000 \text{ g}}{1 \text{ kg}} = 500 \text{ g}$

51. The field has an area of $75 \text{ m} \times 90 \text{ m} = 6750 \text{ m}^2$ and $6750 \text{ m}^2 \times \dfrac{1 \text{ ha}}{10{,}000 \text{ m}^2} = 0.675 \text{ ha}$.

Section 3.5: Time, Current, and Other Units

1. 1 amp

3. 1 second

5. 1 megavolt

7. 43 kW

9. 17 ps

11. 3.2 MW

13. 450 Ω

15. $1 \text{ kW} \times \dfrac{1000 \text{ W}}{1 \text{ kW}} = 1000 \text{ W}$

17. $1 \text{ ns} \times \dfrac{10^{-9} \text{ s}}{1 \text{ ns}} = 0.000000001 \text{ s}$

19. $1 \text{ A} \times \dfrac{1 \text{ } \mu\text{A}}{10^{-6} \text{ A}} = 1{,}000{,}000 \text{ } \mu\text{A}$

21. $1 \text{ V} \times \dfrac{1 \text{ MV}}{10^6 \text{ V}} = 0.000001 \text{ MV}$

23. $0.35 \text{ A} \times \dfrac{1000 \text{ mA}}{1 \text{ A}} = 350 \text{ mA}$

25. $350 \text{ ms} \times \dfrac{1 \text{ s}}{1000 \text{ ms}} = 0.350 \text{ s}$

27.

$$13{,}950 \text{ s} \times \frac{1 \text{ min}}{60 \text{ s}} \times \frac{1 \text{ hr}}{60 \text{ min}} = 3.875 \text{ hr, so there are 3 whole hours.}$$

$$0.875 \text{ hr} \times \frac{60 \text{ min}}{1 \text{ hr}} = 52.5 \text{ min, so there are 52 whole minutes.}$$

$$0.5 \text{ min} \times \frac{60 \text{ s}}{1 \text{ min}} = 30 \text{ s}$$

$$3600 \text{ s} + 1500 \text{ s} + 16 \text{ s} = 5116 \text{ s}$$

$$\text{So, } 13{,}950 \text{ s} = 3 \text{ hr } 52 \text{ min } 30 \text{ s}$$

29. $175 \ \mu\text{F} \times \dfrac{1 \text{ F}}{10^6 \ \mu\text{F}} \times \dfrac{1000 \text{ mF}}{1 \text{ F}} = 0.175 \text{ mF}$

31. $1500 \text{ kHz} \times \dfrac{1 \text{ MHz}}{1000 \text{ kHz}} = 1.5 \text{ MHz}$

Section 3.6: Temperature

1. b

3. c

5. b

7. c

9. d

11.
$$C = \frac{5}{9}\left(F - 32^\circ\right)$$
$$C = \frac{5}{9}\left(77^\circ - 32^\circ\right)$$
$$= \frac{5}{9}\left(45^\circ\right)$$
$$= 25^\circ \text{ C}$$

17.
$$C = \frac{5}{9}\left(F - 32^\circ\right)$$
$$C = \frac{5}{9}\left(\left(-16^\circ\right) - 32^\circ\right)$$
$$= \frac{5}{9}\left(-48^\circ\right)$$
$$= -27^\circ \text{ C}$$

13.
$$F = \frac{9}{5}C + 32^\circ$$
$$F = \frac{9}{5}\left(325^\circ\right) + 32^\circ$$
$$= 585^\circ + 32^\circ$$
$$= 617^\circ \text{ F}$$

19.
$$F = \frac{9}{5}C + 32^\circ$$
$$F = \frac{9}{5}\left(-78^\circ\right) + 32^\circ$$
$$= -140.4^\circ + 32^\circ$$
$$= -108^\circ \text{ F}$$

15.
$$F = \frac{9}{5}C + 32^\circ$$
$$F = \frac{9}{5}\left(-16^\circ\right) + 32^\circ$$
$$= -28.8^\circ + 32^\circ$$
$$= 3.2^\circ \text{ F}$$

Section 3.7: Metric and U.S. Conversion

1. $8 \text{ lb} \times \dfrac{1 \text{ kg}}{2.20 \text{ lb}} = 3.64 \text{ kg}$

3. $38 \text{ cm} \times \dfrac{1 \text{ in.}}{2.54 \text{ cm}} = 15.0 \text{ in.}$

5. $4 \text{ yd} \times \dfrac{1 \text{ m}}{1.09 \text{ yd}} \times \dfrac{100 \text{ cm}}{1 \text{ m}} = 367 \text{ cm}$

7. $30 \text{ kg} \times \dfrac{2.20 \text{ lb}}{1 \text{ kg}} = 66.0 \text{ lb}$

9. $3.2 \text{ in.} \times \dfrac{2.54 \text{ cm}}{1 \text{ in.}} \times \dfrac{10 \text{ mm}}{1 \text{ cm}} = 81.3 \text{ mm}$

15.

$$8 \text{ gal} \times \dfrac{4 \text{ qt}}{1 \text{ gal}} = 32 \text{ qt}$$

$$32 \text{ qt} \times \dfrac{0.946 \text{ L}}{1 \text{ qt}} = 30.3 \text{ L}$$

17.

$$2 \text{ lb} \times \dfrac{16 \text{ oz}}{1 \text{ lb}} + 6 \text{ oz} = 38 \text{ oz}$$

$$38 \text{ oz} \times \dfrac{1 \text{ g}}{0.0353 \text{ oz}} \times \dfrac{1 \text{ kg}}{1000 \text{ g}} = 1.08 \text{ kg}$$

19. a. $100 \text{ yd} \times \dfrac{3 \text{ ft}}{1 \text{ yd}} = 300 \text{ ft}$

b. $100 \text{ yd} \times \dfrac{1 \text{ m}}{1.09 \text{ yd}} = 91.7 \text{ m}$

21. a. $5 \text{ in.} \times \dfrac{2.54 \text{ cm}}{1 \text{ in.}} = 12.7 \text{ cm}$

b. $5 \text{ in.} \times \dfrac{2.54 \text{ cm}}{1 \text{ in.}} \times \dfrac{10 \text{ mm}}{1 \text{ cm}} = 127 \text{ mm}$

23. $3 \text{ yd}^2 \times \left(\dfrac{1 \text{ m}}{1.09 \text{ yd}}\right)^2 = 2.53 \text{ m}^2$

25. $140 \text{ yd}^2 \times \left(\dfrac{3 \text{ ft}}{1 \text{ yd}}\right)^2 = 1260 \text{ ft}^2$

27. $18 \text{ in.}^2 \times \left(\dfrac{2.54 \text{ cm}}{1 \text{ in.}}\right)^2 = 116 \text{ cm}^2$

29.

$$A = (12.6 \text{ m} \times 8.6 \text{ m}) \times \left(\dfrac{1.09 \text{ yd}}{1 \text{ m}}\right)^2 \times \left(\dfrac{3 \text{ ft}}{1 \text{ yd}}\right)^2$$

$$= 1160 \text{ ft}^2$$

11. $75 \text{ km} \times \dfrac{1 \text{ mi}}{1.61 \text{ km}} = 47.2 \text{ mi}$

13. $0.425 \text{ in.} \times \dfrac{2.54 \text{ cm}}{1 \text{ in.}} \times \dfrac{10 \text{ mm}}{1 \text{ cm}} = 10.8 \text{ mm}$

31. $15 \text{ yd}^3 \times \left(\dfrac{1 \text{ m}}{1.09 \text{ yd}}\right)^3 = 11.6 \text{ m}^3$

33.

$$17 \text{ in}^3 \times \left(\dfrac{2.54 \text{ cm}}{1 \text{ in.}}\right)^3 \times \left(\dfrac{10 \text{ mm}}{1 \text{ cm}}\right)^3$$

$$= 279{,}000 \text{ mm}^3$$

35.

$$84 \text{ ft}^3 \times \left(\dfrac{12 \text{ in.}}{1 \text{ ft}}\right)^3 \times \left(\dfrac{2.54 \text{ cm}}{1 \text{ in.}}\right)^3$$

$$= 2{,}380{,}000 \text{ cm}^3$$

37.

$$\dfrac{\$32{,}400}{80 \text{ ft} \times 180 \text{ ft}} = \$2.25/\text{ft}^2$$

$$\dfrac{\$32{,}400}{80 \text{ frontage ft}} = \$405/\text{frontage ft}$$

39.

$$A = (2400 \text{ ft} \times 625 \text{ ft}) \times \dfrac{1 \text{ acre}}{43{,}560 \text{ ft}^2}$$

$$= 34.4 \text{ acre}$$

41. $34.4 \text{ acre} \times \dfrac{1 \text{ ha}}{2.47 \text{ acres}} = 13.9 \text{ ha}$

43. The house lot is $\dfrac{145 \text{ ft} \times 186 \text{ ft}}{43{,}560 \text{ ft}^2} = 0.619$ of an acre.

45. $\dfrac{1}{8} \text{ section} \times \dfrac{640 \text{ acres}}{1 \text{ section}} = 80 \text{ acres}$

47.

$$\dfrac{10{,}550 \text{ kg}}{1 \text{ ha}} \times \dfrac{1 \text{ ha}}{2.47 \text{ acres}} \times \dfrac{2.20 \text{ lb}}{1 \text{ kg}} = 9397 \text{ lb/acre}$$

$$\dfrac{9397 \text{ lb}}{1 \text{ acre}} \times \dfrac{1 \text{ bu}}{56 \text{ lb}} = 168 \text{ bu/acre}$$

49. $8 \times 30 \text{ in.} \times 440 \text{ yd} \times \dfrac{1 \text{ ft}}{12 \text{ in.}} \times \dfrac{3 \text{ ft}}{1 \text{ yd}} \times \dfrac{1 \text{ acre}}{43,560 \text{ ft}^2} = 0.612 \text{ acres}$

51.

$$15 \text{ kg} \times \dfrac{0.0353 \text{ oz}}{1 \text{ g}} \times \dfrac{1000 \text{ g}}{1 \text{ kg}} = 523 \text{ oz}$$

$$\dfrac{523 \text{ oz}}{6 \text{ oz}} = 88.25$$

He can make 88 6-oz brats.

53. $\dfrac{25.6 \text{ kg}}{1 \text{ cm}^2} \times \left(\dfrac{2.54 \text{ cm}}{1 \text{ in.}} \right)^2 \times \dfrac{2.20 \text{ lb}}{1 \text{ kg}} = 363 \dfrac{\text{lb}}{\text{in}^2}$

55. $\dfrac{65 \text{ mi}}{1 \text{ hr}} \times \dfrac{1 \text{ hr}}{60 \text{ min}} \times \dfrac{1 \text{ min}}{60 \text{ s}} \times \dfrac{1.61 \text{ km}}{1 \text{ mi}} \times \dfrac{1000 \text{ m}}{1 \text{ km}} = 29.1 \dfrac{\text{m}}{\text{s}}$

Chapter 3 Review

1. milli
2. kilo
3. M
4. μ
5. 42 mL
6. 8.3 ns
7. 18 kilometers
8. 350 milliamperes
9. 50 microseconds
10. 1 L
11. 1 MW
12. 1 km^2
13. 1 m^3
14. $650 \text{ m} \times \dfrac{1 \text{ km}}{1000 \text{ m}} = 0.650 \text{ km}$
15. $750 \text{ mL} \times \dfrac{1 \text{ L}}{1000 \text{ mL}} = 0.750 \text{ L}$
16. $6.1 \text{ kg} \times \dfrac{1000 \text{ g}}{1 \text{ kg}} = 6100 \text{ g}$
17. $4.2 \text{ A} \times \dfrac{1 \text{ mA}}{10^{-6} \text{ A}} = 4.2 \times 10^6 \text{ mA}$
18. $18 \text{ MW} \times \dfrac{10^6 \text{ W}}{1 \text{ MW}} = 1.8 \times 10^7 \text{ W}$
19. $25 \ \mu\text{s} \times \dfrac{1000 \text{ ns}}{1 \ \mu\text{s}} = 25,000 \text{ ns}$
20. $250 \text{ cm}^2 \times \left(\dfrac{10 \text{ mm}}{1 \text{ cm}} \right)^2 = 25,000 \text{ mm}^2$
21. $25,000 \text{ m}^2 \times \dfrac{1 \text{ ha}}{10,000 \text{ m}^2} = 2.5 \text{ ha}$

22. $0.6 \text{ m}^3 \times \left(\dfrac{100 \text{ cm}}{1 \text{ m}} \right)^3 = 600,000 \text{ cm}^3$
23. $250 \text{ cm}^3 \times \dfrac{1 \text{ mL}}{1 \text{ cm}^3} = 250 \text{ mL}$
24.

$$C = \dfrac{5}{9}\left(F - 32^\circ \right)$$

$$C = \dfrac{5}{9}\left(72^\circ - 32^\circ \right)$$

$$= \dfrac{5}{9}\left(40^\circ \right)$$

$$= 22.2^\circ \text{ C}$$

25.

$$F = \dfrac{9}{5}C + 32^\circ$$

$$F = \dfrac{9}{5}\left(-25^\circ \right) + 32^\circ$$

$$= -45^\circ + 32^\circ$$

$$= -13^\circ \text{ F}$$

26. 0°C
27. 100°C
28. $180 \text{ lb} \times \dfrac{1 \text{ kg}}{2.20 \text{ lb}} = 81.8 \text{ kg}$
29. $126 \text{ ft} \times \dfrac{1 \text{ yd}}{3 \text{ ft}} \times \dfrac{1 \text{ m}}{1.09 \text{ yd}} = 38.5 \text{ m}$
30. $360 \text{ cm} \times \dfrac{1 \text{ in.}}{2.54 \text{ cm}} = 142 \text{ in.}$
31. $275 \text{ in.}^2 \times \left(\dfrac{2.54 \text{ cm}}{1 \text{ in.}} \right)^2 = 1770 \text{ cm}^2$

32. $18 \text{ yd}^2 \times \left(\dfrac{3 \text{ ft}}{1 \text{ yd}}\right)^2 = 162 \text{ ft}^2$

33. $5 \text{ m}^3 \times \left(\dfrac{1.09 \text{ yd}}{1 \text{ m}}\right)^3 \times \left(\dfrac{3 \text{ ft}}{1 \text{ yd}}\right)^3 = 175 \text{ ft}^3$

34. $15.0 \text{ acres} \times \dfrac{1 \text{ ha}}{2.47 \text{ acres}} = 6.07 \text{ ha}$

35. c

36. a

37. d

38. d

39. b

40. b

41. a

42.

Prefix	Symbol	Power of 10	Sample unit	How many?	How many?
tera	T	10^{12}	m	$10^{12} \text{m} = 1 \text{ Tm}$	$1 \text{ m} = 10^{-12} \text{ Tm}$
giga	G	10^{9}	W	$10^{9} \text{ W} = 1 \text{ GW}$	$1 \text{ W} = 10^{-9} \text{ GW}$
mega	M	10^{6}	Hz	$10^{6} \text{ Hz} = 1 \text{ MHz}$	$1 \text{ Hz} = 10{-}6 \text{ MHz}$
kilo	k	10^{3}	g	$10^{3} \text{ g} = 1 \text{ kg}$	$1 \text{ g} = 10^{-3} \text{ kg}$
hecto	h	10^{2}	V	$10^{2} \text{ V} = 1 \text{ hV}$	$1 \text{ V} = 10^{-2} \text{ hV}$
deka	da	10^{1}	L	$10^{1} \text{ L} = 1 \text{ daL}$	$1 \text{ L} = 10^{-1} \text{ daL}$
deci	d	10^{-1}	g	$10^{-1} \text{ g} = 1 \text{ dg}$	$1 \text{ g} = 10^{1} \text{ dg}$
centi	c	10^{-2}	m	$10^{-2} \text{ m} = 1 \text{ cm}$	$1 \text{ m} = 10^{2} \text{ cm}$
milli	m	10^{-3}	A	$10^{-3} \text{ A} = 1 \text{ mA}$	$1 \text{ A} = 10^{3} \text{ mA}$
micro	m	10^{-6}	W	$10^{-6} \text{ W} = 1 \text{ mW}$	$1 \text{ W} = 10^{6} \text{ mW}$
nano	n	10^{-9}	s	$10^{-9} \text{ s} = 1 \text{ ns}$	$1 \text{ s} = 10^{9} \text{ ns}$
pico	p	10^{-12}	s	$10^{-12} \text{ s} = 1 \text{ ps}$	$1 \text{ s} = 10^{12} \text{ ps}$

Chapter 3 Test

1. kilo

3. 1 g

5. 30 hg

7. $4.25 \text{ km} \times \dfrac{1000 \text{ m}}{1 \text{ km}} = 4250 \text{ m}$

9. $72 \text{ m} \times \dfrac{1000 \text{ mm}}{1 \text{ m}} = 72{,}000 \text{ mm}$

11. $12 \text{ dg} \times \dfrac{100 \text{ mg}}{1 \text{ dg}} = 1200 \text{ mg}$

13. $7.236 \text{ metric ton} \times \dfrac{1000 \text{ kg}}{1 \text{ metric ton}} = 7236 \text{ kg}$

15. $72 \text{ hg} \times \dfrac{10^{5} \text{ mg}}{1 \text{ hg}} = 7{,}200{,}000 \text{ mg}$

17. $175 \text{ L} \times \dfrac{1000 \text{ cm}^3}{1 \text{ L}} \times \left(\dfrac{1 \text{ m}}{100 \text{ cm}}\right)^3 = 0.175 \text{ m}^3$

19. $400 \text{ ha} \times \dfrac{10{,}000 \text{ m}^2}{1 \text{ ha}} \times \left(\dfrac{1 \text{ km}}{1000 \text{ m}}\right)^2 = 4 \text{ km}^2$

21. second

23. $280 \text{ W} \times \dfrac{1 \text{ kW}}{1000 \text{ W}} = 0.28 \text{ kW}$

25. $720 \text{ ps} \times \dfrac{1 \text{ ns}}{1000 \text{ ps}} = 0.72 \text{ ns}$

27. 0° C

29.

$$C = \dfrac{5}{9}\left(F - 32^\circ\right)$$
$$C = \dfrac{5}{9}\left(28^\circ - 32^\circ\right)$$
$$= \dfrac{5}{9}\left(-4^\circ\right)$$
$$= -2.22^\circ \text{ C}$$

31. $100 \text{ km} \times \dfrac{1 \text{ mi}}{1.61 \text{ km}} = 62.1 \text{ mi}$

33. $1.8 \text{ ft}^3 \times \left(\dfrac{12 \text{ in.}}{1 \text{ ft}}\right)^3 = 3110 \text{ in}^3$

35. $80.2 \text{ kg} \times \dfrac{2.20 \text{ lb}}{1 \text{ kg}} = 176 \text{ lb}$

Chapter 4: Measurement

Section 4.1: Approximate Numbers and Accuracy

1.	3	**13.**	3	**25.**	2
3.	4	**15.**	3	**27.**	6
5.	4	**17.**	3	**29.**	4
7.	4	**19.**	4	**31.**	6
9.	3	**21.**	4	**33.**	4
11.	4	**23.**	5	**35.**	2

Section 4.2: Precision and Greatest Possible Error

1. a. 0.01 A
   b. 0.005 A

3. a. 0.01 cm
   b. 0.005 cm

5. a. 1 km
   b. 0.5 km

7. a. 0.01 mi
   b. 0.005 mi

9. a. 0.001 A
   b. 0.0005 A

11. a. 0.0001 W
   b. 0.00005 W

13. a. 10 Ω mi
   b. 5 Ω

15. a. 1000 L
   b. 500 L

17. a. 0.1 cm
   b. 0.05 cm

19. a. 10 V
   b. 5 V

21. a. 0.001 m
   b. 0.0005 m

23. a. $\frac{1}{3}$ yd
   b. $\frac{1}{6}$ yd

25. a. $\frac{1}{32}$ in.
   b. $\frac{1}{64}$ in.

27. a. $\frac{1}{16}$ mi
   b. $\frac{1}{32}$ mi

29. a. $\frac{1}{9}$ in^2
   b. $\frac{1}{18}$ in^2

Section 4.3A: The Vernier Caliper

1.	27.20 mm	**13.**	34.60 mm
3.	63.55 mm	**15.**	68.45 mm
5.	8.00 mm	**17.**	5.90 mm
7.	115.90 mm	**19.**	43.55 mm
9.	71.45 mm	**21.**	76.10 mm
11.	10.25 mm	**23.**	12.30 mm

Section 4.3B: The Vernier Caliper

1.	1.362 in.	**5.**	0.234 in.
3.	2.695 in.	**7.**	1.715 in.

9. 2.997 in.

11. 0.483 in.

13. 1.071 in.

15. 2.502 in.

17. 0.316 in.

19. 4.563 in.

21. 2.813 in.

23. 0.402 in.

Section 4.4A: The Micrometer Caliper

1. 4.25 mm

3. 3.90 mm

5. 1.75 mm

7. 7.77 mm

9. 5.81 mm

11. 10.28 mm

13. 7.17 mm

15. 8.75 mm

17. 6.23 mm

19. 5.42 mm

Section 4.4B: The Micrometer Caliper

1. 0.238 in.

3. 0.314 in.

5. 0.147 in.

7. 0.820 in.

9. 0.502 in.

11. 0.200 in.

13. 0.321 in.

15. 0.170 in.

17. 0.658 in.

19. 0.245 in.

Section 4.5: Addition and Subtraction of Measurements

1. a. 14.7 in.

   b. 0.017 in.

3. a. 16.01 mm

   b. 0.737 mm

5. a. 0.0350 A

   b. 0.00050 A

7. a. All have same accuracy.

   b. 0.391 cm

9. a. 205,000 Ω

   b. 45,000 Ω

11. a. 0.04 in.

    b. 15.5 in.

13. a. 0.48 cm

    b. 43.4 cm

15. a. 0.00008 A

    b. 0.91 A

17. a. 0.6 m

    b. All have same precision.

19. a. 500,000 Ω

    b. 500,000 Ω

21. 18.1 m

23. 94.8 cm

25. 97,000 W

27. 840,000 V

29. 19 V

31. 459 mm or 45.9 cm

33. 126.4 cm

35. 8600 mi

37. 35 mm or 3.5 cm

39. 65.4 g

41. 0.330 in.

43. 26.0 mm

45. 12.7 ft

47. 67 lb

49. 1.3 gal

51. 124 gal

53. 10.6666

55. 17.4 lb − 3.6 lb − 5 lb = 9 lb

Section 4.6: Multiplication and Division of Measurements

1. 4400 m^2

3. $1,230,000 \text{ cm}^2$

5. 901 m^2

7. $0.13 \text{ A}\Omega$

9. 7360 cm^3

11. $4.7 \times 10^9 \text{ m}^3$

13. $35 \text{ A}^2\Omega$

15. 2500 in^2

17. 40 m

19. 340 V/A

21. 2.1 km/s

23. $3\bar{0}0 \text{ V}^2 / \Omega$

25. 4.0 g/cm^3

27.
$$V = lwh$$
$$V = (16.4 \text{ ft})(8.6 \text{ ft})(6.4 \text{ ft})$$
$$= 9\bar{0}0 \text{ ft}^3$$

29.
$$s = 4.90t^2$$
$$s = 4.90(2.4)^2$$
$$= 28 \text{ m}$$

31.
$$p = \frac{d^2 n}{2.50}$$
$$p = \frac{(3.00)^2 (8)}{2.50}$$
$$= 28.8 \text{ hp}$$

33.
$$V = \pi r^2 h$$
$$V = \pi (6.2 \text{ m})^2 (8.5 \text{ m})$$
$$= 1\bar{0}00 \text{ m}^3$$

35.
$$\frac{(24 \text{ ft})(14 \text{ ft})(8.0 \text{ ft})}{10 \text{ min}} = 270 \text{ ft}^3/\text{min}$$

37.
$$V = lwh$$
$$V = (13.5 \text{ in.})(17.25 \text{ in.})(2\bar{0} \text{ in.})$$
$$= 4700 \text{ in.}^3$$

39. $\dfrac{32.65 \text{ gal}}{3.4 \text{ hr}} = 9.6 \text{ gal/hr}$

41.
$$A = lw$$
$$A = (55.3 \text{ in.})(28.25 \text{ in.})$$
$$= 1560 \text{ in.}^2$$

43.
$$V = lwh$$
$$V = (3 \text{ ft})(4.2 \text{ ft})(1.5 \text{ ft})$$
$$= 19 \text{ ft}^3$$

45. $\dfrac{52.6 \text{ ft}}{6 \text{ ft}} = 9$

47. $396 \text{ ppm} + 88(2.1 \text{ ppm}) = 580 \text{ ppm}$

49. $9.4 \text{ km}^3 \times \left(\dfrac{1 \text{ mi}}{1.61 \text{ km}}\right)^3 = 2.3 \text{ mi}^3$

Section 4.7: Relative Error and Percent of Error

1.
100 lb; 50 lb;
$$\frac{50 \text{ lb}}{1400 \text{ lb}} = 0.0357; 3.57\%$$

3.
1 rpm; 0.5 rpm;
$$\frac{0.5 \text{ rpm}}{875 \text{ rpm}} = 0.000571; 0.06\%$$

5.
0.001 g; 0.0005 g;
$$\frac{0.0005 \text{ g}}{0.085 \text{ g}} = 0.00588; 0.59\%$$

7.
1 g; 0.5 g;
$$\frac{0.5 \text{ g}}{2 \text{ g}} = 0.25; 25\%$$

9.

0.01 g; 0.005 g;

$$\frac{0.005 \text{ g}}{2.22 \text{ g}} = 0.00225; 0.23\%$$

11.

0.01 kg; 0.005 kg;

$$\frac{0.005 \text{ kg}}{1.00 \text{ kg}} = 0.005; 0.5\%$$

13.

0.001 A; 0.0005 A;

$$\frac{0.0005 \text{ A}}{0.041 \text{ A}} = 0.0122; 1.22\%$$

15.

$\dfrac{1}{8}$ in.; $\dfrac{1}{16}$ in.;

$$\frac{\frac{1}{16} \text{ in.}}{11\frac{7}{8} \text{ in.}} = 0.00526; 0.53\%$$

17.

1 in.; 0.5 in.;

$$\frac{0.5 \text{ in.}}{152 \text{ in.}} = 0.00329; 0.33\%$$

19. $\dfrac{0.05 \text{ cm}}{13.5 \text{ cm}} = 0.37\%$ $\dfrac{\frac{1}{8} \text{ in.}}{8\frac{3}{4} \text{ in.}} = 1.43\%$

So, 13.5 cm is better.

21. $\dfrac{0.5 \text{ mg}}{16 \text{ mg}} = 3.13\%$ $\dfrac{0.05 \text{ g}}{19.7 \text{ g}} = 2.54\%$ $\dfrac{\frac{1}{32} \text{ oz}}{12\frac{3}{16} \text{ oz}} = 2.56\%$; So, 19.7 g is better.

	Upper Limit	Lower Limit	Tolerance
23.	$3\frac{5}{8}$ in.	$3\frac{3}{8}$ in.	$\frac{1}{4}$ in.
25.	$6\frac{21}{32}$ in.	$6\frac{19}{32}$ in.	$\frac{1}{16}$ in.
27.	$3\frac{29}{64}$ in.	$3\frac{27}{64}$ in.	$\frac{1}{32}$ in.
29.	$3\frac{25}{128}$ in.	$3\frac{23}{128}$ in.	$\frac{1}{64}$ in.
31.	1.24 cm	1.14 cm	0.10 cm
33.	0.0185 A	0.0175 A	0.0010 A
35.	26,000 V	22,000 V	4000 V
37.	10.36 kg	10.26 kg	0.10 km

39.

$$10\% \text{ of } \$48,250 = (0.10)(\$48,250) = \$4825$$

Maximum acceptable bid $= \$48,250 + \$4825 = \$53,075$

41.

$$8\% \text{ of } \$1,450,945 = (0.08)(\$1,450,945) = \$116,075.60$$

Maximum acceptable bid $= \$1,450,945 + \$116,075.6 = \$1,567,020.60$

Section 4.8: Color Code of Electrical Resistors

1. $360\,\Omega$; $\pm 10\%$

3. $830,000\,\Omega$; $\pm 20\%$

5. $1,400,000\,\Omega$; $\pm 20\%$

7. $70\,\Omega$; $\pm 5\%$

9. $500,000\,\Omega$; $\pm 20\%$

11. $10,000,000\,\Omega$; $\pm 20\%$

13. yellow, gray, red

25. a. $(0.10)(360\,\Omega) = 36\,\Omega$

 b. $360\,\Omega + 36\,\Omega = 396\,\Omega$

 c. $360\,\Omega - 36\,\Omega = 324\,\Omega$

 d. $2 \times 36\,\Omega = 72\,\Omega$

27. a. $(0.20)(830,000\,\Omega) = 166,000\,\Omega$

 b. $830,000\,\Omega + 166,000\,\Omega = 996,000\,\Omega$

 c. $830,000\,\Omega - 166,000\,\Omega = 664,000\,\Omega$

 d. $2 \times 166,000\,\Omega = 332,000\,\Omega$

15. violet, red, orange

17. blue, green, yellow

19. red, green, silver

21. yellow, green, green

23. violet, blue, gold

29. a. $(0.05)(70\,\Omega) = 3.5\,\Omega$

 b. $70\,\Omega + 3.5\,\Omega = 73.5\,\Omega$

 c. $70\,\Omega - 3.5\,\Omega = 66.5\,\Omega$

 d. $2 \times 3.5\,\Omega = 7\,\Omega$

Section 4.9: Reading Scales

1. $+1.16$ mm

3. -0.67 mm

5. -3.40 mm

7. $+1.74$ mm

9. -4.08 mm

11. -0.056 in.

13. $+0.231$ in.

15. -0.188 in.

17. $+0.437$ in.

19. 6 V

21. 6.4 V

23. 0.4 V

25. 1.4 V

27. 40 V

29. 230 V

31. $7\,\Omega$

33. $12\,\Omega$

35. $11\,\Omega$

37. $35\,\Omega$

39. $85\,\Omega$

41. $300\,\Omega$

Chapter 4 Review

1. 3

2. 2

3. 3

4. 3

5. 2

6. 3

7. 4

8. 5

9. a. 0.01 m

 b. 0.005 m

10. a. 0.1 mi

 b. 0.05 mi

11. a. 100 L

 b. 50 L

12. a. 100 V

 b. 50 V

13. a. 0.01 cm

 b. 0.005 cm

14. a. 10,000 V

 b. 5000 V

15. a. $\dfrac{1}{8}$ in.

 b. $\dfrac{1}{16}$ in.

16. a. $\dfrac{1}{16}$ mi

    b. $\dfrac{1}{32}$ mi

17. 42.35 mm
18. 1.673 in.
19. 11.84 mm
20. 0.438 in.
21. a. 36,500 V

    b. 9.6 V

22. a. 0.0005 A

    b. 0.425 A

23. 720,000 W

24. $4\bar{0}0$ m

25. 400,000 V

32.

    Lower limit: $2000\ \Omega - (0.10)(2000\ \Omega) = 1800\ \Omega$

    Upper limit: $2000\ \Omega + (0.10)(2000\ \Omega) = 2200\ \Omega$

33. $120,000\ \Omega$; $\pm 20\%$

34. $0.85\ \Omega$; $\pm 5\%$

35. -0.563 in.

26. 1900 cm^3
27. 5.88 m^2
28. 1.4 N/m^2
29. 130 V^2/Ω
30.

    Relative error: $\dfrac{\dfrac{1}{32}\ \text{in.}}{5\dfrac{7}{16}\ \text{in.}} = 0.00575$

    Percent error: 0.57%

31.

    Relative error: $\dfrac{0.005\ \text{cm}}{15.60\ \text{cm}} = 0.00032$

    Percent error: 0.03%

36. 8.4 V
37. $2.6\ \Omega$

Chapter 4 Test

1. 4

5. a. $100\ \Omega$

    b. $50\ \Omega$

11. a. 17,060 m

    b. 0.067 m

    c. 0.067 m

    d. 17,060 m

3. 2

7. 63.00 mm

9. 8.32 mm

13. 29 m^3

15. 52.0 g

17. 350 m

19. 1.7 V

Cumulative Review Chapters 1-4

1.

$$1\frac{3}{8} - \frac{1}{2} \times \frac{3}{4} + 1\frac{5}{8} \div \frac{1}{16} = \frac{11}{8} - \frac{1}{2} \times \frac{3}{4} + \frac{13}{8} \div \frac{1}{16}$$

$$= \frac{11}{8} - \frac{1}{2} \times \frac{3}{4} + \frac{13}{8} \times \frac{16}{1}$$

$$= \frac{11}{8} - \frac{3}{8} + 26$$

$$= \frac{11}{8} - \frac{3}{8} + \frac{208}{8}$$

$$= \frac{216}{8}$$

$$= 27$$

2. a. 32,520

b. 32,518.61

3.

$$B = \frac{P}{R}$$

$$B = \frac{18.84}{0.314}$$

$$= 60$$

4.

$$(-4)(5) + (-6)(-4) - 7(-4) \div 2(7)$$

$$= (-20) + 24 + 28 \div 2(7)$$

$$= (-20) + 24 + 14(7)$$

$$= 102$$

5.

$$\frac{\left(10^3 \cdot 10^{-2}\right)^3}{10^3 \cdot 10^5} = \frac{\left(10^1\right)^3}{10^8}$$

$$= \frac{10^3}{10^8}$$

$$= 10^{-5}$$

$$= \frac{1}{10^5}$$

6. 8.70×10^5

7. m

8. 25 kg

9. 250 microseconds

10. 1 mega amp

11. $120 \text{ km} \times \dfrac{1000 \text{ m}}{1 \text{ km}} = 120{,}000 \text{ m}$

12. $250 \text{ cm} \times \dfrac{1 \text{ m}}{100 \text{ cm}} = 2.5 \text{ cm}$

13. $50 \text{ g} \times \dfrac{1 \text{ kg}}{1000 \text{ g}} = 0.05 \text{ kg}$

14. $4060 \text{ kg} \times \dfrac{1 \text{ metric ton}}{1000 \text{ kg}} = 4.06 \text{ metric tons}$

15.

$$F = \frac{9}{5}C + 32°$$

$$F = \frac{9}{5}\left(86°\right) + 32°$$

$$= 154.8° + 32°$$

$$= 187° \text{ F}$$

16.

$$C = \frac{5}{9}\left(F - 32°\right)$$

$$C = \frac{5}{9}\left(50° - 32°\right)$$

$$= \frac{5}{9}\left(18°\right)$$

$$= 10° \text{ C}$$

17. $163 \text{ in}^2 \times \left(\dfrac{2.54 \text{ cm}}{1 \text{ in.}}\right)^2 = 1050 \text{ cm}^2$

18. $120 \text{ m} \times \dfrac{1 \text{ km}}{1000 \text{ m}} = 0.12 \text{ km}$

19. $10 \text{ L} \times \dfrac{1000 \text{ mL}}{1 \text{ L}} = 10{,}000 \text{ L}$

20. 2

21. 4

22. a. 0.01 cm

b. 0.005 cm

23. a. 0.1 lb

b. 0.05 lb

24. 77.75 mm

25. 3.061 in.

26. 7.53 mm

27. 0.537 in.

28. 35,000 km

29. 46.0 L

30. $42{,}000 \text{ cm}^2$

31. 33.7 ft

32. a. 0.001 cm

   b. 0.0005 cm

   c. $\dfrac{0.0005 \text{ cm}}{2.135 \text{ cm}} = 0.000234$

   d. 0.02%

33. 165 V

34. 110 Ω

Chapter 5: Polynomials: An Introduction to Algebra

Section 5.1: Fundamental Operations

1.

$$3(-5)^2 - 4(-2)$$
$$= 3(25) + 8$$
$$= 75 + 8$$
$$= 83$$

3.

$$4(-3) \div (-6) - (-18) \div 3$$
$$= -12 \div (-6) - (-6)$$
$$= -12 \div (-6) - (-6)$$
$$= 2 + 6$$
$$= 8$$

5.

$$(-72) \div (-3) \div (-6) \div (-2) - (-4)(-2)(-5)$$
$$= 24 \div (-6) \div (-2) - (8)(-5)$$
$$= (-4) \div (-2) - (-40)$$
$$= 2 + 40$$
$$= 42$$

7.

$$\left[(-2)(-3) + (-24) \div (-2)\right] \div \left[-10 + 7(-1)^2\right]$$
$$= \left[6 + 12\right] \div \left[-10 + 7(1)\right]$$
$$= 18 \div \left[-10 + 7\right]$$
$$= 18 \div \left[-3\right]$$
$$= -6$$

9.

$$\left[(-2)(-8)^2 \div (-2)^3\right] - \left[-4 + (-2)^4\right]^2$$
$$= \left[(-2)(64) \div (-8)\right] - \left[-4 + 16\right]^2$$
$$= \left[(-128) \div (-8)\right] - \left[12\right]^2$$
$$= 16 - 144$$
$$= -128$$

11.

$$2x - y = 2(2) - (3)$$
$$= 4 - 3$$
$$= 1$$

13.

$$x^2 - y^2 = (2)^2 - (3)^2$$
$$= 4 - 9$$
$$= -5$$

15.

$$\frac{3x + y}{3 + y} = \frac{3(2) + (3)}{3 + (3)}$$
$$= \frac{6 + 3}{3 + 3}$$
$$= \frac{9}{6}$$
$$= \frac{3}{2}$$

17.

$$xy^2 - x = (-1)(5)^2 - (-1)$$
$$= (-1)(25) + 1$$
$$= -25 + 1$$
$$= -24$$

19.

$$\frac{2y}{x} - \frac{2x}{y} = \frac{2(5)}{(-1)} - \frac{2(-1)}{(5)}$$
$$= \frac{10}{-1} - \frac{-2}{5}$$
$$= -\frac{50}{5} + \frac{2}{5}$$
$$= -\frac{48}{5}$$
$$= -5\frac{3}{5}$$

21.

$$3 + 4(x + y) = 3 - 4\left[(-1) + (5)\right]$$
$$= 3 - 4\left[(-1) + (5)\right]$$
$$= 3 - 4\left[4\right]$$
$$= 3 - 16$$
$$= -13$$

62

23.

$$\frac{1}{x} - \frac{1}{y} + \frac{2}{xy} = \frac{1}{(-1)} - \frac{1}{(5)} + \frac{2}{(-1)(5)}$$

$$= -\frac{1}{1} - \frac{1}{5} + \frac{2}{-5}$$

$$= -\frac{5}{5} - \frac{1}{5} - \frac{2}{5}$$

$$= -\frac{8}{5}$$

$$= -1\frac{3}{5}$$

25.

$$\frac{y-4x}{3x-6xy} = \frac{(5)-4(-1)}{3(-1)-6(-1)(5)}$$

$$= \frac{5+4}{(-3)+30}$$

$$= \frac{9}{27}$$

$$= \frac{1}{3}$$

27.

$$\left(2xy^2z\right)^2 = \left[2(-3)(4)^2(6)\right]^2$$

$$= \left[2(-3)(16)(6)\right]^2$$

$$= \left[-576\right]^2$$

$$= 331,776$$

29.

$$\left(y^2 - 2x^2\right)z^2 = \left[(4)^2 - 2(-3)^2\right](6)^2$$

$$= \left[16 - 2(9)\right](36)$$

$$= \left[16 - 18\right](36)$$

$$= \left[-2\right](36)$$

$$= -72$$

31.

$$\frac{(7-x)^2}{z-y} = \frac{\left(7-(-3)\right)^2}{(6)-(4)}$$

$$= \frac{(7+3)^2}{2}$$

$$= \frac{(10)^2}{2}$$

$$= \frac{100}{2}$$

$$= 50$$

33.

$$(2x+6)(3y-4) = \left[2(-1)+6\right]\left[3(2)-4\right]$$

$$= \left[(-2)+6\right]\left[6-4\right]$$

$$= \left[4\right]\left[2\right]$$

$$= 8$$

35.

$$(3x+5)(2y-1)(5z+2)$$

$$= \left[3(-1)+5\right]\left[2(2)-1\right]\left[5(-3)+2\right]$$

$$= \left[(-3)+5\right]\left[4-1\right]\left[(-15)+2\right]$$

$$= \left[2\right]\left[3\right]\left[-13\right]$$

$$= -78$$

37.

$$(x-xy)^2(z-2x)$$

$$= \left[(-1)-(-1)(2)\right]^2\left[(-3)-2(-1)\right]$$

$$= \left[(-1)+2\right]^2\left[(-3)+2\right]$$

$$= \left[1\right]^2\left[-1\right]$$

$$= \left[1\right]\left[-1\right]$$

$$= -1$$

39.

$$\left(x^2+y^2\right)^2 = \left[(-1)^2+(2)^2\right]^2$$

$$= \left[1+4\right]^2$$

$$= \left[5\right]^2$$

$$= 25$$

41.

$$\frac{x^2 + (z - y)^2}{4x^2 + z^2} = \frac{(-1)^2 + \left[(-3) - (2)\right]^2}{4(-1)^2 + (-3)^2}$$

$$= \frac{1 + \left[-5\right]^2}{4(1) + 9}$$

$$= \frac{1 + 25}{4 + 9}$$

$$= \frac{26}{13}$$

$$= 2$$

Section 5.2: Simplifying Algebraic Expressions

1. $a + b + c$

3. $a + b + c$

5. $a - b - c$

7. $x + y - z + 3$

9. $x - y - z - 3$

21.

$$b + b = (1 + 1)b$$
$$= 2b$$

23.

$$x^2 + 2x^2 + 3x + 7x = (1 + 2)x^2 + (3 + 7)x$$
$$= 3x^2 + 10x$$

25.

$$5m - 2m = (5 - 2)m$$
$$= 3m$$

27.

$$3a + 5b - 2a + 7b = (3 - 2)a + (5 + 7)b$$
$$= a + 12b$$

29.

$$6a^2 + a + 1 - 2a = 6a^2 + a - 2a + 1$$
$$= 6a^2 + (1 - 2)a + 1$$
$$6a^2 - a + 1$$

31.

$$2x^2 + 16x + x^2 - 13x$$
$$= 2x^2 + x^2 + 16x - 13x$$
$$= (2 + 1)x^2 + (16 - 13)x$$
$$= 3x^2 + 3x$$

11. $2x + 4 + 3y + 4r$

13. $3x - 5y + 6z - 2w + 11$

15. $-5x - 3y - 6z + 3w + 3$

17. $2x + 3y - z - w + 3r - 2s - 10$

19. $-2x + 3y - z - 4w - 4r + s$

33.

$$1.3x + 5.6x - 13.2x + 4.5x$$
$$= (1.3 + 5.6 - 13.2 + 4.5)x$$
$$= -1.8x$$

35.

$$\frac{5}{9}x + \frac{1}{4}y + \frac{1}{3}x - \frac{3}{8}y$$

$$= \frac{5}{9}x + \frac{1}{3}x + \frac{1}{4}y - \frac{3}{8}y$$

$$= \left(\frac{5}{9} + \frac{1}{3}\right)x + \left(\frac{1}{4} - \frac{3}{8}\right)y$$

$$= \left(\frac{5}{9} + \frac{3}{9}\right)x + \left(\frac{2}{8} - \frac{3}{8}\right)y$$

$$= \frac{8}{9}x - \frac{1}{8}y$$

37.

$$4x^2y - 2xy - y^2 - 3x^2 - 2x^2y + 3y^2$$
$$= 4x^2y - 2x^2y - 2xy - y^2 + 3y^2 - 3x^2$$
$$= (4 - 2)x^2y - 2xy + (-1 + 3)y^2 - 3x^2$$
$$= 2x^2y - 2xy + 2y^2 - 3x^2$$

39.

$$2x^3 + 4x^2 y - 4y^3 + 3x^3 - x^2 y + y - y^3 = 2x^3 + 3x^3 + 4x^2 y - x^2 y - 4y^3 - y^3 + y$$
$$= (2+3)x^3 + (4-1)x^2 y + (-4-1)y^3 + y$$
$$= 5x^3 + 3x^2 y - 5y^3 + y$$

41.

$$y - (y - 1) = y - y + 1$$
$$= 1$$

43.

$$4x + (4 - x) = 4x + 4 - x$$
$$= 3x + 4$$

45.

$$10 - (5 + x) = 10 - 5 - x$$
$$= 5 - x$$

47.

$$2y - (7 - y) = 2y - 7 + y$$
$$= 3y - 7$$

49.

$$(5y + 7) - (y + 2) = 5y + 7 - y - 2$$
$$= 4y + 5$$

51.

$$(4 - 3x) + (3x + 1) = 4 - 3x + 3x + 1$$
$$= 5$$

53.

$$-5y + 9 - (-5y + 3) = -5y + 9 + 5y - 3$$
$$= 6$$

55.

$$0.2x - (0.2x - 28) = 0.2x - 0.2x + 28$$
$$= 28$$

57.

$$\left(\frac{1}{2}x - \frac{2}{3}\right) - \left(2 - \frac{3}{4}x\right)$$
$$= \frac{2}{4}x - \frac{2}{3} - \frac{6}{3} + \frac{3}{4}x$$
$$= \frac{5}{4}x - \frac{8}{3}$$

59.

$$4(3x + 9y) = (4)3x + (4)9y$$
$$= 12x + 36y$$

61.

$$-12(3x^2 - 4y^2) = (-12)3x^2 - (-12)4y^2$$
$$= -36x^2 + 48y^2$$

63.

$$(5x + 13) - 3(x - 2) = 5x + 13 - 3x + 6$$
$$= 2x + 19$$

65.

$$-9y - 0.5(8 - y) = -9y - 4 + 0.5y$$
$$= -8.5y - 4$$

67.

$$2y - 2(y + 21) = 2y - 2y - 42$$
$$= -42$$

69.

$$6n - (2n - 8) = 6n - 2n + 8$$
$$= 4n + 8$$

71.

$$0.8x - (-x + 7) = 0.8x + x - 7$$
$$= 1.8x - 7$$

73.

$$4(2 - 3n) - 2(5 - 3n) = 8 - 12n + 6n - 10$$
$$= -6n - 2$$

75.

$$\frac{2}{3}(6x - 9) - \frac{3}{4}(12x - 16)$$
$$= 4x - 6 - 9x + 12$$
$$= -5x + 6$$

77.

$$0.45(x + 3) - 0.75(2x + 13)$$
$$= 0.45x + 1.35 - 1.5x - 9.75$$
$$= -1.05x - 8.4$$

Section 5.3: Addition and Subtraction of Polynomials

1. binomial

3. binomial

5. monomial

7. trinomial

9. binomial

11. $x^2 - x + 1$; Degree 2.

13. $7x^2 + 4x - 1$; Degree 2.

15. $5x^3 - 4x^2 - 2$; Degree 3.

17. $4y^3 - 6y^2 - 3y + 7$; Degree 3.

19. $-7x^5 - 4x^4 + x^3 + 2x^2 + 5x - 3$; Degree 5.

21. $7a^2 - 10a + 1$

23. $9x^2 - 5x$

25. $9a^3 + 4a^2 + 5a - 5$

27. $4x + 4$

29. $5x^2 + 18x - 22$

31. $5x^3 + 13x^2 - 8x + 7$

33. $8y^2 - 5y + 6$

35. $4a^3 - 3a^2 - 5a$

37.
$$\left(3x^2 + 4x + 7\right) - \left(x^2 - 2x + 5\right)$$
$$= \left(3x^2 + 4x + 7\right) + \left(-x^2 + 2x - 5\right)$$
$$= 2x^2 + 6x + 2$$

39.
$$\left(3x^2 - 5x + 4\right) - \left(6x^2 - 7x + 2\right)$$
$$= \left(3x^2 - 5x + 4\right) + \left(-6x^2 + 7x - 2\right)$$
$$= -3x^2 + 2x + 2$$

41.
$$\left(3a - 4b\right) - \left(2a - 7b\right)$$
$$= \left(3a - 4b\right) + \left(-2a + 7b\right)$$
$$= a + 3b$$

43.
$$\left(7a - 4b\right) - \left(3x - 4y\right)$$
$$= \left(7a - 4b\right) + \left(-3x + 4y\right)$$
$$= 7a - 4b - 3x + 4y$$

45.
$$\left(12x^2 - 3x - 2\right) - \left(11x^2 - 7\right)$$
$$= \left(12x^2 - 3x - 2\right) + \left(-11x^2 + 7\right)$$
$$= x^2 - 3x + 5$$

47.
$$\left(20w^2 - 17w - 6\right) - \left(13w^2 + 7w\right)$$
$$= \left(20w^2 - 17w - 6\right) + \left(-13w^2 - 7w\right)$$
$$= 7w^2 - 24w - 6$$

49.
$$\left(2x^2 - 5x - 2\right) - \left(x^2 - x + 8\right)$$
$$= \left(2x^2 - 5x - 2\right) + \left(-x^2 + x - 8\right)$$
$$= x^2 - 4x - 10$$

51.
$$\left(8x^2 - 2x + 5\right) - \left(4x^2 + 2x - 7\right)$$
$$= \left(8x^2 - 2x + 5\right) + \left(-4x^2 - 2x + 7\right)$$
$$= 4x^2 - 4x + 12$$

53.
$$\left(3x^2 + 2x - 4\right) - \left(9x^2 + 6\right)$$
$$= \left(3x^2 + 2x - 4\right) + \left(-9x^2 - 6\right)$$
$$= -6x^2 + 2x - 10$$

55. $x^3 + 3x - 3$

57. $8x^5 - 18x^4 + 5x^2 + 1$

Section 5.4: Multiplication of Monomials

1.
$$(3a)(-5) = (3)(-5)(a)$$
$$= -15a$$

3.
$$\left(4a^2\right)(7a) = (4)(7)\left(a^2\right)(a)$$
$$= 28a^3$$

5.
$$\left(-9m^2\right)\left(-6m^2\right) = (-9)(-6)\left(m^2\right)\left(m^2\right)$$
$$= 54m^4$$

7.
$$\left(8a^6\right)\left(4a^2\right) = (8)(4)\left(a^6\right)\left(a^2\right)$$
$$= 32a^8$$

9.
$$(13p)(-2pq) = (13)(-2)(p)(p)(q)$$
$$= -26p^2q$$

11.
$$\left(6n\right)\left(5n^2m\right) = (6)(5)(m)(n)\left(n^2\right)$$
$$= 30mn^3$$

13.
$$(-42a)\left(-\frac{1}{2}a^3b\right)$$
$$= (-42)\left(-\frac{1}{2}\right)(a)\left(a^3\right)(b)$$
$$= 21a^4b$$

15.
$$\left(\frac{2}{3}x^2y^2\right)\left(\frac{9}{16}xy^2\right)$$
$$= \left(\frac{2}{3}\right)\left(\frac{9}{16}\right)\left(x^2\right)(x)\left(y^2\right)\left(y^2\right)$$
$$= \frac{3}{8}x^3y^4$$

17.
$$\left(8a^2bc\right)\left(3ab^3c^2\right)$$
$$= (8)(3)\left(a^2\right)(a)(b)\left(b^3\right)(c)\left(c^2\right)$$
$$= 24a^3b^4c^3$$

19.
$$\left(\frac{2}{3}x^2y\right)\left(\frac{9}{32}xy^4z^3\right)$$
$$= \left(\frac{2}{3}\right)\left(\frac{9}{32}\right)\left(x^2\right)(x)(y)\left(y^4\right)\left(z^3\right)$$
$$= \frac{3}{16}x^3y^5z^3$$

21.
$$\left(32.6mnp^2\right)\left(-11.4m^2n\right)$$
$$= (32.6)(-11.4)(m)\left(m^2\right)(n)(n)\left(p^2\right)$$
$$= -371.64m^3n^2p^2$$

23.
$$(5a)\left(-17a^2\right)\left(3a^3b\right)$$
$$= (5)(-17)(3)(a)\left(a^2\right)\left(a^3\right)(b)$$
$$= -255a^6b$$

25.
$$\left(x^3\right)^2 = x^{3(2)}$$
$$= x^6$$

27.
$$\left(x^4\right)^6 = x^{4(6)}$$
$$= x^{24}$$

29.
$$\left(-3x^4\right)^2 = (-3)^2\left(x^4\right)^2$$
$$= 9x^{4(2)}$$
$$= 9x^8$$

31.
$$\left(-x^3\right)^3 = -\left(x^3\right)^3$$
$$= -x^{3(3)}$$
$$= -x^9$$

33.
$$\left(x^2 \cdot x^3\right)^2 = \left(x^5\right)^2$$
$$= x^{5(2)}$$
$$= x^{10}$$

35.
$$\left(x^5\right)^6 = x^{5(6)}$$
$$= x^{30}$$

37.
$$\left(-5x^3y^2\right)^2 = (-5)^2\left(x^3\right)^2\left(y^2\right)^2$$
$$= 25x^6y^4$$

39.
$$\left(15m^2\right)^2 = (15)^2\left(m^2\right)^2$$
$$= 225m^4$$

41.
$$\left(25n^4\right)^3 = (25)^3\left(n^4\right)^3$$
$$= 15,625n^{12}$$

43.

$$\left(3x^2 \cdot x^4\right)^2 = \left(3x^6\right)^2$$
$$= (3)^2\left(x^6\right)^2$$
$$= 9x^{12}$$

45.

$$\left(2x^3 y^4 z\right)^3 = (2)^3\left(x^3\right)^3\left(y^4\right)^3\left(z\right)^3$$
$$= 8x^9 y^{12} z^3$$

47.

$$\left(-2h^3 k^6 m^2\right)^5 = (-2)^5\left(h^3\right)^5\left(k^6\right)^5\left(m^2\right)^5$$
$$= -32h^{15} k^{30} m^{10}$$

49.

$$(4a)(17b) = 68ab$$
$$= 68(2)(-3)$$
$$= -408$$

51.

$$\left(9a^2\right)(-2a) = -18a^3$$
$$= -18(2)^3$$
$$= -18(8)$$
$$= -144$$

53.

$$\left(41a^3\right)\left(-2b^3\right) = -82a^3 b^3$$
$$= -82(2)^3(-3)^3$$
$$= -82(8)(-27)$$
$$= 17,712$$

55.

$$\left(a^2\right)^2 = a^4$$
$$= (2)^4$$
$$= 16$$

57.

$$(4b)^3 = 64b^3$$
$$= 64(-3)^3$$
$$= 64(-27)$$
$$= -1728$$

59.

$$\left(5a^2 b^3\right)^2 = 25a^4 b^6$$
$$= 25(2)^4(-3)^6$$
$$= 25(16)(729)$$
$$= 291,600$$

61.

$$(5ab)\left(a^2 b^2\right) = 5a^3 b^3$$
$$= 5(2)^3(-3)^3$$
$$= 5(8)(-27)$$
$$= -1080$$

63.

$$(9a)\left(ab^2\right) = 9a^2 b^2$$
$$= 9(2)^2(-3)^2$$
$$= 9(4)(9)$$
$$= 324$$

65.

$$-a^2 b^4 = -(2)^2(-3)^4$$
$$= -(4)(81)$$
$$= -324$$

67.

$$(-a)^4 = \left[(-2)\right]^4$$
$$= 16$$

Section 5.5: Multiplication of Polynomials

1.

$$4(a+6) = (4)(a)+(4)(6)$$
$$= 4a + 24$$

3.

$$-6\left(3x^2 + 2y\right) = (-6)\left(3x^2\right)+(-6)(2y)$$
$$= -18x^2 - 12y$$

5.

$$a\left(4x^2 - 6y + 1\right)$$
$$= (a)\left(4x^2\right)+(a)(-6y)+(a)(1)$$
$$= 4ax^2 - 6ay + a$$

7.

$$x\left(3x^2 - 2x + 5\right) = (x)\left(3x^2\right) + (x)(-2x) + (x)(5)$$
$$= 3x^3 - 2x^2 + 5x$$

9.

$$2a\left(3a^2 + 6a - 10\right) = (2a)\left(3a^2\right) + (2a)(+6a) + (2a)(-10)$$
$$= 6a^3 + 12a^2 - 20a$$

11.

$$-3x\left(4x^2 - 7x - 2\right) = (-3x)\left(4x^2\right) + (-3x)(-7x) + (-3x)(-2)$$
$$= -12x^3 + 21x^2 + 6x$$

13.

$$4x\left(-7x^2 - 3y + 2xy\right) = (4x)\left(-7x^2\right) + (4x)(-3y) + (4x)(2xy)$$
$$= -28x^3 - 12xy + 8x^2 y$$

15.

$$(3xy)\left(x^2 y - xy^2 + 4xy\right) = (3xy)\left(x^2 y\right) + (3xy)\left(-xy^2\right) + (3xy)(4xy)$$
$$= 3x^3 y^2 - 3x^2 y^3 + 12x^2 y^2$$

17.

$$-6x^3\left(1 - 6x^2 + 9x^4\right) = \left(-6x^3\right)(1) + \left(-6x^3\right)\left(-6x^2\right) + \left(-6x^3\right)\left(9x^4\right)$$
$$= -54x^7 + 36x^5 - 6x^3$$

19.

$$5ab^2\left(a^3 - b^3 - ab\right) = \left(5ab^2\right)\left(a^3\right) + \left(5ab^2\right)\left(-b^3\right) + \left(5ab^2\right)(-ab)$$
$$= 5a^4 b^2 - 5a^2 b^5 - 5ab^4$$

21.

$$\frac{2}{3}m(14n - 12m) = \left(\frac{2}{3}m\right)(14n) + \left(\frac{2}{3}m\right)(-12m)$$
$$= -8m^2 + \frac{28}{3}mn$$

23.

$$\frac{4}{7}yz^3\left(28y - \frac{2}{5}z\right) = \left(\frac{4}{7}yz^3\right)(28y) + \left(\frac{4}{7}yz^3\right)\left(-\frac{2}{5}z\right)$$
$$= -\frac{8}{35}yz^4 + 16y^2 z^3$$

25.

$$-4a\left(1.3a^5 + 2.5a^2 + 1\right) = (-4a)\left(1.3a^5\right) + (-4a)\left(2.5a^2\right) + (-4a)(1)$$
$$= -5.2a^6 - 10a^3 - 4a$$

27.

$$417a\left(3.2a^2 + 4a\right) = (417a)\left(3.2a^2\right) + (417a)(4a)$$
$$= 1334.4a^3 + 1668a^2$$

29.

$$4x^2y\left(6x^2 - 4xy + 5y^2\right) = \left(4x^2\right)\left(6x^2y\right) + \left(4x^2y\right)(-4xy) + \left(4x^2y\right)\left(5y^2\right)$$

$$= 24x^4y - 16x^3y^2 + 20x^2y^3$$

31.

$$\frac{2}{3}ab^3\left(\frac{3}{4}a^2 - \frac{1}{2}ab^2 + \frac{5}{6}b^3\right) = \left(\frac{2}{3}ab^3\right)\left(\frac{3}{4}a^2\right) + \left(\frac{2}{3}ab^3\right)\left(-\frac{1}{2}ab^2\right) + \left(\frac{2}{3}ab^3\right)\left(\frac{5}{6}b^3\right)$$

$$= \frac{1}{2}a^3b^3 - \frac{2}{3}a^2b^5 + \frac{5}{9}ab^6$$

33.

$$3x(x-4) + 2x(1-5x) - 6x(2x-3) = \left(3x^2 - 12x\right) + \left(2x - 10x^2\right) + \left(-12x^2 + 18x\right)$$

$$= -19x^2 + 8x$$

35.

$$xy\left(3x + 2xy - y^2\right) - 2xy^2\left(2x - xy + 3y\right) = \left(3x^2y + 2x^2y^2 - xy^3\right) + \left(-4x^2y^2 + 2x^2y^3 - 6xy^3\right)$$

$$= 3x^2y + 2x^2y^3 - 7xy^3 - 2x^2y^2$$

37.

$$\begin{array}{r} x+1 \\ \underline{x+6} \\ x^2 + x \\ \underline{\quad x+6} \\ x^2 + 7x + 6 \end{array}$$

39.

$$\begin{array}{r} x+7 \\ \underline{x-2} \\ x^2 + 7x \\ \underline{\quad -2x-14} \\ x^2 + 5x - 14 \end{array}$$

41.

$$\begin{array}{r} x-5 \\ \underline{x-8} \\ x^2 - 5x \\ \underline{\quad -8x+40} \\ x^2 - 13x + 40 \end{array}$$

43.

$$\begin{array}{r} 3a-5 \\ \underline{a-4} \\ 3a^2 - 5a \\ \underline{\quad -12a+20} \\ 3a^2 - 17a + 20 \end{array}$$

45.

$$\begin{array}{r} 6a+4 \\ \underline{2a-3} \\ 12a^2 + 8a \\ \underline{\quad -18a-12} \\ 12a^2 - 10a - 12 \end{array}$$

47.

$$\begin{array}{r} 4a+8 \\ \underline{6a+9} \\ 24a^2 + 48a \\ \underline{\quad 36a+72} \\ 24a^2 + 84a + 72 \end{array}$$

49.

$$\begin{array}{r} 3x-2y \\ \underline{5x+2y} \\ 15x^2 - 10xy \\ \underline{\quad 6xy - 4y^2} \\ 15x^2 - 4xy - 4y^2 \end{array}$$

51.

$$\begin{array}{r} 2x-3 \\ \underline{2x-3} \\ 4x^2 - 6x \\ \underline{\quad -6x+9} \\ 4x^2 - 12x + 9 \end{array}$$

53.

$$2c - 5d$$
$$\underline{2c + 5d}$$
$$4c^2 - 10cd$$
$$\underline{10cd - 25d^2}$$
$$4c^2 - 25d^2$$

55.

$$-7m - 3$$
$$\underline{-13m + 1}$$
$$91m^2 + 39m$$
$$\underline{-7m - 3}$$
$$91m^2 + 32m - 3$$

57.

$$x^5 - x^2$$
$$\underline{x^3 - 1}$$
$$x^8 - x^5$$
$$\underline{-x^5 + x^2}$$
$$x^8 - 2x^5 + x^2$$

59.

$$2y^2 - 4y - 8$$
$$\underline{5y - 2}$$
$$10y^3 - 20y^2 - 40y$$
$$\underline{-4y^2 + 8y + 16}$$
$$10y^3 - 24y^2 - 32y + 16$$

61.

$$4x - 2y - 13$$
$$\underline{6x + 3y}$$
$$24x^2 - 12xy - 78x$$
$$\underline{12xy - 6y^2 - 39y}$$
$$24x^2 - 78x - 6y^2 - 42y$$

63.

$$g + h - 6$$
$$\underline{g - h + 4}$$
$$g^2 + gh - 6g$$
$$- gh - h^2 + 6h$$
$$\underline{4g + 4h - 24}$$
$$g^2 - 2g - h^2 + 10h - 24$$

65. Write polynomials in descending order.

$$2x^4 - x^3 + 0x^2 + 8x - 1$$
$$\underline{5x^3 + x^2 + 2}$$
$$10x^7 - 5x^6 + 0x^5 + 40x^4 - 5x^3$$
$$2x^6 - x^5 + 0x^4 + 8x^3 - x^2$$
$$\underline{4x^4 - 2x^3 + 0x^2 + 16x - 2}$$
$$10x^7 - 3x^6 - x^5 + 44x^4 + x^3 - x^2 + 16x - 2$$

Section 5.6: Division by a Monomial

1. $3x^2$

3. $\dfrac{3x^8}{2}$

5. $\dfrac{6}{x^2}$

7. $\dfrac{2x}{3}$

9. x

11.

$$(15x) \div (6x) = \frac{15x}{6x}$$
$$= \frac{5}{2}$$

13.

$$\left(15a^3b\right) \div \left(3ab^2\right) = \frac{15a^3b}{3ab^2}$$
$$= \frac{5a^2}{b}$$

15.

$$\left(16m^2n\right) \div \left(2m^3n^2\right) = \frac{16m^2n}{2m^3n^2}$$
$$= \frac{8}{mn}$$

17.

$$0 \div \left(113w^2r^3\right) = \frac{0}{113w^2r^3}$$
$$= 0$$

19.

$$\left(207p^3\right) \div \left(9p\right) = \frac{207p^3}{9p}$$
$$= 23p^2$$

21. -2

23. $\dfrac{36}{r^2}$

25. $-\dfrac{23x^2}{7y^2}$

27. $\dfrac{8}{7a^3b^2}$

29.

$$\left(-72x^3yz^4\right) \div \left(-162xy^2\right) = \frac{-72x^3yz^4}{-162xy^2}$$
$$= \frac{4x^2z^4}{9y}$$

31.

$$\left(4x^2 - 8x + 6\right) \div 2 = \frac{4x^2 - 8x + 6}{2}$$
$$= \frac{4x^2}{2} - \frac{8x}{2} + \frac{6}{2}$$
$$= 2x^2 - 4x + 3$$

33.

$$\left(x^4 + x^3 + x^2\right) \div \left(x^2\right) = \frac{x^4 + x^3 + x^2}{x^2}$$
$$= \frac{x^4}{x^2} + \frac{x^3}{x^2} + \frac{x^2}{x^2}$$
$$= x^2 + x + 1$$

35.

$$\left(ax - ay - az\right) \div \left(a\right) = \frac{ax - ay - az}{a}$$
$$= \frac{ax}{a} - \frac{ay}{a} - \frac{az}{a}$$
$$= x - y - z$$

37.

$$\frac{24a^4 - 16a^2 - 8a}{8} = \frac{24a^4}{8} - \frac{16a^2}{8} - \frac{8a}{8}$$
$$= 3a^4 - 2a^2 - a$$

39.

$$\frac{b^{12} - b^9 - b^6}{b^3} = \frac{b^{12}}{b^3} - \frac{b^9}{b^3} - \frac{b^6}{b^3}$$
$$= b^9 - b^6 - b^3$$

41.

$$\frac{bx^4 - bx^3 + bx^2 - 4bx}{-bx}$$
$$= -\frac{bx^4}{bx} + \frac{-bx^3}{-bx} - \frac{bx^2}{bx} + \frac{4bx}{bx}$$
$$= -x^3 + x^2 - x + 4$$

43.

$$\frac{24x^2y^3 + 12x^3y^4 - 6xy^3}{2xy^3}$$
$$= \frac{24x^2y^3}{2xy^3} + \frac{12x^3y^4}{2xy^3} - \frac{6xy^3}{2xy^3}$$
$$= 12x + 6x^2y - 3$$

45.

$$\frac{224x^4y^2z^3 - 168x^3y^3z^4 - 112xy^4z^2}{28xy^2z^2}$$
$$= \frac{224x^4y^2z^3}{28xy^2z^2} - \frac{168x^3y^3z^4}{28xy^2z^2} - \frac{112xy^4z^2}{28xy^2z^2}$$
$$= 8x^3z - 6x^2yz^2 - 4y^2$$

47.

$$\frac{24y^5 - 18y^3 - 12y}{6y^2} = \frac{24y^5}{6y^2} - \frac{18y^3}{6y^2} - \frac{12y}{6y^2}$$
$$= 4y^3 - 3y - \frac{2}{y}$$

49.

$$\frac{1 - 6x^2 - 4x^4}{2x^2} = \frac{1}{2x^2} - \frac{6x^2}{2x^2} - \frac{4x^4}{2x^2}$$
$$= \frac{1}{2x^2} - 3 - 2x^2$$

Section 5.7: Division by a Polynomial

1.

$$
\begin{array}{r}
x + 2 \\
x+1\,\overline{\big)\ x^2 + 3x + 2} \\
\underline{x^2 + x} \\
2x + 2 \\
\underline{2x + 2} \\
0
\end{array}
$$

7.

$$
\begin{array}{r}
y + 2 \\
2y-1\,\overline{\big)\ 2y^2 + 3y - 5} \\
\underline{2y^2 - y} \\
4y - 5 \\
\underline{4y - 2} \\
-3
\end{array}
$$

3.

$$
\begin{array}{r}
3a + 3 \\
2a-3\,\overline{\big)\ 6a^2 - 3a + 2} \\
\underline{6a^2 - 9a} \\
6a + 2 \\
\underline{6a - 9} \\
11
\end{array}
$$

9.

$$
\begin{array}{r}
3b - 4 \\
2b+7\,\overline{\big)\ 6b^2 + 13b - 28} \\
\underline{6b^2 + 21b} \\
-8b - 28 \\
\underline{-8b - 28} \\
0
\end{array}
$$

5.

$$
\begin{array}{r}
4x - 3 \\
3x+2\,\overline{\big)\ 12x^2 - x - 9} \\
\underline{12x^2 + 8x} \\
-9x - 9 \\
\underline{-9x - 6} \\
-3
\end{array}
$$

11.

$$
\begin{array}{r}
6x^2 + x - 1 \\
x+2\,\overline{\big)\ 6x^3 + 13x^2 + x - 2} \\
\underline{6x^3 + 12x^2} \\
x^2 + x \\
\underline{x^2 + 2x} \\
-x - 2 \\
\underline{-x - 2} \\
0
\end{array}
$$

13.

$$
\begin{array}{r}
4x^2 + 7x - 15 \\
2x-7\,\overline{\big)\ 8x^3 - 14x^2 - 79x + 110} \\
\underline{8x^3 - 28x^2} \\
14x^2 - 79x \\
\underline{14x^2 - 49x} \\
-30x + 110 \\
\underline{-30x + 105} \\
5
\end{array}
$$

15.

$$
\begin{array}{r}
2x^2 \quad - \quad 2x \quad - \quad 12 \\
x+1 \overline{\smash{\big)}\ 2x^3 \ + \ 0x^2 \ - \ 14x \ - \ 12} \\
\underline{2x^3 \ + \ 2x^2} \\
-2x^2 \ - \ 14x \\
\underline{-2x^2 \ - \ 2x} \\
-12x \ - \ 12 \\
\underline{-12x \ - \ 12} \\
0
\end{array}
$$

17.

$$
\begin{array}{r}
2x^2 \quad - \quad 16x \quad + \quad 32 \\
2x+4 \overline{\smash{\big)}\ 4x^3 \ - \ 24x^2 \ + \ 0x \ + \ 128} \\
\underline{4x^3 \ + \ 8x^2} \\
-32x^2 \ + \ 0x \\
\underline{-32x^2 \ - \ 64x} \\
64x \ + \ 128 \\
\underline{64x \ + \ 128} \\
0
\end{array}
$$

19.

$$
\begin{array}{r}
3x^2 \quad + \quad 10x \quad + \quad 20 \\
x-2 \overline{\smash{\big)}\ 3x^3 \ + \ 4x^2 \ + \ 0x \ - \ 6} \\
\underline{3x^3 \ - \ 6x^2} \\
10x^2 \ + \ 0x \\
\underline{10x^2 \ - \ 20x} \\
20x \ - \ 6 \\
\underline{20x \ - \ 40} \\
34
\end{array}
$$

21.

$$
\begin{array}{r}
2x^2 \quad + \quad 6x \quad + \quad 30 \\
2x-5 \overline{\smash{\big)}\ 4x^3 \ + \ 2x^2 \ + \ 30x \ + \ 20} \\
\underline{4x^3 \ - \ 10x^2} \\
12x^2 \ + \ 30x \\
\underline{12x^2 \ - \ 30x} \\
60x \ + \ 20 \\
\underline{60x \ - \ 150} \\
170
\end{array}
$$

23.

$$
\begin{array}{r}
2x^3 \quad\quad\quad + \quad 4x \quad + \quad 6 \\
4x-5 \overline{\smash{\big)}\ 8x^4 \ - \ 10x^3 \ + \ 16x^2 \ + \ 4x \ - \ 30} \\
\underline{8x^4 \ - \ 10x^3} \\
+ \ 16x^2 \ + \ 4x \\
\underline{16x^2 \ - \ 20x} \\
24x \ - \ 30 \\
\underline{24x \ - \ 30} \\
0
\end{array}
$$

25.

$$
\begin{array}{r}
4x^2 \;-\; 2x \;+\; 1 \\
2x+1\,\big)\overline{\,8x^3 + 0x^2 + 0x - 1\,} \\
\underline{8x^3 + 4x^2} \\
-4x^2 + 0x \\
\underline{-4x^2 - 2x} \\
2x - 1 \\
\underline{2x + 1} \\
-2
\end{array}
$$

27.

$$
\begin{array}{r}
x^3 - 2x^2 + 4x - 8 \\
x+2\,\big)\overline{\,x^4 + 0x^3 + 0x^2 + 0x - 16\,} \\
\underline{x^4 + 2x^3} \\
-2x^3 + 0x^2 \\
\underline{-2x^3 - 4x^2} \\
4x^2 + 0x \\
\underline{4x^2 + 8x} \\
-8x - 16 \\
\underline{-8x - 16} \\
0
\end{array}
$$

29.

$$
\begin{array}{r}
3x^2 - 4x + 1 \\
x^2+3x-2\,\big)\overline{\,3x^4 + 5x^3 - 17x^2 + 11x - 2\,} \\
\underline{3x^4 + 9x^3 - 6x^2} \\
-4x^3 - 11x^2 + 11x \\
\underline{-4x^3 - 12x^2 + 8x} \\
x^2 + 3x - 2 \\
\underline{x^2 + 3x - 2} \\
0
\end{array}
$$

Chapter 5 Review

1. a

2. 0

3. 1

4.

$$10 - 4(3) = 10 - 12$$
$$= -2$$

5.

$$2 + 3 \cdot 4^2 = 2 + 3 \cdot 16$$
$$= 2 + 48$$
$$= 50$$

6.

$$(4)(12) \div 6 - 2^3 + 18 \div 3^2$$
$$= 48 \div 6 - 8 + 18 \div 9$$
$$= 8 - 8 + 2$$
$$= 2$$

7.

$$x + y = (3) + (-2)$$
$$= 1$$

8.

$$x - 3y = (3) - 3(-2)$$
$$= 3 + 6$$
$$= 9$$

9.

$$5xy = 5(3)(-2)$$
$$= -30$$

10.

$$\frac{x^2}{y} = \frac{(3)^2}{(-2)}$$
$$= \frac{9}{-2}$$
$$= -4\frac{1}{2}$$

11.

$$y^3 - y^2 = (-2)^3 - (-2)^2$$
$$= (-8) - 4$$
$$= -12$$

12.

$$\frac{2x^3 - 3y}{xy^2} = \frac{2(3)^3 - 3(-2)}{(3)(-2)^2}$$
$$= \frac{2(27) + 6}{(3)(4)}$$
$$= \frac{54 + 6}{12}$$
$$= \frac{60}{12}$$
$$= 5$$

13.

$$(5y - 3) - (2 - y) = 5y - 3 - 2 + y$$
$$= (5 + 1)y + (-3 - 2)$$
$$= 6y - 5$$

14.

$$(7 - 3x) - (5x + 1) = 7 - 3x - 5x - 1$$
$$= (-3 - 5)x + (7 - 1)$$
$$= -8x + 6$$

15.

$$11(2x + 1) - 4(3x - 4)$$
$$= 22x + 11 - 12x + 16$$
$$= 10x + 27$$

16.

$$\left(x^3 + 2x^2 y\right) - \left(3y^3 - 2x^3 + x^2 y + y\right)$$
$$= \left(x^3 + 2x^2 y\right) + \left(-3y^3 + 2x^3 - x^2 y - y\right)$$
$$= 3x^3 + x^2 y - 3y^3 - y$$

17. binomial

18. 4

19.

$$\left(3a^2 + 7a - 2\right) + \left(5a^2 - 2a + 4\right) = (3 + 5)a^2 + (7 - 2)a + (-2 + 4)$$
$$= 8a^2 + 5a + 2$$

20.

$$\left(6x^3 + 3x^2 + 1\right) - \left(-3x^3 - x^2 - x - 1\right) = \left(6x^3 + 3x^2 + 1\right) + \left(3x^3 + x^2 + x + 1\right)$$
$$= 9x^3 + 4x^2 + x + 2$$

21.

$$\left(3x^2 + 5x + 2\right) + \left(9x^2 - 6x - 2\right) - \left(2x^2 + 6x - 4\right) = \left(3x^2 + 5x + 2\right) + \left(9x^2 - 6x - 2\right) + \left(-2x^2 - 6x + 4\right)$$
$$= (3 + 9 - 2)x^2 + (5 - 6 - 6)x + (2 - 2 + 4)$$
$$= 10x^2 - 7x + 4$$

22.

$$\left(6x^2\right)\left(4x^3\right) = (6)(4)\left(x^2\right)\left(x^3\right)$$
$$= 24x^5$$

23.

$$\left(-7x^2 y\right)\left(8x^3 y^2\right)$$
$$= (-7)(8)\left(x^2\right)\left(x^3\right)\left(y\right)\left(y^2\right)$$
$$= -56x^5 y^3$$

24.

$$\left(3x^2\right)^3 = \left(3\right)^3\left(x^2\right)^3$$
$$= 27x^6$$

25.

$$5a\left(3a+4b\right) = \left(5a\right)\left(3a\right)+\left(5a\right)\left(4b\right)$$
$$= 15a^2 + 20ab$$

26.

$$-4x^2\left(8-2x+3x^2\right)$$
$$= \left(-4x^2\right)\left(8\right)+\left(-4x^2\right)\left(-2x\right)+\left(-4x^2\right)\left(3x^2\right)$$
$$= -12x^4 + 8x^3 - 32x^2$$

27.

$$5x+3$$
$$3x-4$$
$$\overline{15x^2+9x}$$
$$\underline{-20x-12}$$
$$15x^2-11x-12$$

28.

$$3x^2-6x+1$$
$$2x-4$$
$$\overline{6x^3-12x^2+2x}$$
$$\underline{-12x^2+24x-4}$$
$$6x^3-24x^2+26x-4$$

29.

$$\left(49x^2\right)\div\left(7x^3\right) = \frac{49x^2}{7x^3}$$
$$= \frac{7}{x}$$

30. $5x$

31.

$$\frac{36a^3-27a^2+9a}{9a} = \frac{36a^3}{9a}-\frac{27a^2}{9a}+\frac{9a}{9a}$$
$$= 4a^2-3a+1$$

32.

$$
\begin{array}{r}
3x \quad - \quad 4 \\
2x+3 \overline{\smash{\big)}\ 6x^2 \ + \ x \ - \ 12} \\
\underline{6x^2 \ + \ 9x } \\
-8x \ - \ 12 \\
\underline{-8x \ - \ 12} \\
0
\end{array}
$$

33.

$$
\begin{array}{r}
3x^2 \quad - \quad 4x \quad + \quad 2 \\
x+2 \overline{\smash{\big)}\ 3x^3 \ + \ 2x^2 \ - \ 6x \ + \ 4} \\
\underline{3x^3 \ + \ 6x^2 } \\
-4x^2 \ - \ 6x \\
\underline{-4x^2 \ - \ 8x } \\
2x \ + \ 4 \\
\underline{2x \ + \ 4} \\
0
\end{array}
$$

Chapter 5 Test

1.

$$3\cdot5-2\cdot4^2 = 15-2\cdot16$$
$$= 15-32$$
$$= -17$$

3.

$$\frac{3x^2y - 4x}{2y} = \frac{3(4)^2(-1) - 4(4)}{2(-1)}$$

$$= \frac{3(16)(-1) - 16}{-2}$$

$$= \frac{-64}{-2}$$

$$= 32$$

5.

$$5(2+x) - 2(x+4) = 10 + 5x - 2x - 8$$

$$= 5x - 2x + 10 - 8$$

$$= 3x + 2$$

7.

$$(5a - 5b + 7) - (2a - 5b - 3)$$

$$= (5a - 5b + 7) + (-2a + 5b + 3)$$

$$= 3a + 10$$

9.

$$\left(6x^4y^2\right)^3 = (6)^3\left(x^4\right)^3\left(y^2\right)^3$$

$$= 216x^{12}y^6$$

11. $\dfrac{5x^2}{y^3}$

13.

$$x + y - 5$$
$$\underline{x - y}$$
$$x^2 + xy - 5x$$
$$\underline{-xy \qquad -y^2 + 5y}$$
$$x^2 - 5x - y^2 + 5y$$

15.

$$\frac{9x^4y^3 - 12x^2y + 18y^2}{3x^2y^3}$$

$$= \frac{9x^4y^3}{3x^2y^3} - \frac{12x^2y}{3x^2y^3} + \frac{18y^2}{3x^2y^3}$$

$$= 3x^2 - \frac{4}{y^2} + \frac{6}{x^2y}$$

17.

$$\begin{array}{r}
3x + 2 \\
x - 5 \overline{\smash{\big)}\ 3x^2 - 13x - 10} \\
\underline{3x^2 - 15x} \\
2x - 10 \\
2x - 10 \\
\underline{} \\
0
\end{array}$$

19.

$$\begin{array}{r}
2x + 1 \\
3x - 5 \overline{\smash{\big)}\ 6x^2 - 7x - 6} \\
\underline{6x^2 - 10x} \\
3x - 6 \\
3x - 5 \\
\underline{} \\
-1
\end{array}$$

Chapter 6: Equations and Formulas

Section 6.1: Equations

1.
$$x + 2 = 8$$
$$x + 2 - 2 = 8 - 2$$
$$x = 6$$

3.
$$y - 5 = 12$$
$$y - 5 + 5 = 12 + 5$$
$$y = 17$$

5.
$$w - 7\frac{1}{2} = 3$$
$$w - \frac{15}{2} + \frac{15}{2} = \frac{6}{2} + \frac{15}{2}$$
$$w = \frac{21}{2} = 10\frac{1}{2}$$

7.
$$\frac{x}{13} = 1.5$$
$$13\left(\frac{x}{13}\right) = (1.5)13$$
$$x = 19.5$$

9.
$$3b = 15.6$$
$$\frac{3b}{3} = \frac{15.6}{3}$$
$$b = 5.2$$

11.
$$17x = 5117$$
$$\frac{17x}{17} = \frac{5117}{17}$$
$$x = 301$$

13.
$$2 = x - 5$$
$$2 + 5 = x - 5 + 5$$
$$x = 7$$

15.
$$17 = -3 + w$$
$$17 + 3 = -3 + w + 3$$
$$20 = w$$

17.
$$14b = 57$$
$$\frac{14b}{14} = \frac{57}{14}$$
$$b = 4\frac{1}{14}$$

19.
$$5m = 0$$
$$\frac{5m}{5} = \frac{0}{5}$$
$$m = 0$$

21.
$$x + 5 = 5$$
$$x + 5 - 5 = 5 - 5$$
$$x = 0$$

23.
$$4x = 64$$
$$\frac{4x}{4} = \frac{64}{4}$$
$$x = 16$$

25.
$$\frac{x}{7} = 56$$
$$7\left(\frac{x}{7}\right) = (56)7$$
$$x = 392$$

27.
$$-48 = 12y$$
$$\frac{-48}{12} = \frac{12y}{12}$$
$$-4 = y$$

29.
$$-x = 2$$
$$\frac{-x}{-1} = \frac{2}{-1}$$
$$x = -2$$

31.
$$5y + 3 = 13$$
$$5y + 3 - 3 = 13 - 3$$
$$5y = 10$$
$$\frac{5y}{5} = \frac{10}{5}$$
$$y = 2$$

33.
$$10 - 3x = 16$$
$$10 - 3x - 10 = 16 - 10$$
$$-3x = 6$$
$$\frac{-3x}{-3} = \frac{6}{-3}$$
$$x = -2$$

35.
$$\frac{x}{4} - 5 = 3$$
$$\frac{x}{4} - 5 + 5 = 3 + 5$$
$$\frac{x}{4} = 8$$
$$4\left(\frac{x}{4}\right) = (8)4$$
$$x = 32$$

37.
$$2 - x = 6$$
$$2 - x - 2 = 6 - 2$$
$$-x = 4$$
$$\frac{-x}{-1} = \frac{4}{-1}$$
$$x = -4$$

39.
$$\frac{2}{3}y - 4 = 8$$
$$\frac{2}{3}y - 4 + 4 = 8 + 4$$
$$\frac{2y}{3} = 12$$
$$\frac{3}{2}\left(\frac{2y}{3}\right) = (12)\frac{3}{2}$$
$$y = 18$$

41.
$$3x - 5 = 12$$
$$3x - 5 + 5 = 12 + 5$$
$$3x = 17$$
$$\frac{3x}{3} = \frac{17}{3}$$
$$x = 5\frac{2}{3}$$

43.
$$\frac{m}{3} - 6 = 8$$
$$\frac{m}{3} - 6 + 6 = 8 + 6$$
$$\frac{m}{3} = 14$$
$$3\left(\frac{m}{3}\right) = (14)3$$
$$m = 42$$

45.
$$\frac{2x}{3} = 7$$
$$\frac{3}{2}\left(\frac{2x}{3}\right) = (7)\frac{3}{2}$$
$$x = \frac{21}{2}$$
$$x = 10\frac{1}{2}$$

47.
$$-3y - 7 = -6$$
$$-3y - 7 + 7 = -6 + 7$$
$$-3y = 1$$
$$\frac{-3y}{-3} = \frac{1}{-3}$$
$$y = -\frac{1}{3}$$

49.
$$5 - x = 6$$
$$5 - x - 5 = 6 - 5$$
$$-x = 1$$
$$\frac{-x}{-1} = \frac{1}{-1}$$
$$x = -1$$

51.

$$54y - 13 = 17.8$$
$$54y - 13 + 13 = 17.8 + 13$$
$$54y = 30.8$$
$$\frac{54y}{54} = \frac{30.8}{54}$$
$$y = \frac{308}{540}$$
$$y = \frac{77}{135}$$

53.

$$28w - 56 = -8$$
$$28w - 56 + 56 = -8 + 56$$
$$28w = 48$$
$$\frac{28w}{28} = \frac{48}{28}$$
$$x = 1\frac{5}{7}$$

55.

$$29r - 13 = 57$$
$$29r - 13 + 13 = 57 + 13$$
$$29r = 70$$
$$\frac{29r}{29} = \frac{70}{29}$$
$$t = 2\frac{12}{29}$$

57.

$$31 - 3y = 41$$
$$31 - 3y - 31 = 41 - 31$$
$$-3y = 10$$
$$\frac{-3y}{-3} = \frac{10}{-3}$$
$$y = -3\frac{1}{3}$$

59.

$$-83 = 17 - 4x$$
$$-83 - 17 = 17 - 4x - 17$$
$$-100 = -4x$$
$$\frac{-100}{-4} = \frac{-4x}{-4}$$
$$25 = x$$

Section 6.2: Equations with Variables in Both Members

1.

$$4y + 9 = 7y - 15$$
$$4y + 9 - 4y = 7y - 15 - 4y$$
$$9 = 3y - 15$$
$$9 + 15 = 3y - 15 + 15$$
$$24 = 3y$$
$$\frac{24}{3} = \frac{3y}{3}$$
$$8 = y$$

3.

$$5x + 3 = 7x - 5$$
$$5x + 3 - 5x = 7x - 5 - 5x$$
$$3 = 2x - 5$$
$$3 + 5 = 2x - 5 + 5$$
$$8 = 2x$$
$$\frac{8}{2} = \frac{2x}{2}$$
$$4 = x$$

5.
$$-2x + 7 = 5x - 21$$
$$-2x + 7 + 2x = 5x - 21 + 2x$$
$$7 = 7x - 21$$
$$7 + 21 = 7x - 21 + 21$$
$$28 = 7x$$
$$\frac{28}{7} = \frac{7x}{7}$$
$$4 = x$$

7.
$$3y + 5 = 5y - 1$$
$$3y + 5 - 3y = 5y - 1 - 3y$$
$$5 = 2y - 1$$
$$5 + 1 = 2y - 1 + 1$$
$$6 = 2y$$
$$\frac{6}{2} = \frac{2y}{2}$$
$$3 = y$$

9.
$$-3x + 17 = 6x - 37$$
$$-3x + 17 + 3x = 6x - 37 + 3x$$
$$17 = 9x - 37$$
$$17 + 37 = 9x - 37 + 37$$
$$54 = 9x$$
$$\frac{54}{9} = \frac{9x}{9}$$
$$6 = x$$

11.
$$7x + 9 = 9x - 3$$
$$7x + 9 - 7x = 9x - 3 - 7x$$
$$9 = 2x - 3$$
$$9 + 3 = 2x - 3 + 3$$
$$12 = 2x$$
$$\frac{12}{2} = \frac{2x}{2}$$
$$6 = x$$

13.
$$3x - 2 = 5x + 8$$
$$3x - 2 - 3x = 5x + 8 - 3x$$
$$-2 = 2x + 8$$
$$-2 - 8 = 2x + 8 - 8$$
$$-10 = 2x$$
$$\frac{-10}{2} = \frac{2x}{2}$$
$$-5 = x$$

15.
$$-4x + 25 = 6x - 45$$
$$-4x + 25 + 4x = 6x - 45 + 4x$$
$$25 = 10x - 45$$
$$25 + 45 = 10x - 45 + 45$$
$$70 = 10x$$
$$\frac{70}{10} = \frac{10x}{10}$$
$$7 = x$$

17.
$$5x + 4 = 10x - 7$$
$$5x + 4 - 5x = 10x - 7 - 5x$$
$$4 = 5x - 7$$
$$4 + 7 = 5x - 7 + 7$$
$$11 = 5x$$
$$\frac{11}{5} = \frac{5x}{5}$$
$$2\frac{1}{5} = x$$

19.
$$27 + 5x = 9 + 3x$$
$$27 + 5x - 3x = 9 + 3x - 3x$$
$$2x + 27 = 9$$
$$2x + 27 - 27 = 9 - 27$$
$$2x = -18$$
$$\frac{2x}{2} = \frac{-18}{2}$$
$$x = -9$$

21.

$$-7x + 18 = 11x - 36$$
$$-7x + 18 + 7x = 11x - 36 + 7x$$
$$18 = 18x - 36$$
$$18 + 36 = 18x - 36 + 36$$
$$54 = 18x$$
$$\frac{54}{18} = \frac{18x}{18}$$
$$3 = x$$

23.

$$4y + 11 = 7y - 28$$
$$4y + 11 - 4y = 7y - 28 - 4y$$
$$11 = 3y - 28$$
$$11 + 28 = 3y - 28 + 28$$
$$39 = 3y$$
$$\frac{39}{3} = \frac{3y}{3}$$
$$13 = y$$

25.

$$-4x + 2 = 8x - 7$$
$$-4x + 2 + 4x = 8x - 7 + 4x$$
$$2 = 12x - 7$$
$$2 + 7 = 12x - 7 + 7$$
$$9 = 12x$$
$$\frac{9}{12} = \frac{12x}{12}$$
$$\frac{3}{4} = x$$

29.

$$3x + 1 = 17 - x$$
$$3x + 1 + x = 17 - x + x$$
$$4x + 1 = 17$$
$$4x + 1 - 1 = 17 - 1$$
$$4x = 16$$
$$\frac{4x}{4} = \frac{16}{4}$$
$$x = 4$$

27.

$$13x + 6 = 6x - 1$$
$$13x + 6 - 6x = 6x - 1 - 6x$$
$$7x + 6 = -1$$
$$7x + 6 - 6 = -1 - 6$$
$$7x = -7$$
$$\frac{7x}{7} = \frac{-7}{7}$$
$$x = -1$$

Section 6.3: Equations with Parentheses

1.

$$2(x + 3) - 6 = 10$$
$$2x + 6 - 6 = 10$$
$$2x = 10$$
$$\frac{2x}{2} = \frac{10}{2}$$
$$x = 5$$

3.

$$3n + (2n + 4) = 6$$
$$3n + 2n + 4 = 6$$
$$5n + 4 = 6$$
$$5n + 4 - 4 = 6 - 4$$
$$5n = 2$$
$$\frac{5n}{5} = \frac{2}{5}$$
$$n = \frac{2}{5}$$

5.
$$16 = -3(x-4)$$
$$16 = -3x + 12$$
$$16 - 12 = -3x + 12 - 12$$
$$4 = -3x$$
$$\frac{4}{-3} = \frac{-3x}{-3}$$
$$-1\frac{1}{3} = x$$

7.
$$5a - (3a+4) = 8$$
$$5a - 3a - 4 = 8$$
$$2a - 4 = 8$$
$$2a - 4 + 4 = 8 + 4$$
$$2a = 12$$
$$\frac{2a}{2} = \frac{12}{2}$$
$$a = 6$$

9.
$$5a - 4(a-3) = 7$$
$$5a - 4a + 12 = 7$$
$$a + 12 = 7$$
$$a + 12 - 12 = 7 - 12$$
$$a = -5$$

11.
$$5(x-3) = 21$$
$$5x - 15 = 21$$
$$5x - 15 + 15 = 21 + 15$$
$$5x = 36$$
$$\frac{5x}{5} = \frac{36}{5}$$
$$x = 7\frac{1}{5}$$

13.
$$2a - (5a - 7) = 22$$
$$2a - 5a + 7 = 22$$
$$-3a + 7 = 22$$
$$-3a + 7 - 7 = 22 - 7$$
$$-3a = 15$$
$$\frac{-3a}{-3} = \frac{15}{-3}$$
$$a = -5$$

15.
$$2(w-3) + 6 = 0$$
$$2w - 6 + 6 = 0$$
$$2w = 0$$
$$\frac{2w}{2} = \frac{0}{2}$$
$$w = 0$$

17.
$$3x - 7 + 17(1-x) = -6$$
$$3x - 7 + 17 - 17x = -6$$
$$10 - 14x = -6$$
$$10 - 14x - 10 = -6 - 10$$
$$-14x = -16$$
$$\frac{-14x}{-14} = \frac{-16}{-14}$$
$$x = 1\frac{1}{7}$$

19.
$$6b = 27 + 3b$$
$$6b - 3b = 27 + 3b - 3b$$
$$3b = 27$$
$$\frac{3b}{3} = \frac{27}{3}$$
$$b = 9$$

21.
$$4(25 - x) = 3x + 2$$
$$100 - 4x = 3x + 2$$
$$100 - 4x + 4x = 3x + 2 + 4x$$
$$100 = 7x + 2$$
$$100 - 2 = 7x + 2 - 2$$
$$98 = 7x$$
$$\frac{98}{7} = \frac{7x}{7}$$
$$x = 14$$

23.

$$x + 3 = 4(57 - x)$$
$$x + 3 = 228 - 4x$$
$$x + 3 + 4x = 228 - 4x + 4x$$
$$5x + 3 = 228$$
$$5x + 3 - 3 = 228 - 3$$
$$5x = 225$$
$$\frac{5x}{5} = \frac{225}{5}$$
$$x = 45$$

25.

$$6x + 2 = 2(17 - x)$$
$$6x + 2 = 34 - 2x$$
$$6x + 2 + 2x = 34 - 2x + 2x$$
$$8x + 2 = 34$$
$$8x + 2 - 2 = 34 - 2$$
$$8x = 32$$
$$\frac{8x}{8} = \frac{32}{8}$$
$$x = 4$$

27.

$$5(x - 8) - 3x - 4 = 0$$
$$5x - 40 - 3x - 4 = 0$$
$$2x - 44 = 0$$
$$2x - 44 + 44 = 0 + 44$$
$$2x = 44$$
$$\frac{2x}{2} = \frac{44}{2}$$
$$x = 22$$

29.

$$3(x + 4) + 3x = 6$$
$$3x + 12 + 3x = 6$$
$$6x + 12 = 6$$
$$6x + 12 - 12 = 6 - 12$$
$$6x = -6$$
$$\frac{6x}{6} = \frac{-6}{6}$$
$$x = -1$$

31.

$$y - 4 = 2(y - 7)$$
$$y - 4 = 2y - 14$$
$$y - 4 - y = 2y - 14 - y$$
$$-4 = y - 14$$
$$-4 + 14 = y - 14 + 14$$
$$10 = y$$

33.

$$9m - 3(m - 5) = 7m - 3$$
$$9m - 3m + 15 = 7m - 3$$
$$6m + 15 = 7m - 3$$
$$6m + 15 - 6m = 7m - 3 - 6m$$
$$15 = m - 3$$
$$15 + 3 = m - 3 + 3$$
$$18 = m$$

35.

$$3(2x + 7) = 13 + 2(4x + 2)$$
$$6x + 21 = 13 + 8x + 4$$
$$6x + 21 = 8x + 17$$
$$6x + 21 - 6x = 8x + 17 - 6x$$
$$21 = 2x + 17$$
$$21 - 17 = 2x + 17 - 17$$
$$4 = 2x$$
$$\frac{4}{2} = \frac{2x}{2}$$
$$2 = x$$

37.

$$8(x - 5) = 13x$$
$$8x - 40 = 13x$$
$$8x - 40 - 8x = 13x - 8x$$
$$-40 = 5x$$
$$\frac{-40}{5} = \frac{5x}{5}$$
$$-8 = x$$

39.

$$5 + 3(x+7) = 26 - 6(5x+11)$$
$$5 + 3x + 21 = 26 - 30x - 66$$
$$3x + 26 = -30x - 40$$
$$3x + 26 + 30x = -30x - 40 + 30x$$
$$33x + 26 = -40$$
$$33x + 26 - 26 = -40 - 26$$
$$33x = -66$$
$$\frac{33x}{33} = \frac{-66}{33}$$
$$x = -2$$

41.

$$5(2y-3) = 3(7y-6) + 19(y+1) + 14$$
$$10y - 15 = 21y - 18 + 19y + 19 + 14$$
$$10y - 15 = 40y + 15$$
$$10y - 15 - 10y = 40y + 15 - 10y$$
$$-15 = 30y + 15$$
$$-15 - 15 = 30y + 15 - 15$$
$$-30 = 30y$$
$$\frac{-30}{30} = \frac{30y}{30}$$
$$-1 = y$$

43.

$$16(x+3) = 7(x-5) - 9(x+4) - 7$$
$$16x + 48 = 7x - 35 - 9x - 36 - 7$$
$$16x + 48 = -2x - 78$$
$$16x + 48 + 2x = -2x - 78 + 2x$$
$$18x + 48 = -78$$
$$18x + 48 - 48 = -78 - 48$$
$$18x = -126$$
$$\frac{18x}{18} = \frac{-126}{18}$$
$$x = -7$$

45.

$$4(y+2) = 8(y-4) + 7$$
$$4y + 8 = 8y - 32 + 7$$
$$4y + 8 = 8y - 25$$
$$4y + 8 - 4y = 8y - 25 - 4y$$
$$8 = 4y - 25$$
$$8 + 25 = 4y - 25 + 25$$
$$33 = 4y$$
$$\frac{33}{4} = \frac{4y}{4}$$
$$8\frac{1}{4} = y$$

47.

$$4(5y-2) + 3(2y+6) = 25(3y+2) - 19y$$
$$20y - 8 + 6y + 18 = 75y + 50 - 19y$$
$$26y + 10 = 56y + 50$$
$$26y + 10 - 26y = 56y + 50 - 26y$$
$$10 = 30y + 50$$
$$10 - 50 = 30y + 50 - 50$$
$$-40 = 30y$$
$$\frac{-40}{30} = \frac{30y}{30}$$
$$-1\frac{1}{3} = y$$

49.

$$12 + 8(2y+3) = (y+7) - 16$$
$$12 + 16y + 24 = y + 7 - 16$$
$$16y + 36 = y - 9$$
$$16y + 36 - y = y - 9 - y$$
$$15y + 36 = -9$$
$$15y + 36 - 36 = -9 - 36$$
$$15y = -45$$
$$\frac{15y}{15} = \frac{-45}{15}$$
$$y = -3$$

51.

$$5x - 10(3x - 6) = 3(24 - 9x)$$
$$5x - 30x + 60 = 72 - 27x$$
$$60 - 25x = 72 - 27x$$
$$60 - 25x + 27x = 72 - 27x + 27x$$
$$2x + 60 = 72$$
$$2x + 60 - 60 = 72 - 60$$
$$2x = 12$$
$$\frac{2x}{2} = \frac{12}{2}$$
$$x = 6$$

53.

$$6(y - 4) - 4(5y + 1) = 3(y - 2) - 4(2y + 1)$$
$$6y - 24 - 20y - 4 = 3y - 6 - 8y - 4$$
$$-14y - 28 = -5y - 10$$
$$-14y - 28 + 14y = -5y - 10 + 14y$$
$$-28 = 9y - 10$$
$$-28 + 10 = 9y - 10 + 10$$
$$-18 = 9y$$
$$\frac{-18}{9} = \frac{9y}{9}$$
$$-2 = y$$

55.

$$-6(x - 5) + 3x = 6x - 10(-3 + x)$$
$$-6x + 30 + 3x = 6x + 30 - 10x$$
$$30 - 3x = 30 - 4x$$
$$30 - 3x + 4x = 30 - 4x + 4x$$
$$x + 30 = 30$$
$$x + 30 - 30 = 30 - 30$$
$$x = 0$$

57.

$$2.3x - 4.7 + 0.6(3x + 5) = 0.7(3 - x)$$
$$2.3x - 4.7 + 1.8x + 3 = 2.1 - 0.7x$$
$$4.1x - 1.7 = 2.1 - 0.7x$$
$$4.1x - 1.7 + 0.7x = 2.1 - 0.7x + 0.7x$$
$$4.8x - 1.7 = 2.1$$
$$4.8x - 1.7 + 1.7 = 2.1 + 1.7$$
$$4.8x = 3.8$$
$$\frac{4.8x}{4.8} = \frac{3.8}{4.8}$$
$$x = \frac{19}{24}$$

59.

$$0.089x - 0.32 + 0.001(5 - x) = 0.231$$
$$0.089x - 0.32 + 0.005 - 0.001x = 0.231$$
$$0.088x - 0.315 = 0.231$$
$$0.088x - 0.315 + 0.315 = 0.231 + 0.315$$
$$0.088x = 0.546$$
$$\frac{0.088}{0.088}x = \frac{0.546}{0.088}$$
$$x = \frac{546}{88}$$
$$x = 6\frac{9}{44}$$

Section 6.4: Equations with Fractions

1.

$$\frac{2x}{3} = \frac{32}{6}$$
$$6\left(\frac{2x}{3}\right) = \left(\frac{32}{6}\right)6$$
$$4x = 32$$
$$\frac{4x}{4} = \frac{32}{4}$$
$$x = 8$$

3.

$$\frac{3}{8}y = 1\frac{14}{16}$$
$$16\left(\frac{3}{8}y\right) = \left(\frac{30}{16}\right)16$$
$$6y = 30$$
$$\frac{6y}{6} = \frac{30}{6}$$
$$x = 5$$

5.

$$2\frac{1}{2}x = 7\frac{1}{2}$$

$$2\left(\frac{5}{2}x\right) = \left(\frac{15}{2}\right)2$$

$$5x = 15$$

$$\frac{5x}{5} = \frac{15}{5}$$

$$x = 3$$

7.

$$\frac{2}{3} + \frac{x}{4} = \frac{28}{6}$$

$$12\left(\frac{2}{3} + \frac{x}{4}\right) = \left(\frac{28}{6}\right)12$$

$$12\left(\frac{2}{3}\right) + 12\left(\frac{x}{4}\right) = \left(\frac{28}{6}\right)12$$

$$4(2) + 3(x) = 2(28)$$

$$3x + 8 = 56$$

$$3x + 8 - 8 = 56 - 8$$

$$3x = 48$$

$$\frac{3x}{3} = \frac{48}{3}$$

$$x = 16$$

9.

$$\frac{3}{4} - \frac{x}{3} = \frac{5}{12}$$

$$12\left(\frac{3}{4} - \frac{x}{3}\right) = \left(\frac{5}{12}\right)12$$

$$12\left(\frac{3}{4}\right) - 12\left(\frac{x}{3}\right) = \left(\frac{5}{12}\right)12$$

$$3(3) - 4(x) = 1(5)$$

$$9 - 4x = 5$$

$$9 - 4x - 9 = 5 - 9$$

$$-4x = -4$$

$$\frac{-4x}{-4} = \frac{-4}{-4}$$

$$x = 1$$

11.

$$\frac{3}{5}x - 25 = \frac{x}{10}$$

$$10\left(\frac{3}{5}x - 25\right) = \left(\frac{x}{10}\right)10$$

$$10\left(\frac{3}{5}x\right) - 10(25) = \left(\frac{x}{10}\right)10$$

$$2(3x) - 10(25) = 1(x)$$

$$6x - 250 = x$$

$$6x - 250 - 6x = x - 6x$$

$$-250 = -5x$$

$$\frac{-250}{-5} = \frac{-5x}{-5}$$

$$50 = x$$

13.

$$\frac{y}{3} - 1 = \frac{y}{6}$$

$$6\left(\frac{y}{3} - 1\right) = \left(\frac{y}{6}\right)6$$

$$6\left(\frac{y}{3}\right) - 6(1) = \left(\frac{y}{6}\right)6$$

$$2(y) - 6(1) = 1(y)$$

$$2y - 6 = y$$

$$2y - 6 - 2y = y - 2y$$

$$-6 = -y$$

$$\frac{-6}{-1} = \frac{-y}{-1}$$

$$6 = y$$

15.

$$\frac{3x}{4} - \frac{7}{20} = \frac{2}{5}x$$

$$20\left(\frac{3x}{4} - \frac{7}{20}\right) = \left(\frac{2}{5}x\right)20$$

$$20\left(\frac{3x}{4}\right) - 20\left(\frac{7}{20}\right) = \left(\frac{2}{5}x\right)20$$

$$5(3x) - 1(7) = 4(2x)$$

$$15x - 7 = 8x$$

$$15x - 8x - 7 = 8x - 8x$$

$$7x - 7 = 0$$

$$7x - 7 + 7 = 0 + 7$$

$$7x = 7$$

$$\frac{7x}{7} = \frac{7}{7}$$

$$x = 1$$

17.

$$\frac{x}{2} + \frac{2}{3}(2x + 3) = 46$$

$$6\left[\frac{x}{2} + \frac{2}{3}(2x + 3)\right] = (46)6$$

$$6\left[\frac{x}{2}\right] + 6\left[\frac{2}{3}(2x + 3)\right] = (46)6$$

$$3(x) + 4(2x + 3) = 6(46)$$

$$3x + 8x + 12 = 276$$

$$11x + 12 = 276$$

$$11x + 12 - 12 = 276 - 12$$

$$11x = 264$$

$$\frac{11x}{11} = \frac{264}{11}$$

$$x = 24$$

19.

$$0.96 = 0.06(12 + x)$$

$$100(0.96) = \left[0.06(12 + x)\right]100$$

$$96 = 6(12 + x)$$

$$96 = 72 + 6x$$

$$96 - 72 = 72 + 6x - 72$$

$$24 = 6x$$

$$\frac{24}{6} = \frac{6x}{6}$$

$$4 = x$$

21.

$$\frac{3x - 24}{16} - \frac{3x - 12}{12} = 3$$

$$48\left(\frac{3x - 24}{16} - \frac{3x - 12}{12}\right) = (3)48$$

$$48\left(\frac{3x - 24}{16}\right) - 48\left(\frac{3x - 12}{12}\right) = (3)48$$

$$3(3x - 24) - 4(3x - 12) = 48(3)$$

$$9x - 72 - 12x + 48 = 144$$

$$-3x - 24 = 144$$

$$-3x - 24 + 24 = 144 + 24$$

$$-3x = 168$$

$$\frac{-3x}{-3} = \frac{168}{-3}$$

$$x = -56$$

23.

$$5x + \frac{6x - 8}{14} + \frac{10x + 6}{6} = 43$$

$$42\left(5x + \frac{6x - 8}{14} + \frac{10x + 6}{6}\right) = (43)42$$

$$42(5x) + 42\left(\frac{6x - 8}{14}\right) + 42\left(\frac{10x + 6}{6}\right) = (43)42$$

$$42(5x) + 3(6x - 8) + 7(10x + 6) = (43)42$$

$$210x + 18x - 24 + 70x + 42 = 1806$$

$$298x + 18 = 1806$$

$$298x + 18 - 18 = 1806 - 18$$

$$298x = 1788$$

$$\frac{298x}{298} = \frac{1788}{298}$$

$$x = 6$$

25.

$$\frac{4x}{6} - \frac{x+5}{2} = \frac{6x-6}{8}$$

$$24\left(\frac{4x}{6} - \frac{x+5}{2}\right) = \left(\frac{6x-6}{8}\right)24$$

$$24\left(\frac{4x}{6}\right) - 24\left(\frac{x+5}{2}\right) = \left(\frac{6x-6}{8}\right)24$$

$$4(4x) - 12(x+5) = 3(6x-6)$$

$$16x - 12x - 60 = 18x - 18$$

$$4x - 60 = 18x - 18$$

$$4x - 60 - 4x = 18x - 18 - 4x$$

$$-60 = 14x - 18$$

$$-60 + 18 = 14x - 18 + 18$$

$$-42 = 14x$$

$$\frac{-42}{14} = \frac{14x}{14}$$

$$-3 = x$$

27.

$$\frac{4}{x} = 6$$

$$x\left(\frac{4}{x}\right) = (6)x$$

$$4 = 6x$$

$$\frac{4}{6} = \frac{6x}{6}$$

$$\frac{2}{3} = x$$

29.

$$5 - \frac{1}{x} = 7$$

$$x\left(5 - \frac{1}{x}\right) = (7)x$$

$$x(5) - x\left(\frac{1}{x}\right) = (7)x$$

$$5x - 1 = 7x$$

$$5x - 1 - 5x = 7x - 5x$$

$$-1 = 2x$$

$$\frac{-1}{2} = \frac{2x}{2}$$

$$-\frac{1}{2} = x$$

31.

$$\frac{5}{y} - 1 = 4$$

$$y\left(\frac{5}{y} - 1\right) = (4)y$$

$$y\left(\frac{5}{y}\right) - y(1) = (4)y$$

$$5 - y = 4y$$

$$5 - y + y = 4y + y$$

$$5 = 5y$$

$$\frac{5}{5} = \frac{5y}{5}$$

$$1 = y$$

33.

$$\frac{3}{x} - 8 = 7$$

$$x\left(\frac{3}{x} - 8\right) = (7)x$$

$$x\left(\frac{3}{x}\right) - x(8) = (7)x$$

$$3 - 8x = 7x$$

$$3 - 8x + 8x = 7x + 8x$$

$$3 = 15x$$

$$\frac{3}{15} = \frac{15x}{15}$$

$$\frac{1}{5} = x$$

35.

$$7 - \frac{6}{x} = 5$$

$$x\left(7 - \frac{6}{x}\right) = (5)x$$

$$x(7) - x\left(\frac{6}{x}\right) = (5)x$$

$$7x - 6 = 5x$$

$$7x - 6 - 5x = 5x - 5x$$

$$2x - 6 = 0$$

$$2x - 6 + 6 = 0 + 6$$

$$2x = 6$$

$$\frac{2x}{2} = \frac{6}{2}$$

$$x = 3$$

37.

$$\frac{6}{x} + 5 = 14$$

$$x\left(\frac{6}{x} + 5\right) = (14)x$$

$$x\left(\frac{6}{x}\right) + x(5) = (14)x$$

$$6 + 5x = 14x$$

$$6 + 5x - 5x = 14x - 5x$$

$$6 = 9x$$

$$\frac{6}{9} = \frac{9x}{9}$$

$$\frac{2}{3} = x$$

39.

$$1 - \frac{2}{x} = \frac{14}{3x} - \frac{1}{3}$$

$$3x\left(1 - \frac{2}{x}\right) = \left(\frac{14}{3x} - \frac{1}{3}\right)3x$$

$$3x(1) - 3x\left(\frac{2}{x}\right) = \left(\frac{14}{3x}\right)3x - \left(\frac{1}{3}\right)3x$$

$$3x - 6 = 14 - x$$

$$3x - 6 + x = 14 - x + x$$

$$4x - 6 = 14$$

$$4x - 6 + 6 = 14 + 6$$

$$4x = 20$$

$$\frac{4x}{4} = \frac{20}{4}$$

$$x = 5$$

41.

$$\frac{7}{2x} + 14\frac{1}{2} = \frac{7}{x} - 10$$

$$\frac{7}{2x} + \frac{29}{2} = \frac{7}{x} - 10$$

$$2x\left(\frac{7}{2x} + \frac{29}{2}\right) = \left(\frac{7}{x} - 10\right)2x$$

$$2x\left(\frac{7}{2x}\right) + 2x\left(\frac{29}{2}\right) = \left(\frac{7}{x}\right)2x - (10)2x$$

$$29x + 7 = 14 - 20x$$

$$29x + 7 + 20x = 14 - 20x + 20x$$

$$49x + 7 = 14$$

$$49x + 7 - 7 = 14 - 7$$

$$49x = 7$$

$$\frac{49x}{49} = \frac{7}{49}$$

$$x = \frac{1}{7}$$

Section 6.5: Translating Words into Algebraic Symbols

1. $x - 20$

3. $\frac{x}{6}$

5. $x + 16$

7. $26 - x$

9. $2x$

11. $6x + 28 = 40$

13. $\frac{x}{6} = 5$

15. $5(x + 28) = 150$

17. $\frac{x}{6} - 7 = 2$

19. $30 - 2x = 4$

21. $(x - 7)(x + 5) = 13$

23. $6x - 17 = 7$

25. $4x - 17 = 63$

Section 6.6: Applications Involving Equations

1.

Let x = space between shelves.

8 ft 4 in. is $8 \text{ ft} \times \dfrac{12 \text{ in.}}{1 \text{ ft}} + 2 \text{ in.} = 98 \text{ in.}$ and there will be 8 shelves and 9 spaces, so:

$$9(x) + 9(1) = 98$$
$$9x + 9 = 98$$
$$9x + 8 - 8 = 98 - 8$$
$$9x = 90$$
$$\dfrac{9x}{9} = \dfrac{90}{9}$$
$$x = 10$$

The shelves should be 10 in. apart.

3.

Let x = number of incandescent lights.

$2x - 20$ = number of flourescent lights

The order contains 256 light fixtures, so:

$$x + (2x - 20) = 256$$
$$3x - 20 = 256$$
$$3x - 20 + 20 = 256 + 20$$
$$3x = 276$$
$$\dfrac{3x}{3} = \dfrac{276}{3}$$
$$x = 92$$

The order will consist of 92 incandescent light fixtures and 184 fluorescent light fixtures.

5.

Let x = John's share.

$2x$ = Maria's share.

$3x$ = Betsy's share.

$4950 is being distributed, so:

$$x + 2x + 3x = 4950$$
$$6x = 4950$$
$$\dfrac{6x}{6} = \dfrac{4950}{6}$$
$$x = 825$$

John should receive $825, Maria should receive $1650, and Betsy should receive $2475.

7.

Let x = width.

$2x$ = length.

The perimeter is 60 cm, so:

$$P = 2l + 2w$$
$$60 = 2(x) + 2(2x)$$
$$60 = 2x + 4x$$
$$60 = 6x$$
$$60 = 6x$$
$$\dfrac{60}{6} = \dfrac{6x}{6}$$
$$10 = x$$

The width is 10 cm and the length is 20 cm.

9.

Let x = lengths of fence touching house.

$x - 15$ = length of fence opposite house.

The total length is 345 ft, so:

$$x + x + (x - 15) = 345$$
$$3x - 15 = 345$$
$$3x - 15 + 15 = 345 + 15$$
$$3x = 360$$
$$\dfrac{3x}{3} = \dfrac{360}{3}$$
$$x = 120$$

The two equal sides are 120 ft and the side opposite the house is 105 ft.

11.

Let x = length of first side.

x = length of second equal side.

$x - 4$ = length of third side.

The perimeter is 122 ft, so:

$$x + x + (x - 4) = 122$$
$$3x - 4 = 122$$
$$3x - 4 + 4 = 122 + 4$$
$$3x = 126$$
$$\dfrac{3x}{3} = \dfrac{126}{3}$$
$$x = 42$$

The two equal sides are 42 ft in length and the third side is 38 ft in length.

13. Note: 18 inches is $18 \text{ in.} \times \dfrac{1 \text{ ft}}{12 \text{ in.}} = 1.5 \text{ ft.}$

Let x = length of first piece.

$x + 1.5$ = length of second piece.

The combined length of the two pieces is 12 feet, so:

$$x + (x + 1.5) = 12$$
$$2x + 1.5 = 12$$
$$2x + 1.5 - 1.5 = 12 - 1.5$$
$$2x = 10.5$$
$$\frac{2x}{2} = \frac{10.5}{2}$$
$$x = 5.25$$

The pieces are 5.25 ft and 6.75 ft long.

15.

Let x = number of \$6.50 boards.

$20 - x$ = number of \$9.50 boards.

Their total cost was \$166, so:

$$6.5x + 9.5(20 - x) = 166$$
$$6.5x + 190 - 9.5x = 166$$
$$190 - 3x = 166$$
$$190 - 3x - 190 = 166 - 190$$
$$-3x = -24$$
$$\frac{-3x}{-3} = \frac{-24}{-3}$$
$$x = 8$$

Eight \$6.50 boards and 12 \$9.50 boards were purchased.

17.

Let x = amount in 4% account.

$7500 - x$ = amount in 2.5% account.

The total interest was \$232.50, so:

$$0.04x + 0.025(7500 - x) = 232.50$$
$$0.04x + 187.5 - 0.025x = 232.50$$
$$0.015x + 187.5 = 232.5$$
$$0.015x + 187.5 - 187.5 = 232.5 - 187.5$$
$$0.015x = 45$$
$$\frac{0.015x}{0.015} = \frac{45}{0.015}$$
$$x = 3000$$

\$3000 was in the 4% account and \$4500 was in the 2.5% account.

19.

Let x = litres of 4% milk.

You need 2% milk, so:

$$0.04x + 0.01(40) = 0.02(x + 40)$$
$$0.04x + 0.4 = 0.02x + 0.8$$
$$0.04x + 0.4 - 0.02x = 0.02x + 0.8 - 0.02x$$
$$0.02x + 0.4 = 0.8$$
$$0.02x + 0.4 - 0.4 = 0.8 - 0.4$$
$$0.02x = 0.4$$
$$\frac{0.02x}{0.02} = \frac{0.4}{0.02}$$
$$x = 20$$

20 L of 4% milk should be added.

21.

Let x = millilitres of 30% alcohol solution.

$800 - x$ = millilitres of 80% alcohol solution.

You need a 60% alcohol solution, so:

$$0.30x + 0.80(800 - x) = 0.60(800)$$
$$0.30x + 640.0 - 0.8x = 0.60(800)$$
$$640 - 0.5x = 480$$
$$640 - 0.5x - 640 = 480 - 640$$
$$-0.5x = -160$$
$$\frac{-0.5x}{-0.5} = \frac{-160}{-0.5}$$
$$x = 320$$

320 mL of 30% alcohol solution should be combined with 480 mL of 80% alcohol solution.

23.

Let x = quarts of pure antifreeze.

$12 - x$ = quarts of 40% gasoline mixture.

You need a 60% antifreeze mixture, so:

$$1.00x + 0.40(12 - x) = 0.60(12)$$
$$1.00x + 4.8 - 0.4x = 0.60(12)$$
$$0.6x + 4.8 = 7.2$$
$$0.6x + 4.8 - 4.8 = 7.2 - 4.8$$
$$0.6x = 2.4$$
$$\frac{0.6x}{0.6} = \frac{2.4}{0.6}$$
$$x = 4$$

4 quarts of solution should be drained and replaced with pure antifreeze.

25.

Let x = cubic feet of sea water.

You need 125 lb of sea salt, so:

$$\frac{64\text{ lb}}{1\text{ ft}^3}(0.0035)x = 125$$

$$0.224x = 125$$

$$\frac{0.224x}{0.224} = \frac{125}{0.224}$$

$$x = 558$$

558 cubic feet of sea water is required.

27.

Let x = additional investment.

The interest on the total amount invested should be $6250, so:

$$0.05(30,000 + 40,000 + x) = 6250$$

$$0.05(70,000 + x) = 6250$$

$$0.05x + 3500 = 6250$$

$$0.05x + 3500 - 3500 = 6250 - 3500$$

$$0.05x = 2750$$

$$\frac{0.05x}{0.05} = \frac{2750}{0.05}$$

$$x = 55,000$$

An additional $55,000 should be invested.

29.

Let x = the width of the table.

$3x$ = the length of the table.

$$P = 2l + 2w$$

$$240 = 2(3x) + 2(x)$$

$$240 = 8x$$

$$\frac{240}{8} = \frac{8x}{8}$$

$$30 = x$$

The tables are 30 in. by 90 in.

31.

Let x = cost of 80% lean ground beef.

$1.25x$ = cost of 90% lean ground beef.

$$x(1) + 1.25x(1) = 5.40$$

$$2.25x = 5.4$$

$$\frac{2.25x}{2.25} = \frac{5.4}{2.25}$$

$$x = 2.40$$

80% lean ground beef costs $2.40 per pound and 90% lean ground beef costs $3.00 per pound.

Section 6.7: Formulas

1.

$$E = Ir$$

$$\frac{E}{I} = \frac{Ir}{I}$$

$$\frac{E}{I} = r,\text{ so } r = \frac{E}{I}$$

3.

$$F = ma$$

$$\frac{F}{m} = \frac{ma}{m}$$

$$\frac{F}{m} = a,\text{ so } a = \frac{F}{m}$$

5.

$$C = \pi d$$

$$\frac{C}{\pi} = \frac{\pi d}{\pi}$$

$$\frac{C}{\pi} = d,\text{ so } d = \frac{C}{\pi}$$

7.

$$V = lwh$$

$$\frac{V}{lh} = \frac{lwh}{lh}$$

$$\frac{V}{hl} = w,\text{ so } w = \frac{V}{hl}$$

9.

$$A = 2\pi rh$$

$$\frac{A}{2\pi r} = \frac{2\pi rh}{2\pi r}$$

$$\frac{A}{2\pi r} = h,\text{ so } h = \frac{A}{2\pi r}$$

11.

$$v^2 = 2gh$$

$$\frac{v^2}{2g} = \frac{2gh}{2g}$$

$$\frac{v^2}{2g} = h,\text{ so } h = \frac{v^2}{2g}$$

13.

$$I = \frac{Q}{t}$$

$$t(I) = \left(\frac{Q}{t}\right)t$$

$$tI = Q$$

$$\frac{tI}{I} = \frac{Q}{I}$$

$$t = \frac{Q}{I}$$

15.

$$v = \frac{s}{t}$$

$$t(v) = \left(\frac{s}{t}\right)t$$

$$vt = s, \text{ so } s = vt$$

17.

$$I = \frac{V}{R}$$

$$R(I) = \left(\frac{V}{R}\right)R$$

$$RI = V$$

$$\frac{RI}{I} = \frac{V}{I}$$

$$R = \frac{V}{I}$$

19.

$$E = \frac{I}{4\pi r^2}$$

$$4\pi r^2(E) = \left(\frac{I}{4\pi r^2}\right)4\pi r^2$$

$$4\pi r^2 E = I, \text{ so } I = 4\pi r^2 E$$

21.

$$X_c = \frac{1}{2\pi fC}$$

$$2\pi fC(X_c) = \left(\frac{1}{2\pi fC}\right)2\pi fC$$

$$2\pi fCX_c = 1$$

$$\frac{2\pi fCX_c}{2\pi CX_c} = \frac{1}{2\pi CX_c}$$

$$f = \frac{1}{2\pi CX_c}$$

23.

$$A = \frac{1}{2}bh$$

$$2(A) = \left(\frac{1}{2}bh\right)2$$

$$2A = bh$$

$$\frac{2A}{h} = \frac{bh}{h}$$

$$\frac{2A}{h} = b, \text{ so } b = \frac{2A}{h}$$

25.

$$Q = \frac{I^2 Rt}{J}$$

$$J(Q) = \left(\frac{I^2 Rt}{J}\right)J$$

$$JQ = RtI^2$$

$$\frac{JQ}{tI^2} = \frac{RtI^2}{tI^2}$$

$$\frac{JQ}{tI^2} = R, \text{ so } R = \frac{JQ}{tI^2}$$

27.

$$F = \frac{9}{5}C + 32$$

$$F - 32 = \frac{9}{5}C + 32 - 32$$

$$\frac{5}{9}(F - 32) = \left(\frac{9}{5}C\right)\frac{5}{9}$$

$$\frac{5}{9}(F - 32) = C, \text{ so } C = \frac{5}{9}(F - 32)$$

29.

$$C_T = C_1 + C_2 + C_3 + C_4$$

$$C_T - C_1 - C_3 - C_4 = C_1 + C_2 + C_3 + C_4 - C_1 - C_3 - C_4$$

$$C_T - C_1 - C_3 - C_4 = C_2$$

$$\text{So } C_2 = C_T - C_1 - C_3 - C_4$$

31.

$$Ax + By + C = 0$$

$$Ax + By + C - By - C = 0 - By - C$$

$$Ax = -By - C$$

$$\frac{Ax}{A} = \frac{-By - C}{A}$$

$$x = \frac{-By - C}{A}$$

33.

$$Q_1 = P(Q_2 - Q_1)$$

$$Q_1 = PQ_2 - PQ_1$$

$$Q_1 + PQ_1 = PQ_2 - PQ_1 + PQ_1$$

$$Q_1 + PQ_1 = PQ_2$$

$$\frac{Q_1 + PQ_1}{P} = \frac{PQ_2}{P}$$

$$\frac{Q_1 + PQ_1}{P} = Q_2$$

$$\frac{Q_1}{P} + \frac{PQ_1}{P} = Q_2$$

$$\frac{Q_1}{P} + Q_1 = Q_2$$

$$\text{So } Q_2 = \frac{Q_1 + PQ_1}{P} \text{ or } \frac{Q_1}{P} + Q_1$$

35.

$$A = \left(\frac{a+b}{2}\right)h$$

$$2(A) = \left[\left(\frac{a+b}{2}\right)h\right]2$$

$$2A = (a+b)h$$

$$\frac{2A}{a+b} = \frac{h(a+b)}{a+b}$$

$$\frac{2A}{a+b} = h, \text{ so } h = \frac{2A}{a+b}$$

37.

$$l = a + (n-1)d$$

$$l - a = a + (n-1)d - a$$

$$l - a = d(n-1)$$

$$\frac{l-a}{n-1} = \frac{d(n-1)}{n-1}$$

$$\frac{l-a}{n-1} = d, \text{ so } d = \frac{l-a}{n-1}$$

39.

$$Ft = m(V_2 - V_1)$$

$$\frac{Ft}{V_2 - V_1} = \frac{m(V_2 - V_1)}{V_2 - V_1}$$

$$\frac{Ft}{V_2 - V_1} = m, \text{ so } m = \frac{Ft}{V_2 - V_1}$$

41.

$$Q = wc(T_1 - T_2)$$

$$\frac{Q}{w(T_1 - T_2)} = \frac{wc(T_1 - T_2)}{w(T_1 - T_2)}$$

$$\frac{Q}{w(T_1 - T_2)} = c, \text{ so } c = \frac{Q}{w(T_1 - T_2)}$$

43.

$$V = \frac{2\pi(3690 + h)}{P}$$

$$P(V) = \left[\frac{2\pi(3690 + h)}{P}\right]P$$

$$PV = 7380\pi + 2\pi h$$

$$PV - 7380\pi = 7380\pi + 2\pi h - 7380\pi$$

$$PV - 7380\pi = 2\pi h$$

$$\frac{PV - 7380\pi}{2\pi} = \frac{2\pi h}{2\pi}$$

$$\frac{PV}{2\pi} - \frac{7380\pi}{2\pi} = h$$

$$\frac{PV}{2\pi} - 3690 = h, \text{ so } h = \frac{PV}{2\pi} - 3690$$

Section 6.8: Substituting Data into Formulas

1.

a.

$$A = lw$$

$$\frac{A}{w} = \frac{lw}{w}$$

$$\frac{A}{w} = l, \text{ so } l = \frac{A}{w}$$

b.

$$l = \frac{A}{w}$$

$$l = \frac{414}{18.0}$$

$$= 23.0$$

3.

a.

$$V = \frac{\pi r^2 h}{3}$$

$$3(V) = \left(\frac{\pi r^2 h}{3}\right) 3$$

$$3V = \pi h r^2$$

$$\frac{3V}{\pi r^2} = \frac{\pi h r^2}{\pi r^2}$$

$$\frac{3V}{\pi r^2} = h, \text{ so } h = \frac{3V}{\pi r^2}$$

b.

$$h = \frac{3V}{\pi r^2}$$

$$h = \frac{3(753.6)}{\pi (6.00)^2}$$

$$= 20.0$$

5.

a.

$$E = \frac{mv^2}{2}$$

$$2(E) = \left(\frac{mv^2}{2}\right) 2$$

$$2E = mv^2$$

$$\frac{2E}{v^2} = \frac{mv^2}{v^2}$$

$$\frac{2E}{v^2} = m, \text{ so } m = \frac{2E}{v^2}$$

b.

$$m = \frac{2E}{v^2}$$

$$m = \frac{2(484,000)}{(22.0)^2}$$

$$= 2000$$

7.

a.

$$v_f = v_i + at$$

$$v_f - v_i = v_i + at - v_i$$

$$v_f - v_i = at$$

$$\frac{v_f - v_i}{a} = \frac{at}{a}$$

$$\frac{v_f - v_i}{a} = t, \text{ so } t = \frac{v_f - v_i}{a}$$

b.

$$t = \frac{v_f - v_i}{a}$$

$$t = \frac{(193.1) - (14.9)}{18.0}$$

$$= 9.90$$

9.

a.
$$v_f^2 = v_i^2 + 2gh$$
$$v_f^2 - v_i^2 = v_i^2 + 2gh - v_i^2$$
$$v_f^2 - v_i^2 = 2gh$$
$$\frac{v_f^2 - v_i^2}{2g} = \frac{2gh}{2g}$$
$$\frac{v_f^2 - v_i^2}{2g} = h, \text{ so } h = \frac{v_f^2 - v_i^2}{2g}$$

b.
$$h = \frac{v_f^2 - v_i^2}{2g}$$
$$h = \frac{(192)^2 - (0)^2}{2(32.0)}$$
$$= 576$$

11.

a.
$$L = \pi(r_1 + r_2) + 2d$$
$$L = \pi r_1 + \pi r_2 + 2d$$
$$L - \pi r_2 - 2d = \pi r_1 + \pi r_2 + 2d - \pi r_2 - 2d$$
$$L - 2d - \pi r_2 = \pi r_1$$
$$\frac{L - 2d - \pi r_2}{\pi} = \frac{\pi r_1}{\pi}$$
$$\frac{L - 2d - \pi r_2}{\pi} = r_1, \text{ so } r_1 = \frac{L - 2d - \pi r_2}{\pi}$$

b.
$$r_1 = \frac{L - 2d - \pi r_2}{\pi}$$
$$r_1 = \frac{(37.68) - 2(6.28) - \pi(5.00)}{\pi}$$
$$= 3.00$$

13.

a.
$$Fgr = Wv^2$$
$$\frac{Fgr}{Fg} = \frac{Wv^2}{Fg}$$
$$r = \frac{Wv^2}{Fg}$$

b.
$$r = \frac{Wv^2}{Fg}$$
$$r = \frac{(24,000)(176)^2}{(12,000)(32)}$$
$$= 1900$$

15.

a.
$$A = \frac{h}{2}(a + b)$$
$$2(A) = \left[\frac{h}{2}(a + b)\right]2$$
$$2A = h(a + b)$$
$$2A = ha + hb$$
$$2A - ha = ha + hb - ha$$
$$2A - ah = bh$$
$$\frac{2A - ah}{h} = \frac{bh}{h}$$
$$\frac{2A - ah}{h} = b, \text{ so } b = \frac{2A - ah}{h}$$

b.
$$b = \frac{2A - ah}{h}$$
$$b = \frac{2(1160) - (56.5)(22.0)}{22.0}$$
$$= 49.0$$

17. a.

$$V = \frac{1}{2}lw(D+d)$$

$$2(V) = \left[\frac{1}{2}lw(D+d)\right]2$$

$$2V = lwD + lwd$$

$$2V - lwD = lwD + lwd - lwD$$

$$2V - lwD = dlw$$

$$\frac{2V - lwD}{lw} = \frac{dlw}{lw}$$

$$\frac{2V - lwD}{lw} = d$$

$$\frac{2V - lwD}{lw} = d$$

$$\text{so } d = \frac{2V - lwD}{lw}$$

b.

$$d = \frac{2V - lwD}{lw}$$

$$d = \frac{2(226.8) - (9.00)(6.30)(5.00)}{(9.00)(6.30)}$$

$$= 3.00$$

21.

$$P = I^2R$$

$$P = (4.50 \text{ A})^2 (16.0 \text{ }\Omega)$$

$$= 324 \text{ W}$$

23. First, solve for w.

$$A = lw$$

$$\frac{A}{l} = \frac{lw}{l}$$

$$\frac{A}{l} = w$$

Then, substitute the data.

$$w = \frac{A}{l}$$

$$w = \frac{84.0 \text{ ft}^2}{12.5 \text{ ft}}$$

$$= 6.72 \text{ ft}$$

19. a.

$$S = \frac{n}{2}(a+l)$$

$$2(S) = \left[\frac{n}{2}(a+l)\right]2$$

$$2S = n(a+l)$$

$$2S = na + nl$$

$$2S - nl = na + nl - nl$$

$$2S - nl = na$$

$$\frac{2S - nl}{n} = \frac{na}{n}$$

$$\frac{2S - nl}{n} = a$$

$$\text{so } a = \frac{2S - nl}{n}$$

b.

$$a = \frac{2S - nl}{n}$$

$$a = \frac{2(147.9) - (14.5)(3.80)}{14.5}$$

$$= 16.6$$

25. First, solve for h.

$$V = \pi r^2 h$$

$$\frac{V}{\pi r^2} = \frac{\pi r^2 h}{\pi r^2}$$

$$\frac{V}{\pi r^2} = h$$

Then, substitute the data.

$$h = \frac{V}{\pi r^2}$$

$$h = \frac{8550 \text{ m}^3}{\pi (15.0 \text{ m})^2} = 12.1 \text{ m}$$

27.

$$\frac{1}{R} = \frac{1}{R_1} + \frac{1}{R_2}$$

$$\frac{1}{R} = \frac{1}{20.0\ \Omega} + \frac{1}{60.0\ \Omega}$$

$$\frac{1}{R} = \frac{1}{15.0\ \Omega}$$

$$R \cdot 15.0\ \Omega \left(\frac{1}{R}\right) = \left(\frac{1}{15.0\ \Omega}\right) R \cdot 15.0\ \Omega$$

$$15.0\ \Omega = R$$

29.

$$\Delta l = \alpha l \Delta T$$

$$\Delta l = \left(6.5 \times 10^{-6}/{}^{\circ}\text{F}\right)\left(50.0\ \text{ft}\right)\left(110{}^{\circ}\text{F}\right) \times \frac{12\ \text{in.}}{1\ \text{ft}}$$

$$= 0.43\ \text{in.}$$

Section 6.9: Reciprocal Formulas Using a Calculator

1.

$$\frac{1}{R} = \frac{1}{R_1} + \frac{1}{R_2}$$

$$\frac{1}{R} = \frac{1}{8.00\ \Omega} + \frac{1}{12.0\ \Omega}$$

$$R = 4.8\ \Omega$$

3.

$$\frac{1}{R} = \frac{1}{R_1} + \frac{1}{R_2}$$

$$\frac{1}{R_1} = \frac{1}{R} - \frac{1}{R_2}$$

$$\frac{1}{R_1} = \frac{1}{12.0\ \Omega} - \frac{1}{36.0\ \Omega}$$

$$R_1 = 18.0\ \Omega$$

5.

$$\frac{1}{R} = \frac{1}{R_1} + \frac{1}{R_2}$$

$$\frac{1}{R_1} = \frac{1}{R} - \frac{1}{R_2}$$

$$\frac{1}{R_1} = \frac{1}{15.0\ \Omega} - \frac{1}{24.0\ \Omega}$$

$$R_1 = 40.0\ \Omega$$

7.

$$\frac{1}{f} = \frac{1}{s_0} + \frac{1}{s_i}$$

$$\frac{1}{f} = \frac{1}{3.00\ \text{cm}} + \frac{1}{15.0\ \text{cm}}$$

$$f = 2.50\ \text{cm}$$

9.

$$\frac{1}{f} = \frac{1}{s_0} + \frac{1}{s_i}$$

$$\frac{1}{s_i} = \frac{1}{f} - \frac{1}{so}$$

$$\frac{1}{s_i} = \frac{1}{14.5\ \text{cm}} - \frac{1}{21.5\ \text{cm}}$$

$$f = 44.5\ \text{cm}$$

11.

$$\frac{1}{R} = \frac{1}{R_1} + \frac{1}{R_2} + \frac{1}{R_3}$$

$$\frac{1}{R} = \frac{1}{30.0\ \Omega} + \frac{1}{18.0\ \Omega} + \frac{1}{45.0\ \Omega}$$

$$R = 9.00\ \Omega$$

13.

$$\frac{1}{R} = \frac{1}{R_1} + \frac{1}{R_2} + \frac{1}{R_3}$$

$$\frac{1}{R_3} = \frac{1}{R} - \frac{1}{R_1} - \frac{1}{R_2}$$

$$\frac{1}{R_3} = \frac{1}{80.0\ \Omega} - \frac{1}{175\ \Omega} - \frac{1}{275\ \Omega}$$

$$R_3 = 318\ \Omega$$

15.

$$\frac{1}{R} = \frac{1}{R_1} + \frac{1}{R_2} + \frac{1}{R_3}$$

$$\frac{1}{R_2} = \frac{1}{R} - \frac{1}{R_1} - \frac{1}{R_3}$$

$$\frac{1}{R_2} = \frac{1}{1250\ \Omega} - \frac{1}{3750\ \Omega} - \frac{1}{4450\ \Omega}$$

$$R_2 = 3240\ \Omega$$

17.

$$\frac{1}{C} = \frac{1}{C_1} + \frac{1}{C_2} + \frac{1}{C_3}$$

$$\frac{1}{C} = \frac{1}{12.0\ \mu F} + \frac{1}{24.0\ \mu F} + \frac{1}{24.0\ \mu F}$$

$$C = 6.00\ \mu F$$

19.

$$\frac{1}{C} = \frac{1}{C_1} + \frac{1}{C_2} + \frac{1}{C_3}$$

$$\frac{1}{C_3} = \frac{1}{C} - \frac{1}{C_1} - \frac{1}{C_2}$$

$$\frac{1}{C_3} = \frac{1}{1.25 \times 10^{-6}\ F} - \frac{1}{8.75 \times 10^{-6}\ F} - \frac{1}{6.15 \times 10^{-6}\ F}$$

$$C_3 = 1.91 \times 10^{-6}\ F$$

21.

$$\frac{1}{C} = \frac{1}{C_1} + \frac{1}{C_2} + \frac{1}{C_3}$$

$$\frac{1}{C} = \frac{1}{6.56 \times 10^{-7}\ F} + \frac{1}{5.05 \times 10^{-6}\ F} + \frac{1}{1.79 \times 10^{-8}\ F}$$

$$C = 1.74 \times 10^{-8}\ F$$

23.

$$\frac{1}{R} = \frac{1}{R_1} + \frac{1}{R_2} + \frac{1}{R_3} + \frac{1}{R_4}$$

$$\frac{1}{R} = \frac{1}{655\ \Omega} + \frac{1}{775\ \Omega} + \frac{1}{1050\ \Omega} + \frac{1}{1250\ \Omega}$$

$$R = 219\ \Omega$$

Chapter 6 Review

1.

$$2x + 4 = 7$$
$$2x + 4 - 4 = 7 - 4$$
$$2x = 3$$
$$\frac{2x}{2} = \frac{3}{2}$$
$$x = 1\frac{1}{2}$$

2.

$$11 - 3x = 23$$
$$11 - 3x - 11 = 23 - 11$$
$$-3x = 12$$
$$\frac{-3x}{-3} = \frac{12}{-3}$$
$$x = -4$$

3.

$$\frac{x}{3} - 7 = 12$$
$$3\left(\frac{x}{3} - 7\right) = (12)3$$
$$3\left(\frac{x}{3}\right) - 3(7) = 3(12)$$
$$x - 21 = 36$$
$$x - 21 + 21 = 36 + 21$$
$$x = 57$$

4.

$$5 - \frac{x}{6} = 1$$

$$6\left(5 - \frac{x}{6}\right) = (1)6$$

$$6(5) - 6\left(\frac{x}{6}\right) = 6(1)$$

$$30 - x = 6$$

$$30 - x - 30 = 6 - 30$$

$$-x = -24$$

$$\frac{-x}{-1} = \frac{-24}{-1}$$

$$x = 24$$

5.

$$78 - 16y = 190$$

$$78 - 16y - 78 = 190 - 78$$

$$-16y = 112$$

$$\frac{-16y}{-16} = \frac{112}{-16}$$

$$y = -7$$

6.

$$25 = 3x - 2$$

$$25 + 2 = 3x - 2 + 2$$

$$27 = 3x$$

$$\frac{27}{3} = \frac{3x}{3}$$

$$9 = x$$

7.

$$2x + 9 = 5x - 15$$

$$2x + 9 - 2x = 5x - 15 - 2x$$

$$9 = 3x - 15$$

$$9 + 15 = 3x - 15 + 15$$

$$24 = 3x$$

$$\frac{24}{3} = \frac{3x}{3}$$

$$8 = x$$

8.

$$-6x + 5 = 2x - 19$$

$$-6x + 5 + 6x = 2x - 19 + 6x$$

$$5 = 8x - 19$$

$$5 + 19 = 8x - 19 + 19$$

$$24 = 8x$$

$$\frac{24}{8} = \frac{8x}{8}$$

$$3 = x$$

9.

$$3 - 2x = 9 - 3x$$

$$3 - 2x + 3x = 9 - 3x + 3x$$

$$x + 3 = 9$$

$$x + 3 - 3 = 9 - 3$$

$$x = 6$$

10.

$$4x + 1 = 4 - x$$

$$4x + 1 + x = 4 - x + x$$

$$5x + 1 = 4$$

$$5x + 1 - 1 = 4 - 1$$

$$5x = 3$$

$$\frac{5x}{5} = \frac{3}{5}$$

$$x = \frac{3}{5}$$

11.

$$7 - (x - 5) = 11$$

$$7 - x + 5 = 11$$

$$12 - x = 1$$

$$12 - x - 12 = 1 - 12$$

$$-x = -11$$

$$\frac{-x}{-1} = \frac{-11}{-1}$$

$$x = 11$$

12.

$$4x + 2(x + 3) = 42$$

$$4x + 2x + 6 = 42$$

$$6x + 6 = 42$$

$$6x + 6 - 6 = 42 - 6$$

$$6x = 36$$

$$\frac{6x}{6} = \frac{36}{6}$$

$$x = 6$$

13.

$$3y - 5(2 - y) = 22$$
$$3y - 10 + 5y = 22$$
$$8y - 10 = 22$$
$$8y - 10 + 10 = 22 + 10$$
$$8y = 32$$
$$\frac{8y}{8} = \frac{32}{8}$$
$$y = 4$$

14.

$$6(x + 7) - 5(x + 8) = 0$$
$$6x + 42 - 5x - 40 = 0$$
$$x + 2 = 0$$
$$x + 2 - 2 = 0 - 2$$
$$x = -2$$

15.

$$3x - 4(x - 3) = 3(x - 4)$$
$$3x - 4x + 12 = 3x - 12$$
$$12 - x = 3x - 12$$
$$12 - x + x = 3x - 12 + x$$
$$12 = 4x - 12$$
$$12 + 12 = 4x - 12 + 12$$
$$24 = 4x$$
$$\frac{24}{4} = \frac{4x}{4}$$
$$6 = x$$

16.

$$4(x + 3) - 9(x - 2) = x + 27$$
$$4x + 12 - 9x + 18 = x + 27$$
$$30 - 5x = x + 27$$
$$30 - 5x + 5x = x + 27 + 5x$$
$$30 = 6x + 27$$
$$30 - 27 = 6x + 27 - 27$$
$$3 = 6x$$
$$\frac{3}{6} = \frac{6x}{6}$$
$$\frac{1}{2} = x$$

17.

$$\frac{2x}{3} = \frac{16}{9}$$
$$9\left(\frac{2x}{3}\right) = \left(\frac{16}{9}\right)9$$
$$6x = 16$$
$$\frac{6x}{6} = \frac{16}{6}$$
$$x = 2\frac{2}{3}$$

18.

$$\frac{x}{3} - 2 = \frac{3x}{5}$$
$$15\left(\frac{x}{3} - 2\right) = \left(\frac{3x}{5}\right)15$$
$$15\left(\frac{x}{3}\right) - 15(2) = \left(\frac{3x}{5}\right)15 = \left(\frac{3x}{5}\right)15$$
$$5x - 30 = 9x$$
$$5x - 30 - 5x = 9x - 5x$$
$$-30 = 4x$$
$$\frac{-30}{4} = \frac{4x}{4}$$
$$-7\frac{1}{2} = x$$

19.

$$\frac{3x}{4} - \frac{x - 1}{5} = \frac{3 + x}{2}$$
$$20\left(\frac{3x}{4} - \frac{x - 1}{5}\right) = \left(\frac{3 + x}{2}\right)20$$
$$20\left(\frac{3x}{4}\right) - 20\left(\frac{x - 1}{5}\right) = \left(\frac{3 + x}{2}\right)20$$
$$5(3x) - 4(x - 1) = 10(3 + x)$$
$$15x - 4x + 4 = 10x + 30$$
$$11x + 4 = 10x + 30$$
$$11x + 4 - 10x = 10x + 30 - 10x$$
$$x + 4 = 30$$
$$x + 4 - 4 = 30 - 4$$
$$x = 26$$

20.

$$\frac{7}{x} - 3 = \frac{1}{x}$$

$$x\left(\frac{7}{x} - 3\right) = \left(\frac{1}{x}\right)x$$

$$x\left(\frac{7}{x}\right) - x(3) = \left(\frac{1}{x}\right)x$$

$$7 - 3x = 1$$

$$7 - 3x - 7 = 1 - 7$$

$$-3x = -6$$

$$\frac{-3x}{-3} = \frac{-6}{-3}$$

$$x = 2$$

21.

$$5 - \frac{7}{x} = 3\frac{3}{5}$$

$$5x\left(5 - \frac{7}{x}\right) = \left(\frac{18}{5}\right)5x$$

$$5x(5) - 5x\left(\frac{7}{x}\right) = \left(\frac{18}{5}\right)5x$$

$$25x - 35 = 18x$$

$$25x - 35 - 18x = 18x - 18x$$

$$7x - 35 = 0$$

$$7x - 35 + 35 = 0 + 35$$

$$7x = 35$$

$$\frac{7x}{7} = \frac{35}{7}$$

$$x = 5$$

22.

Let x = width.

$2x + 6$ = length.

The perimeter is 48 m, so:

$$P = 2l + 2w$$

$$48 = 2(2x + 6) + 2(x)$$

$$48 = 4x + 12 + 2x$$

$$48 = 6x + 12$$

$$48 - 12 = 6x + 12 - 12$$

$$36 = 6x$$

$$\frac{36}{6} = \frac{6x}{6}$$

$$6 = x$$

The width is 6 m and the length is 18 m.

23.

Let x = litres of 60% acid solution.

$12 - x$ = litres of 100% acid solution.

You need a 75% acid solution, so:

$$0.60x + 1.00(12 - x) = 0.75(12)$$

$$0.6x + 12 - x = 9$$

$$12 - 0.4x = 9$$

$$12 - 0.4x - 12 = 9 - 12$$

$$-0.4x = -3$$

$$\frac{-0.4x}{-0.4} = \frac{-3}{-0.4}$$

$$x = 7.5$$

7.5 L of 60% acid solution should be combined with 4.5 L of 100% acid solution.

24.

$$F = Wg$$

$$\frac{F}{W} = \frac{Wg}{W}$$

$$\frac{F}{W} = g, \text{ so } g = \frac{F}{W}$$

25.

$$P = \frac{W}{A}$$

$$A(P) = \left(\frac{W}{A}\right)A$$

$$AP = W$$

$$\frac{AP}{P} = \frac{W}{P}$$

$$A = \frac{W}{P}$$

26.

$$L = A + B + \frac{1}{2}t$$

$$2(L) = \left(A + B + \frac{1}{2}t\right)2$$

$$2L = 2A + 2B + t$$

$$2L - 2A - 2B = 2A + 2B + t - 2A - 2B$$

$$2L - 2B - 2A = t, \text{ so } t = 2L - 2B - 2A$$

27.

$$k = \frac{1}{2}mv^2$$

$$2(k) = \left(\frac{1}{2}mv^2\right)2$$

$$2k = mv^2$$

$$\frac{2k}{v^2} = \frac{mv^2}{v^2}$$

$$\frac{2k}{v^2} = m, \text{ so } m = \frac{2k}{v^2}$$

28.

$$P_2 = \frac{P_1 T_2}{T_1}$$

$$T_1(P_2) = \left(\frac{P_1 T_2}{T_1}\right)T_1$$

$$P_2 T_1 = P_1 T_2$$

$$\frac{P_2 T_1}{P_2} = \frac{P_1 T_2}{P_2}$$

$$T_1 = \frac{P_1 T_2}{P_2}$$

29.

$$v = \frac{v_f + v_0}{2}$$

$$2(v) = \left(\frac{v_f + v_0}{2}\right)2$$

$$2v = v_0 + v_f$$

$$2v - v_f = v_0 + v_f - v_f$$

$$2v - v_f = v_0, \text{ so } v_0 = 2v - v_f$$

30. First, solve for F.

$$C = \frac{5}{9}(F - 32)$$

$$\frac{9}{5}(C) = \left[\frac{5}{9}(F - 32)\right]\frac{9}{5}$$

$$\frac{9}{5}C = F - 32$$

$$\frac{9}{5}C + 32 = F - 32 + 32$$

$$\frac{9}{5}C + 32 = F$$

Then, substitute the data.

$$F = \frac{9}{5}C + 32$$

$$F = \frac{9}{5}(175) + 32$$

$$= 347°C$$

31. First, solve for w.

$$P = 2(l + w)$$

$$P = 2l + 2w$$

$$P - 2l = 2l + 2w - 2l$$

$$P - 2l = 2w$$

$$\frac{P - 2l}{2} = \frac{2w}{2}$$

$$\frac{P - 2l}{2} = w$$

Then, substitute the data.

$$w = \frac{P - 2l}{2}$$

$$w = \frac{112.8 - 2(36.9)}{2}$$

$$= 19.5$$

32. First, solve for m.

$$k = \frac{1}{2}mv^2$$

$$2(k) = \left(\frac{1}{2}mv^2\right)2$$

$$2k = mv^2$$

$$\frac{2k}{v^2} = \frac{mv^2}{v^2}$$

$$\frac{2k}{v^2} = m$$

Then, substitute the data.

$$m = \frac{2k}{v^2}$$

$$m = \frac{2(460)}{(5.0)^2}$$

$$= 37$$

33.

$$\frac{1}{R} = \frac{1}{R_1} + \frac{1}{R_2}$$

$$\frac{1}{R} = \frac{1}{50.0\,\Omega} + \frac{1}{75.0\,\Omega}$$

$$R = 30.0\,\Omega$$

34.

$$\frac{1}{C} = \frac{1}{C_1} + \frac{1}{C_2} + \frac{1}{C_3}$$

$$\frac{1}{C_2} = \frac{1}{C} - \frac{1}{C_1} - \frac{1}{C_3}$$

$$\frac{1}{C_2} = \frac{1}{25.0\,\mu F} - \frac{1}{75.0\,\mu F} - \frac{1}{80.0\,\mu F}$$

$$C_2 = 70.6\,\mu F$$

Chapter 6 Test

1.

$$x - 8 = -6$$

$$x - 8 + 8 = -6 + 8$$

$$x = 2$$

3.

$$10 - 2x = 42$$

$$10 - 2x - 10 = 42 - 10$$

$$-2x = 32$$

$$\frac{-2x}{-2} = \frac{32}{-2}$$

$$x = -16$$

5.

$$7x - 20 = 5x + 4$$

$$7x - 20 - 5x = 5x + 4 - 5x$$

$$2x - 20 = 4$$

$$2x - 20 + 20 = 4 + 20$$

$$2x = 24$$

$$\frac{2x}{2} = \frac{24}{2}$$

$$x = 12$$

7.

$$\frac{1}{2}(3x - 6) = 3(x - 2)$$

$$2\left[\frac{1}{2}(3x - 6)\right] = [3(x - 2)]2$$

$$3x - 6 = 6(x - 2)$$

$$3x - 6 = 6x - 12$$

$$3x - 6 - 3x = 6x - 12 - 3x$$

$$-6 = 3x - 12$$

$$-6 + 12 = 3x - 12 + 12$$

$$6 = 3x$$

$$\frac{6}{3} = \frac{3x}{3}$$

$$2 = x$$

9.

$$\frac{3x}{5} - 2 = \frac{x}{5} - \frac{x}{10}$$

$$10\left(\frac{3x}{5} - 2\right) = \left(\frac{x}{5} - \frac{x}{10}\right)10$$

$$10\left(\frac{3x}{5}\right) - 10(2) = \left(\frac{x}{5}\right)10 - \left(\frac{x}{10}\right)10$$

$$6x - 20 = x$$

$$6x - 20 - 6x = x - 6x$$

$$-20 = -5x$$

$$\frac{-20}{-5} = \frac{-5x}{-5}$$

$$4 = x$$

11.

$$\frac{x}{2} - \frac{2}{5} = \frac{2x}{5} - \frac{3}{4}$$

$$20\left(\frac{x}{2} - \frac{2}{5}\right) = \left(\frac{2x}{5} - \frac{3}{4}\right)20$$

$$20\left(\frac{x}{2}\right) - 20\left(\frac{2}{5}\right) = \left(\frac{2x}{5}\right)20 - \left(\frac{3}{4}\right)20$$

$$10x - 8 = 8x - 15$$

$$10x - 8 - 8x = 8x - 15 - 8x$$

$$2x - 8 = -15$$

$$2x - 8 + 8 = -15 + 8$$

$$2x = -7$$

$$\frac{2x}{2} = \frac{-7}{2}$$

$$x = -3\frac{1}{2}$$

13.

Let x = litres of pure antifreeze.

You need a 80% antifreeze mixture, so:

$$1.00x + 0.60(20) = 0.80(20 + x)$$

$$x + 12 = 0.8x + 16$$

$$x + 12 - 0.8x = 0.8x + 16 - 0.8x$$

$$0.2x + 12 = 16$$

$$0.2x + 12 - 12 = 16 - 12$$

$$0.2x = 4$$

$$\frac{0.2x}{0.2} = \frac{4}{0.2}$$

$$x = 20$$

20 L of pure antifreeze should be added.

15.

$$C_T = C_1 + C_2 + C_3$$

$$C_T - C_1 - C_3 = C_1 + C_2 + C_3 - C_1 - C_3$$

$$C_T - C_1 - C_3 = C_2$$

$$\text{So } C_2 = C_T - C_1 - C_3$$

17. First, solve for R.

$$P = I^2 R$$

$$\frac{P}{I^2} = \frac{I^2 R}{I^2}$$

$$\frac{P}{I^2} = R, \text{ so } R = \frac{P}{I^2}$$

Then, substitute the data.

$$R = \frac{P}{I^2}$$

$$R = \frac{480}{(5.0)^2}$$

$$= 19$$

19.

$$\frac{1}{C} = \frac{1}{C_1} + \frac{1}{C_2}$$

$$\frac{1}{C_1} = \frac{1}{C} - \frac{1}{C_2}$$

$$\frac{1}{C_1} = \frac{1}{20.0 \ \mu F} - \frac{1}{30.0 \ \mu F}$$

$$C_1 = 60.0 \ \mu F$$

Cumulative Review: Chapters 1-6

1. $2 \cdot 2 \cdot 2 \cdot 3 \cdot 29$

2. 8.1%

5. $5 \text{ ha} \times \dfrac{10,000 \text{ m}^2}{1 \text{ ha}} = 50,000 \text{ m}^2$

6.

$$C = \frac{5}{9}\left(F - 32°\right)$$

$$C = \frac{5}{9}\left(101° - 32°\right)$$

$$= \frac{5}{9}\left(69°\right)$$

$$= 38.3° \ C$$

3. 0.0003015

4. 2.85×10^4

7. $6250 \text{ in}^2 \times \left(\dfrac{1 \text{ ft}}{12 \text{ in.}}\right)^2 = 43.4 \text{ ft}^2$

8. a. 2

b. 1

c. 5

9. a. 55.60 mm

b. 2.189 in.

10. 0.428 in.

11. 494,000 W

12.

$$(2x-5y)+(3y-4x)-2(3x-5y)$$
$$=2x-5y-4x+3y-6x+10y$$
$$=(2-4-6)x+(-5+3+10)y$$
$$=-8x+8y$$

13.

$$\left(4y^3+3y-5\right)-\left(2y^3-4y^2-2y+6\right)$$
$$=\left(4y^3+3y-5\right)+\left(-2y^3+4y^2+2y-6\right)$$
$$=(4-2)y^3+4y^2+(3+2)y+(-5-6)$$
$$=2y^3+4y^2+5y-11$$

14.

$$\left(3y^3\right)^3=(3)^3\left(y^3\right)^3$$
$$=27y^9$$

15.

$$-2x\left(x^2-3x+4\right)$$
$$=(-2x)\left(x^2\right)+(-2x)(-3x)+(-2x)(4)$$
$$=-2x^3+6x^2-8x$$

16.

$$
\begin{array}{l}
6y^3-5y^2-y+2 \\
\underline{\qquad 2y-1 \qquad} \\
12y^4-10y^3-2y^2+4y \\
\underline{\quad -6y^3+5y^2+y-2} \\
12y^4-16y^3+3y^2+5y-2
\end{array}
$$

17.

$$
\begin{array}{l}
4x-3y \\
\underline{5x+2y} \\
20x^2-15xy \\
\underline{\qquad 8xy-6y^2} \\
20x^2-7xy-6y^2
\end{array}
$$

18. $\dfrac{43}{9xy^2}$

19.

$$\left(16x^2y^3\right)\left(-5x^4y^5\right)$$
$$=(16)(-5)\left(x^2\right)\left(x^4\right)\left(y^3\right)\left(y^5\right)$$
$$=-80x^6y^8$$

20.

$$
\begin{array}{r}
x^2 \quad - \quad 3x \quad + \quad 4 \\
x+5 \,\overline{\smash{\big)}\, x^3 \;+\; 2x^2 \;-\; 11x \;-\; 20} \\
\underline{x^3 \;+\; 5x^2 \qquad\qquad\qquad} \\
-3x^2 \;-\; 11x \qquad \\
\underline{-3x^2 \;-\; 15x \qquad} \\
4x \;-\; 20 \\
\underline{4x \;+\; 20} \\
-40
\end{array}
$$

21.

$$3x^2-4xy+5y^2-\left(-3x^2\right)+(-7xy)+10y^2$$
$$=3x^2-4xy+5y^2+3x^2+(-7xy)+10y^2$$
$$=(3+3)x^2+(-4-7)xy+(5+10)y^2$$
$$=6x^2-11xy+15y^2$$

22.

$$4x-2=12$$
$$4x-2+2=12+2$$
$$4x=14$$
$$\frac{4x}{4}=\frac{14}{4}$$
$$x=3\frac{1}{2}$$

23.

$$\frac{x}{4}-5=9$$
$$\frac{x}{4}-5+5=9+5$$
$$\frac{x}{4}=14$$
$$4\left(\frac{x}{4}\right)=(14)4$$
$$x=56$$

24.

$$4x-3=7x+15$$
$$4x-3-4x=7x+15-4x$$
$$-3=3x+15$$
$$-3-15=3x+15-15$$
$$-18=3x$$
$$\frac{-18}{3}=\frac{3x}{3}$$
$$-6=x$$

25.

$$\frac{5x}{8} = \frac{3}{2}$$

$$8\left(\frac{5x}{8}\right) = \left(\frac{3}{2}\right)8$$

$$5x = 12$$

$$\frac{5x}{5} = \frac{12}{5}$$

$$x = 2\frac{2}{5}$$

26.

$$5 - (x - 3) = (2 + x) - 5$$

$$5 - x + 3 = 2 + x - 5$$

$$8 - x = x - 3$$

$$8 - x + x = x - 3 + x$$

$$8 = 2x - 3$$

$$8 + 3 = 2x - 3 + 3$$

$$11 = 2x$$

$$\frac{11}{2} = \frac{2x}{2}$$

$$5\frac{1}{2} = x$$

27.

$$C = \frac{1}{2}(a + b + c)$$

$$2(C) = 2\left[\frac{1}{2}(a + b + c)\right]$$

$$2C = a + b + c$$

$$2C - b - c = a + b + c - b - c$$

$$2C - b - c = a$$

$$a = 2C - b - c$$

28. First, solve for w.

$$A = lw$$

$$\frac{A}{l} = \frac{lw}{l}$$

$$\frac{A}{l} = w$$

Then, substitute the data.

$$w = \frac{A}{l}$$

$$w = \frac{91.3 \text{ m}^2}{8.20 \text{ m}}$$

$$= 11.1 \text{ m}$$

29. $7x = 250$

30.

Let x = length.

$$\frac{x}{2} = \text{width.}$$

The perimeter is 30 ft, so:

$$P = 2l + 2w$$

$$30 = 2\left(\frac{x}{2}\right) + 2(x)$$

$$30 = 3x$$

$$\frac{30}{3} = \frac{3x}{3}$$

$$10 = x$$

The width is 5 ft and the length is 10 ft.

Chapter 7: Ratio and Proportion

Section 7.1: Ratio

1. $3 \text{ to } 15 = \dfrac{3}{15} = \dfrac{1}{5}$

3. $7 : 21 = \dfrac{7}{21} = \dfrac{1}{3}$

5. $\dfrac{80}{48} = \dfrac{5}{3}$

7. $3 \text{ in} : 15 \text{ in} = \dfrac{3 \text{ in.}}{15 \text{ in.}} = \dfrac{1}{5}$

9.
$$3 \text{ cm} : 15 \text{ mm}$$
$$= \dfrac{3 \text{ cm}}{15 \text{ mm}}$$
$$= \dfrac{30 \text{ mm}}{15 \text{ mm}}$$
$$= \dfrac{2}{1} \text{ or } 2$$

11. $2 \text{ ft}^2 \times \left(\dfrac{12 \text{ in.}}{1 \text{ ft}} \right)^2 = 288 \text{ in}^2$

    $9 \text{ in}^2 : 2 \text{ in}^2 = \dfrac{9 \text{ in}^2}{2 \text{ ft}^2} = \dfrac{9 \text{ in}^2}{288 \text{ in}^2} = \dfrac{1}{32}$

13. $\dfrac{3}{4} \text{ to } \dfrac{7}{6} = \dfrac{3}{4} \div \dfrac{7}{6} = \dfrac{3}{4} \times \dfrac{6}{7} = \dfrac{9}{14}$

15. $2\dfrac{3}{4} : 4 = \dfrac{11}{4} \div \dfrac{4}{1} = \dfrac{11}{4} \times \dfrac{1}{4} = \dfrac{11}{16}$

17. $\dfrac{5\dfrac{1}{3}}{2\dfrac{2}{3}} = \dfrac{\dfrac{16}{3}}{\dfrac{8}{3}} = \dfrac{16}{3} \div \dfrac{8}{3} = \dfrac{16}{3} \times \dfrac{3}{8} = \dfrac{2}{1} \text{ or } 2$

19. $10 \text{ to } 2\dfrac{1}{2} = \dfrac{10}{1} \div \dfrac{5}{2} = \dfrac{10}{1} \times \dfrac{2}{5} = \dfrac{4}{1} \text{ or } 4$

21. $3\dfrac{1}{2} \text{ to } 2\dfrac{1}{2} = \dfrac{7}{2} \div \dfrac{5}{2} = \dfrac{7}{2} \times \dfrac{2}{5} = \dfrac{7}{5}$

23. $1\dfrac{3}{4} \text{ to } 7 = \dfrac{7}{4} \div \dfrac{7}{1} = \dfrac{7}{4} \times \dfrac{1}{7} = \dfrac{1}{4}$

25. 30 mi/gal

27. 46 gal/h

29. $\dfrac{625 \text{ mi}}{12\dfrac{1}{2} \text{ h}} = \dfrac{625 \text{ mi} \times 2}{12\dfrac{1}{2} \text{ h} \times 2} = \dfrac{1250 \text{ mi}}{25 \text{ h}} = 50 \text{ mi/h}$

31. $\dfrac{2\dfrac{1}{4} \text{ lb}}{6 \text{ gal}} = \dfrac{2\dfrac{1}{4} \text{ lb} \times 4}{6 \text{ gal} \times 4} = \dfrac{9 \text{ lb}}{24 \text{ gal}} = \dfrac{3}{8} \text{ lb/gal}$

33. $96 \text{ lb to } 15 \text{ lb} = \dfrac{96 \text{ lb}}{15 \text{ lb}} = \dfrac{32}{5}$

35. $90 \text{ gal to } 5 \text{ min} = \dfrac{90 \text{ gal}}{5 \text{ min}} = 18 \text{ gal/min}$

37. $18 \text{ V to } 4950 \text{ V} = \dfrac{18 \text{ V}}{4950 \text{ V}} = \dfrac{1}{275}$

39. $540 \text{ turns to } 45 \text{ turns} = \dfrac{540 \text{ turns}}{45 \text{ turns}} = \dfrac{12}{1} \text{ or } 12$

41. $2.7 \text{ tons} \times \dfrac{2000 \text{ lb}}{1 \text{ ton}} = 5400 \text{ lb}$

    $\dfrac{2.7 \text{ tons}}{150 \text{ bu}} = \dfrac{5400 \text{ lb}}{150 \text{ bu}} = 36 \text{ lb/bu}$

43. $\dfrac{350 \text{ gal}}{14 \text{ acres}} = 25 \text{ gal/acre}$

45. $\dfrac{2 \text{ ft}^3}{2 \text{ ft}^3 + 6 \text{ ft}^3} = \dfrac{2 \text{ ft}^3}{8 \text{ ft}^3} = \dfrac{1}{4}$

47. $\dfrac{\$27.04}{16 \text{ ft}} = \$1.69 \text{ /ft}$

49. $\dfrac{\$225,750}{2150 \text{ ft}^2} = \$105 \text{ /ft}$

51. $\dfrac{9}{12} = \dfrac{3}{4}$

53. $\dfrac{40 \text{ hr}}{250 \text{ hr}} = \dfrac{4}{25}$

55. $\dfrac{50 \text{ g}}{1000 \text{ mL}} = \dfrac{1}{20} \text{ g/mL or } 0.05 \text{ g/mL}$

57.
$$5 \text{ hr} \times \dfrac{60 \text{ min}}{1 \text{ hr}} = 300 \text{ min}$$
$$750 \text{ mL} \times \dfrac{15 \text{ drops}}{1 \text{ mL}} = 11,250 \text{ drops}$$
$$\dfrac{11,250 \text{ drops}}{300 \text{ min}} = 38 \text{ drops/min}$$

59.

$$\frac{280 \text{ male cougars}}{45,000 \text{ mi}^2 - 3000 \text{ mi}^2}$$

$$= \frac{280 \text{ male cougars}}{42,000 \text{ mi}^2}$$

$$= 1 \text{ male cougar}/150\text{mi}^2$$

61.

$$6 \text{ hr} \times \frac{60 \text{ min}}{1 \text{ hr}} = 360 \text{ min}$$

$$1200 \text{ mL} \times \frac{15 \text{ drops}}{1 \text{ mL}} = 18,000 \text{ drops}$$

$$\frac{18,000 \text{ drops}}{360 \text{ min}} = 50 \text{ drops/min}$$

63.

$$5.5 \text{ hr} \times \frac{60 \text{ min}}{1 \text{ hr}} = 330 \text{ min}$$

$$1000 \text{ mL} \times \frac{15 \text{ drops}}{1 \text{ mL}} = 15,000 \text{ drops}$$

$$\frac{15,000 \text{ drops}}{330 \text{ min}} = 45 \text{ drops/min}$$

65.

$$1000 \text{ mL} \times \frac{10 \text{ drops}}{1 \text{ mL}} = 10,000 \text{ drops}$$

$$\frac{10,000 \text{ drops}}{50 \text{ drops/min}} = 200 \text{ min}$$

$$200 \text{ min} \times \frac{1 \text{ hr}}{60 \text{ min}} = 3\frac{1}{3} \text{ hr}$$

67.

$$2000 \text{ mL} \times \frac{10 \text{ drops}}{1 \text{ mL}} = 20,000 \text{ drops}$$

$$\frac{20,000 \text{ drops}}{40 \text{ drops/min}} = 500 \text{ min}$$

$$500 \text{ min} \times \frac{1 \text{ hr}}{60 \text{ min}} = 8\frac{1}{3} \text{ hr}$$

69. $\dfrac{\$13,225}{2300 \text{ ft}^2} = \$5.75/\text{ft}^2$

71.

$$\frac{\$5.88}{12 \text{ dozen pencils}}$$

$$= \frac{\$5.88}{144 \text{ pencils}}$$

$$= \$0.04/\text{pencil}$$

73. $\dfrac{1 \text{ gal}}{3 \text{ qt}} = \dfrac{4 \text{ qt}}{3 \text{ qt}} = \dfrac{4}{3}$

75. $\dfrac{\$125}{25 \text{ lb}} = \$5/\text{lb}$

Section 7.2: Proportion

1. a. $2, 3$ b. $1, 6$

    c. $2 \times 3 = 6$ d. $1 \times 6 = 6$

3. a. $9, 28$ b. $7, 36$

    c. $9 \times 28 = 252$ d. $7 \times 36 = 252$

5. a. $7, w$ b. x, z

    c. $7 \times w = 7w$ d. $x \times z = xz$

7. $3 \times 10 = 30 = 2 \times 15$; The ratios are equal.

9. $5 \times 18 = 90 \neq 60 = 3 \times 20$; The ratios are not equal.

11. $3 \times 4 = 12 = 1 \times 12$; The ratios are equal.

13.

$$\frac{x}{4} = \frac{9}{12}$$

$$12x = 36$$

$$x = 3$$

15.

$$\frac{5}{7} = \frac{4}{y}$$

$$5y = 28$$

$$y = \frac{28}{5} = 5\frac{3}{5}$$

17.

$$\frac{2}{x} = \frac{4}{28}$$
$$4x = 56$$
$$x = 14$$

19.

$$\frac{5}{7} = \frac{3x}{14}$$
$$21x = 70$$
$$x = \frac{10}{3} = 3\frac{1}{3}$$

21.

$$\frac{5}{7} = \frac{25}{y}$$
$$5y = 175$$
$$y = 35$$

23.

$$\frac{-5}{x} = \frac{2}{3}$$
$$2x = -15$$
$$x = -7.5$$

25.

$$\frac{8}{0.04} = \frac{700}{x}$$
$$8x = 28$$
$$x = \frac{7}{2} = 3.5$$

27.

$$\frac{3x}{27} = \frac{0.5}{9}$$
$$27x = 13.5$$
$$x = 0.5$$

29.

$$\frac{17}{28} = \frac{153}{2x}$$
$$34x = 4284$$
$$x = 126$$

31.

$$\frac{12}{y} = \frac{84}{144}$$
$$84y = 1728$$
$$y = 20.6$$

33.

$$\frac{x}{48} = \frac{56}{72}$$
$$72x = 2688$$
$$x = \frac{112}{3} = 37\frac{1}{3}$$

35.

$$\frac{472}{x} = \frac{793}{64.2}$$
$$793x = (472)(64.2)$$
$$x = \frac{(472)(64.2)}{793}$$
$$x = 38.2$$

37.

$$\frac{30.1}{442} = \frac{55.7}{x}$$
$$30.1x = (442)(55.7)$$
$$x = \frac{(442)(55.7)}{30.1}$$
$$x = 818$$

39.

$$\frac{36.9}{104} = \frac{3210}{x}$$
$$36.9x = (104)(3210)$$
$$x = \frac{(104)(3210)}{36.9}$$
$$x = 9050$$

41.

$$\frac{x}{4.2} = \frac{19.6}{3.87}$$
$$3.87x = (4.2)(19.6)$$
$$x = \frac{(4.2)(19.6)}{3.87}$$
$$x = 21.3$$

43.

$$\frac{2\frac{3}{4} \text{ ft}^3}{8 \text{ ft}^3} = \frac{x}{128 \text{ ft}^3}$$

$$8x = \left(\frac{11}{4}\right)(128)$$

$$x = \frac{\left(\frac{11}{4}\right)(128)}{8}$$

$$x = 44 \text{ ft}^3$$

45.

$$\frac{\$171,000}{1800 \text{ ft}^2} = \frac{x}{2400 \text{ ft}^2}$$

$$1800x = (2400)(171,000)$$

$$x = \frac{(2400)(171,000)}{1800}$$

$$x = \$228,000$$

47.

$$\frac{\$120}{75 \text{ yd}} = \frac{x}{90 \text{ yd}}$$

$$75x = (90)(120)$$

$$x = \frac{(90)(120)}{75}$$

$$x = \$144$$

49.

$$\frac{25 \text{ gal}}{3 \text{ h}} = \frac{x}{1.2 \text{ h}}$$

$$3x = (25)(1.2)$$

$$x = \frac{(25)(1.2)}{3}$$

$$x = 10 \text{ gal}$$

51.

$$\frac{3 \text{ lb}}{20 \text{ gal}} = \frac{x}{350 \text{ gal}}$$

$$20x = 1050$$

$$x = 52.5 \text{ lb}$$

53.

$$\frac{90 \text{ lb}}{100 \text{ bu/acre}} = \frac{x}{120 \text{ bu/acre}}$$

$$100x = (90)(120)$$

$$x = \frac{(90)(120)}{100}$$

$$x = 108 \text{ lb}$$

55. a. $\dfrac{180 \text{ bu}}{15 \text{ trees}} = 12 \text{ bu/tree}$

 b. $180 \text{ bu} \times \dfrac{\$13}{\frac{1}{2} \text{ bu}} = \$4680$

57.

$$\frac{750 \text{ ft}}{1.563 \, \Omega} = \frac{x}{2.605 \, \Omega}$$

$$1.563x = (750)(2.605)$$

$$x = \frac{(750)(2.605)}{1.563}$$

$$x = 1250 \text{ ft}$$

59.

$$\frac{35 \text{ turns}}{4 \text{ turns}} = \frac{x}{68 \text{ turns}}$$

$$4x = 2380$$

$$x = 595 \text{ turns}$$

61.

$$\frac{270 \text{ hp}}{2 \text{ L}} = \frac{x}{1.6 \text{ L}}$$

$$2x = 432$$

$$x = 216 \text{ hp}$$

63.

$$\frac{18 \text{ gal}}{560 \text{ mi}} = \frac{x}{820 \text{ mi}}$$

$$560x = (18)(820)$$

$$x = \frac{(18)(820)}{560}$$

$$x = 26.4 \text{ gal}$$

65.

$$\frac{1000 \text{ mg}}{2.5 \text{ mL}} = \frac{800 \text{ mg}}{x}$$

$$1000x = 2000$$

$$x = 2 \text{ mL}$$

67.

$$\frac{250 \text{ mg}}{10 \text{ mL}} = \frac{150 \text{ mg}}{x}$$

$$250x = 1500$$

$$x = 6 \text{ mL}$$

69. The reduction is $20,400 - $19,200 = $1200.

$$\frac{\$1200}{\$20,400} = \frac{x}{100}$$

$$20,400x = 120,000$$

$$x = 5.9\%$$

71. a.

$$\frac{18 \text{ g}}{100 \text{ g}} = \frac{x}{100}$$

$$100x = 1800$$

$$x = 18\%$$

b.

$$\frac{x}{650 \text{ lb}} = \frac{18}{100}$$

$$100x = (18)(650)$$

$$x = \frac{(18)(650)}{100} = 117 \text{ lb}$$

73.

$$\frac{5.7 \text{ hL}}{x} = \frac{30}{100}$$

$$30x = 570$$

$$x = 19\%$$

75. There is a total of $1 + 2.5 + 4 = 7.5$ parts.

a.

$$\frac{1}{7.5} = \frac{x}{100}$$

$$7.5x = 100$$

$$x = 13.3\%$$

b.

$$\frac{2.5}{7.5} = \frac{x}{100}$$

$$7.5x = 250$$

$$x = 33.3\%$$

c.

$$\frac{4}{7.5} = \frac{x}{100}$$

$$7.5x = 400$$

$$x = 53.3\%$$

77.

$$\frac{25 \text{ ft}}{3 \text{ min}} = \frac{x}{10 \text{ min}}$$

$$3x = 250$$

$$x = 83\frac{1}{3} \text{ ft or 8 ft 4 in.}$$

79.

$$\frac{35 \text{ lb salt}}{1000 \text{ lb sea water}} = \frac{x}{2000 \text{ lb sea water}}$$

$$1000x = 70,000$$

$$x = 70 \text{ lb salt}$$

81.

$$\frac{20 \text{ turns}}{100 \text{ ft}} = \frac{x}{175 \text{ ft}}$$

$$100x = 3500$$

$$x = 35 \text{ turns}$$

83.

$$\frac{3}{100} = \frac{x}{150 \text{ mL}}$$

$$100x = 450$$

$$x = 4.5 \text{ mL}$$

85.

$$\frac{1}{100} = \frac{x}{500}$$

$$100x = 500$$

$$x = 5 \text{ g}$$

87.

$$\frac{1}{1000} = \frac{x}{1500}$$

$$1000x = 1500$$

$$x = 1.5 \text{ g}$$

89.

$$\frac{1}{10} = \frac{x}{300}$$

$$10x = 300$$

$$x = 30 \text{ g}$$

91.

$$\frac{5 \text{ lb beef}}{2 \text{ lb pork}} = \frac{6 \text{ lb beef}}{x}$$

$$5x = 12$$

$$x = 2.4 \text{ lb pork}$$

93.

a.

$$\frac{125 \text{ mL}}{25 \text{ mL}} = 5; \quad \frac{50 \text{ mL}}{25 \text{ mL}} = 2; \quad \frac{50 \text{ mL}}{25 \text{ mL}} = 1$$

The kitchen ratios are 5 : 2 : 1.

b.

vegetable oil: salt:

$$\frac{2}{5} = \frac{150 \text{ mL}}{x} \qquad \frac{2}{1} = \frac{150 \text{ mL}}{x}$$

$$2x = 750 \text{ mL} \qquad 2x = 150$$

$$x = 375 \text{ mL} \qquad x = 75 \text{ mL}$$

You would need 375 mL of vegetable oil, 150 mL of sherry, and 75 mL of salt.

95. You will need a total of $5 \times 8 = 40$ crepes and $\dfrac{40}{16} = 2.5$.

$$\frac{1}{4} \text{ cup margarine} \times 2.5 = \frac{5}{8} \text{ cup margarine}$$

$$3 \text{ tbs sugar} \times 2.5 = 7\frac{1}{2} \text{ tbs sugar}$$

$$\frac{1}{2} \text{ cup orange juice} \times 2.5 = 1\frac{1}{4} \text{ cups orange juice}$$

$$2 \text{ tsp lemon juice} \times 2.5 = 5 \text{ tsp lemon juice}$$

$$1 \text{ tbs orange rind} \times 2.5 = 2\frac{1}{2} \text{ tbs orange rind}$$

$$\frac{1}{2} \text{ tsp lemon rind} \times 2.5 = 1\frac{1}{4} \text{ tsp lemon rind}$$

$$\frac{1}{2} \text{ cup almonds} \times 2.5 = 1\frac{1}{4} \text{ cups almonds}$$

97. $\dfrac{18}{4} = 4.5$

$$\frac{1}{4} \text{ cup butter} \times 4.5 = 1\frac{1}{8} \text{ cup butter}$$

$$\frac{1}{4} \text{ cup flour} \times 4.5 = 1\frac{1}{8} \text{ cup flour}$$

$$1 \text{ cup milk} \times 4.5 = 4\frac{1}{2} \text{ cups milk}$$

$$\frac{1}{2} \text{ tsp salt} \times 4.5 = 2\frac{1}{4} \text{ tsp salt}$$

$$8 \text{ oz cheese} \times 4.5 = 36 \text{ oz cheese}$$

$$4 \text{ egg yolks} \times 4.5 = 18 \text{ egg yolks}$$

$$4 \text{ egg whites} \times 4.5 = 18 \text{ egg whites}$$

99.

pie crust:	butter:	eggs:	salt:	water:	pepper:
$\frac{3}{3} = \frac{x}{12}$	$\frac{2}{3} = \frac{x}{12}$	$\frac{1}{3} = \frac{x}{12}$	$\frac{1}{3} = \frac{x}{12}$	$\frac{1}{3} = \frac{x}{12}$	$\frac{\frac{1}{2}}{3} = \frac{x}{12}$
$3x = 12$	$3x = 12$	$3x = 12$	$3x = 12$	$3x = 12$	$3x = 6$
$x = 12$ sticks	$x = 4$ tbs	$x = 4$ eggs	$x = 4$ tsp	$x = 4$ tbs	$x = 2$ tsp

12 sticks of pie crust, 8 tbs soft butter, 4 eggs, 4 tsp salt, 4 tbs water, and 2 tsp pepper are required.

Section 7.3: Direct Variation

1. $d = 2\left(1\frac{1}{2}\right) = 32\left(\dfrac{3}{2}\right) = 48$ mi

3. $d = 32\left(1\frac{7}{8}\right) = 32\left(\dfrac{15}{8}\right) = 60$ mi

5. $d = 32\left(2\frac{11}{16}\right) = 32\left(\frac{43}{16}\right) = 86$ mi

7. $d = 32\left(4\frac{1}{4}\right) = 32\left(\frac{13}{4}\right) = 104$ mi

9. $d = 32(3) = 96$ mi

11. $d = 85(5) = 425$ km

13. $d = 85(3) = 70$ km

15. $d = 85\left(3\frac{1}{5}\right) = 85\left(\frac{16}{5}\right) = 272$ km

17.
$$\frac{x}{25 \text{ spaces}} = \frac{0.5 \text{ cm}}{1 \text{ space}}$$
$$(1 \text{ space})x = (25 \text{ spaces})(0.5 \text{ cm})$$
$$x = \frac{(25 \text{ spaces})(0.5 \text{ cm})}{1 \text{ space}}$$
$$x = 12.5 \text{ cm}$$

19. $A = lw = (12.5 \text{ cm})(8 \text{ cm}) = 100 \text{ cm}^2$

21.
$$\frac{x}{2 \text{ spaces}} = \frac{0.5 \text{ cm}}{1 \text{ space}}$$
$$(1 \text{ space})x = (2 \text{ spaces})(0.5 \text{ cm})$$
$$x = \frac{(2 \text{ spaces})(0.5 \text{ cm})}{1 \text{ space}}$$
$$x = 1 \text{ cm}$$
The square is $1 \text{ cm} \times 5 \text{ cm}$

23.
$$\frac{x}{13 \text{ spaces}} = \frac{0.5 \text{ cm}}{1 \text{ space}}$$
$$(1 \text{ space})x = (13 \text{ spaces})(0.5 \text{ cm})$$
$$x = \frac{(13 \text{ spaces})(0.5 \text{ cm})}{1 \text{ space}}$$
$$x = 6.5 \text{ cm}$$

25.
$$\frac{x}{13 \text{ spaces}} = \frac{0.5 \text{ cm}}{1 \text{ space}}$$
$$(1 \text{ space})x = (13 \text{ spaces})(0.5 \text{ cm})$$
$$x = \frac{(13 \text{ spaces})(0.5 \text{ cm})}{1 \text{ space}}$$
$$x = 6.5 \text{ cm}$$

27.

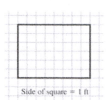

Side of square = 1 ft

29.

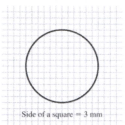

Side of a square = 3 mm

31. a. yes; the rectangle is much larger.

b. no; the rectangle would be 8×6 squares, but the circle would have a diameter of 12 squares.

33. $\text{MA} = \dfrac{F_l}{F_s} = \dfrac{4000}{200} = \dfrac{20}{1}$ or 20 : 1

35. $\text{MA} = \dfrac{F_l}{F_s} = \dfrac{3600}{400} = \dfrac{9}{1}$ or 9 : 1

37.
$$\text{MA} = \frac{F_l}{F_s}$$
$$\frac{75}{1} = \frac{5100 \text{ lb}}{x}$$
$$75x = 1(5100 \text{ lb})$$
$$x = \frac{1(5100 \text{ lb})}{75}$$
$$x = 68 \text{ lb}$$

39.
$$\text{MA} = \frac{F_l}{F_s}$$
$$90 = \frac{x}{2650 \text{ lb}}$$
$$x = 90(2650 \text{ lb})$$
$$x = 238,500 \text{ lb}$$

41. $\text{MA} = \dfrac{r_l^2}{r_s^2} = \dfrac{(27 \text{ in.})^2}{(3 \text{ in.})^2} = \dfrac{81}{1}$ or 81 : 1

43. $\text{MA} = \dfrac{r_l^2}{r_s^2} = \dfrac{(15 \text{ in.})^2}{(3 \text{ in.})^2} = \dfrac{25}{1}$ or 25 : 1

45.

$$\frac{F_l}{F_s} = \frac{r_l^2}{r_s^2}$$

$$\frac{x}{25 \text{ lb}} = \frac{(8 \text{ in.})^2}{(2 \text{ in.})^2}$$

$$(x)(2 \text{ in.})^2 = (25 \text{ lb})(8 \text{ in.})^2$$

$$x = \frac{(25 \text{ lb})(8 \text{ in.})^2}{(2 \text{ in.})^2}$$

$$x = 400 \text{ lb}$$

47.

$$\frac{F_s}{F_l} = \frac{r_s^2}{r_l^2}$$

$$\frac{x}{6400 \text{ lb}} = \frac{(4 \text{ in.})^2}{(16 \text{ in.})^2}$$

$$(x)(16 \text{ in.})^2 = (6400 \text{ lb})(4 \text{ in.})^2$$

$$x = \frac{(6400 \text{ lb})(4 \text{ in.})^2}{(16 \text{ in.})^2}$$

$$x = 400 \text{ lb}$$

49.

$$\frac{F_l}{F_s} = \frac{r_l^2}{r_s^2}$$

$$\frac{x}{40 \text{ lb}} = \frac{(28 \text{ in.})^2}{(7 \text{ in.})^2}$$

$$(x)(7 \text{ in.})^2 = (40 \text{ lb})(28 \text{ in.})^2$$

$$x = \frac{(40 \text{ lb})(28 \text{ in.})^2}{(7 \text{ in.})^2}$$

$$x = 640 \text{ lb}$$

51.

$$\frac{x}{3\frac{3}{4} \text{ h}} = \frac{90 \text{ mi}}{1\frac{1}{2} \text{ h}}$$

$$\left(1\frac{1}{2} \text{ h}\right)(x) = \left(3\frac{3}{4} \text{ h}\right)(90 \text{ mi})$$

$$x = \frac{\left(3\frac{3}{4} \text{ h}\right)(90 \text{ mi})}{1\frac{1}{2} \text{ h}}$$

$$x = 225 \text{ mi}$$

53.

$$\frac{x}{188 \text{ ft}^2} = \frac{93 \text{ blocks}}{124 \text{ ft}^2}$$

$$(124 \text{ ft}^2)(x) = (188 \text{ ft}^2)(93 \text{ blocks})$$

$$x = \frac{(188 \text{ ft}^2)(93 \text{ blocks})}{124 \text{ ft}^2}$$

$$x = 141 \text{ blocks}$$

Section 7.4: Inverse Variation

1.

$$(d_A)(\text{rpm}_A) = (d_B)(\text{rpm}_B)$$

$$(25 \text{ cm})(72 \text{ rpm}) = (50 \text{ cm})(x)$$

$$x = \frac{(25 \text{ cm})(72 \text{ rpm})}{50 \text{ cm}}$$

$$x = 36 \text{ rpm}$$

3.

$$(d_A)(\text{rpm}_A) = (d_B)(\text{rpm}_B)$$

$$(10 \text{ cm})(120 \text{ rpm}) = (15 \text{ cm})(x)$$

$$x = \frac{(10 \text{ cm})(120 \text{ rpm})}{15 \text{ cm}}$$

$$x = 80 \text{ rpm}$$

5.

$$(d_A)(\text{rpm}_A) = (d_B)(\text{rpm}_B)$$
$$(34 \text{ cm})(440 \text{ rpm}) = (x)(680 \text{ rpm})$$
$$x = \frac{(34 \text{ cm})(440 \text{ rpm})}{680 \text{ rpm}}$$
$$x = 22 \text{ cm}$$

7.

$$(d_A)(\text{rpm}_A) = (d_B)(\text{rpm}_B)$$
$$(x)(225 \text{ rpm}) = (15 \text{ in.})(465 \text{ rpm})$$
$$x = \frac{(15 \text{ in.})(465 \text{ rpm})}{225 \text{ rpm}}$$
$$x = 31 \text{ in.}$$

13.

$$(d_A)(\text{rpm}_A) = (d_B)(\text{rpm}_B)$$
$$(x)(680 \text{ rpm}) = (4 \text{ in.})(1870 \text{ rpm})$$
$$x = \frac{(4 \text{ in.})(1870 \text{ rpm})}{680 \text{ rpm}}$$
$$x = 11 \text{ in.}$$

15.

$$(d_A)(\text{rpm}_A) = (d_B)(\text{rpm}_B)$$
$$(x+7)(80 \text{ rpm}) = (x)(136 \text{ rpm})$$
$$80x + 560 = 136x$$
$$80x + 560 - 80x = 136x - 80x$$
$$560 = 56x$$
$$\frac{560}{56} = \frac{56x}{56}$$
$$x = 10$$

The larger pulley is 17 cm in diameter and the smaller pulley is 10 cm in diameter.

17.

$$(d_A)(\text{rpm}_A) = (d_B)(\text{rpm}_B)$$
$$(6.50 \text{ cm})(x) = (9.00 \text{ cm})(15\overline{0}0 \text{ rpm})$$
$$x = \frac{(9.00 \text{ cm})(15\overline{0}0 \text{ rpm})}{6.50 \text{ cm}}$$
$$x = 2080 \text{ rpm}$$

9.

$$(d_A)(\text{rpm}_A) = (d_B)(\text{rpm}_B)$$
$$(13 \text{ in.})(720 \text{ rpm}) = (18 \text{ in.})(x)$$
$$x = \frac{(13 \text{ in.})(720 \text{ rpm})}{18 \text{ in.}}$$
$$x = 520 \text{ rpm}$$

11.

$$(d_A)(\text{rpm}_A) = (d_B)(\text{rpm}_B)$$
$$(x)(48 \text{ rpm}) = (8 \text{ in.})(300 \text{ rpm})$$
$$x = \frac{(8 \text{ in.})(300 \text{ rpm})}{48 \text{ rpm}}$$
$$x = 50 \text{ in.}$$

19.

$$(d_B)(\text{rpm}_B) = (d_A)(\text{rpm}_A)$$
$$(5.00 \text{ cm})(x) = (18.0 \text{ cm})(12\overline{0}0 \text{ rpm})$$
$$x = \frac{(18.0 \text{ cm})(12\overline{0}0 \text{ rpm})}{5.00 \text{ cm}}$$
$$x = 4320 \text{ rpm}$$

$$(d_D)(\text{rpm}_D) = (d_C)(\text{rpm}_C)$$
$$(5.50 \text{ cm})(x) = (15.0 \text{ cm})(4320 \text{ rpm})$$
$$x = \frac{(15.0 \text{ cm})(4320 \text{ rpm})}{5.50 \text{ cm}}$$
$$x = 11,800 \text{ rpm}$$

21.

$$(n_A)(\text{rpm}_A) = (n_B)(\text{rpm}_B)$$
$$(x)(125 \text{ rpm}) = (50 \text{ teeth})(400 \text{ rpm})$$
$$x = \frac{(50 \text{ teeth})(400 \text{ rpm})}{125 \text{ rpm}}$$
$$x = 160 \text{ teeth}$$

23.

$$(n_A)(\text{rpm}_A) = (n_B)(\text{rpm}_B)$$
$$(25 \text{ teeth})(x) = (42 \text{ teeth})(600 \text{ rpm})$$
$$x = \frac{(42 \text{ teeth})(600 \text{ rpm})}{25 \text{ teeth}}$$
$$x = 1008 \text{ rpm}$$

25.

$$(n_A)(\text{rpm}_A) = (n_B)(\text{rpm}_B)$$

$$(x)\left(6\frac{1}{4}\ \text{rpm}\right) = (120\ \text{teeth})(30\ \text{rpm})$$

$$x = \frac{(120)(30\ \text{rpm})}{6\frac{1}{4}\ \text{rpm}}$$

$$x = 576\ \text{teeth}$$

27.

$$(n_A)(\text{rpm}_A) = (n_B)(\text{rpm}_B)$$

$$(25\ \text{teeth})(x) = (75\ \text{teeth})(32\ \text{rpm})$$

$$x = \frac{(75\ \text{teeth})(32\ \text{rpm})}{25\ \text{teeth}}$$

$$x = 96\ \text{rpm}$$

29.

$$(n_A)(\text{rpm}_A) = (n_B)(\text{rpm}_B)$$

$$(30\ \text{teeth})(x) = (60\ \text{teeth})(72\ \text{rpm})$$

$$x = \frac{(60\ \text{teeth})(72\ \text{rpm})}{30\ \text{teeth}}$$

$$x = 144\ \text{rpm}$$

31.

$$(n_A)(\text{rpm}_A) = (n_B)(\text{rpm}_B)$$

$$(x)(90\ \text{rpm}) = (120\ \text{teeth})(30\ \text{rpm})$$

$$x = \frac{(120\ \text{teeth})(30\ \text{rpm})}{90\ \text{rpm}}$$

$$x = 40\ \text{teeth}$$

33.

$$F_1 d_1 = F_2 d_2$$

$$(18\ \text{lb})(5\ \text{in.}) = (9\ \text{lb})(x)$$

$$x = \frac{(18\ \text{lb})(5\ \text{in.})}{9\ \text{lb}}$$

$$x = 10\ \text{in.}$$

35.

$$F_1 d_1 = F_2 d_2$$

$$(40\ \text{lb})(9\ \text{in.}) = (x)(3\ \text{in.})$$

$$x = \frac{(40\ \text{lb})(9\ \text{in.})}{3\ \text{in.}}$$

$$x = 120\ \text{lb}$$

37.

$$F_1 d_1 = F_2 d_2$$

$$(x)(8\ \text{ft}) = (180\ \text{lb})(6\ \text{ft})$$

$$x = \frac{(180\ \text{lb})(68\ \text{ft})}{8\ \text{ft}}$$

$$x = 135\ \text{lb}$$

39.

$$F_1 d_1 = F_2 d_2$$

$$(x)(8\ \text{ft}) = (1\ \text{ton})(4\ \text{ft})$$

$$x = \frac{(1\ \text{ton})(4\ \text{ft})}{8\ \text{ft}}$$

$$x = \frac{1}{2}\ \text{ton or } 1000\ \text{lb}$$

41.

$$F_1 d_1 = F_2 d_2$$

$$(1350\ \text{g})(x) = (1200\ \text{g})(72\ \text{cm})$$

$$x = \frac{(1200\ \text{g})(72\ \text{cm})}{1350\ \text{g}}$$

$$x = 64\ \text{cm}$$

43.

$$F_1 d_1 = F_2 d_2$$

$$(210\ \text{lb})(x) = (190\ \text{lb})(28\ \text{in.})$$

$$x = \frac{(190\ \text{lb})(28\ \text{in.})}{210\ \text{lb}}$$

$$x = 25\frac{1}{3}\ \text{in.}$$

45.

$$(x)(8\ \text{cm}) = (18\ \text{cm})(12\ \text{cm})$$

$$x = \frac{(18\ \text{cm})(12\ \text{cm})}{8\ \text{cm}}$$

$$x = 27\ \text{cm}$$

47.

$$(x)(80\ \Omega) = (8\ \text{A})(30\ \Omega)$$

$$x = \frac{(8\ \text{A})(30\ \Omega)}{80\ \Omega}$$

$$x = 3\ \text{A}$$

Chapter 7 Review

1. $7 \text{ to } 28 = \dfrac{7}{28} = \dfrac{1}{4}$

2. $60 : 40 = \dfrac{60}{40} = \dfrac{3}{2}$

3. $1 \text{ g to } 500 \text{ mg} = \dfrac{1000 \text{ g}}{500 \text{ g}} = \dfrac{2}{1}$

4.
$$5 \text{ ft} \times \dfrac{12 \text{ in.}}{1 \text{ ft}} + 6 \text{ in.} = 66 \text{ in.}$$
$$9 \text{ ft} \times \dfrac{12 \text{ in.}}{1 \text{ ft}} = 108 \text{ in.}$$
$$\dfrac{5 \text{ ft } 6 \text{ in.}}{9 \text{ ft}} = \dfrac{66 \text{ in.}}{108 \text{ in.}} = \dfrac{11}{18}$$

5. $2 \times 35 = 70 = 7 \times 10$; The ratios are equal.

6. $18 \times 30 = 540 \neq 575 = 5 \times 115$; The ratios are not equal.

7.
$$\dfrac{x}{4} = \dfrac{5}{20}$$
$$20x = 20$$
$$x = 1$$

8.
$$\dfrac{10}{25} = \dfrac{x}{75}$$
$$25x = 750$$
$$x = 30$$

9.
$$\dfrac{3}{x} = \dfrac{8}{64}$$
$$8x = 192$$
$$x = 24$$

10.
$$\dfrac{72}{96} = \dfrac{30}{x}$$
$$72x = 2880$$
$$x = 40$$

11.
$$\dfrac{73.4}{x} = \dfrac{25.9}{37.4}$$
$$25.9x = (73.4)(37.4)$$
$$x = \dfrac{(73.4)(37.4)}{25.9}$$
$$x = 106$$

12.
$$\dfrac{x}{19.7} = \dfrac{144}{68.7}$$
$$68.7x = (19.7)(144)$$
$$x = \dfrac{(19.7)(144)}{68.7}$$
$$x = 41.3$$

13.
$$\dfrac{61.1}{81.3} = \dfrac{592}{x}$$
$$61.1x = (81.3)(592)$$
$$x = \dfrac{(81.3)(592)}{61.1}$$
$$x = 788$$

14.
$$\dfrac{243}{58.3} = \dfrac{x}{127}$$
$$58.3x = (243)(127)$$
$$x = \dfrac{(243)(127)}{58.3}$$
$$x = 529$$

15.
$$\dfrac{x}{500 \text{ ft}} = \dfrac{\$67.50}{180 \text{ ft}}$$
$$180x = (500)(67.50)$$
$$x = \dfrac{(500)(67.50)}{180}$$
$$x = \$187.50$$

16.
$$\dfrac{x}{3.15 \ \Omega} = \dfrac{750 \text{ ft}}{1.89 \ \Omega}$$
$$(1.89 \ \Omega)x = (3.15 \ \Omega)(750 \text{ ft})$$
$$x = \dfrac{(3.15 \ \Omega)(750 \text{ ft})}{1.89 \ \Omega}$$
$$x = 1250 \text{ ft}$$

17.

$$\frac{x}{9 \text{ houses}} = \frac{144 \text{ h}}{6 \text{ houses}}$$

$$(6 \text{ houses})x = (9 \text{ houses})(144 \text{ h})$$

$$x = \frac{(9 \text{ houses})(144 \text{ h})}{6 \text{ houses}}$$

$$x = 216 \text{ h}$$

18.

$$\frac{x}{25 \text{ lb}} = \frac{12}{1}$$

$$x = (12)(25 \text{ lb})$$

$$x = 300 \text{ lb}$$

19. The total amount invested was
$6380 + $4620 = $11,000$.

Jones:

$$\frac{\$6380}{\$11,000} = \frac{x}{100}$$

$$11,000x = 638,000$$

$$x = 58\%$$

Hernandez:

$$\frac{\$4620}{\$11,000} = \frac{x}{100}$$

$$11,000x = 462,000$$

$$x = 42\%$$

20. $7 \text{ lb} \times \dfrac{16 \text{ oz}}{1 \text{ lb}} + 13 \text{ oz.} = 125 \text{ oz}$;

$$\frac{11 \text{ oz}}{125 \text{ oz}} = \frac{x}{100}$$

$$125x = 1100$$

$$x = 808\%$$

21. a. direct variation

b. indirect variation

22. direct variation

23.

$$\frac{x}{3\frac{5}{8} \text{ in.}} = \frac{25 \text{ mi}}{\frac{1}{4} \text{ in.}}$$

$$\left(\frac{1}{4} \text{ in.}\right)x = \left(3\frac{5}{8} \text{ in.}\right)(25 \text{ mi})$$

$$x = \frac{\left(3\frac{5}{8} \text{ in.}\right)(25 \text{ mi})}{\frac{1}{4} \text{ in.}}$$

$$x = 362.5 \text{ mi}$$

24. $2 \text{ mi} \times \dfrac{5280 \text{ ft}}{1 \text{ mi}} = 10,596 \text{ ft}$

$$\frac{x}{10,596 \text{ ft}} = \frac{1 \text{ in.}}{600 \text{ ft}}$$

$$(600 \text{ ft})x = (10,596 \text{ ft})(1 \text{ in.})$$

$$x = \frac{(10,596 \text{ ft})(1 \text{ in.})}{600 \text{ ft}}$$

$$x = 17.6 \text{ in.}$$

25.

$$(d_A)(\text{rpm}_A) = (d_B)(\text{rpm}_B)$$

$$(40 \text{ cm})(x) = (25 \text{ cm})(900 \text{ rpm})$$

$$x = \frac{(25 \text{ cm})(900 \text{ rpm})}{40 \text{ cm}}$$

$$x = 562.5 \text{ rpm}$$

26.

$$(n_A)(\text{rpm}_A) = (n_B)(\text{rpm}_B)$$

$$(14 \text{ teeth})(x) = (42 \text{ teeth})(25 \text{ rpm})$$

$$x = \frac{(42 \text{ teeth})(25 \text{ rpm})}{14 \text{ teeth}}$$

$$x = 75 \text{ rpm}$$

27.

$$\frac{F_s}{F_l} = \frac{r_s^2}{r_l^2}$$

$$\frac{x}{6050 \text{ lb}} = \frac{(2 \text{ in.})^2}{(22 \text{ in.})^2}$$

$$(x)(22 \text{ in.})^2 = (6050 \text{ lb})(2 \text{ in.})^2$$

$$x = \frac{(6050 \text{ lb})(2 \text{ in.})^2}{(22 \text{ in.})^2} = 50 \text{ lb}$$

28.
$$F_1 d_1 = F_2 d_2$$
$$(x)(12 \text{ ft}) = (240 \text{ lb})(9 \text{ ft})$$
$$x = \frac{(240 \text{ lb})(9 \text{ ft})}{12 \text{ ft}}$$
$$x = 180 \text{ lb}$$

29.
$$\frac{I}{100 \text{ V}} = \frac{0.6 \text{ A}}{30 \text{ V}}$$
$$(30 \text{ V})(I) = (100 \text{ V})(0.6 \text{ A})$$
$$I = \frac{(100 \text{ V})(0.6 \text{ A})}{30 \text{ V}}$$
$$I = 2 \text{ A}$$

30.
$$(8 \text{ workers})(x) = (12 \text{ workers})(72 \text{ h})$$
$$x = \frac{(12 \text{ workers})(72 \text{ h})}{8 \text{ workers}}$$
$$x = 108 \text{ h}$$

Chapter 7 Test

1. $16 \text{ m to } 64 \text{ m} = \dfrac{16 \text{ m}}{64 \text{ m}} = \dfrac{1}{4}$

3. $400 \text{ mL to } 5 \text{ L} = \dfrac{400 \text{ mL}}{5000 \text{ mL}} = \dfrac{2}{25}$

5.
$$\frac{8}{x} = \frac{24}{5}$$
$$24x = 40$$
$$x = \frac{5}{3} = 1\frac{2}{3}$$

11.
$$(n_A)(\text{rpm}_A) = (n_B)(\text{rpm}_B)$$
$$(36 \text{ teeth})(x) = (48 \text{ teeth})(150 \text{ rpm})$$
$$x = \frac{(48 \text{ teeth})(150 \text{ rpm})}{36 \text{ teeth}}$$
$$x = 200 \text{ rpm}$$

7.
$$\frac{x}{80 \text{ ft}} = \frac{\$138}{60 \text{ ft}}$$
$$60x = (80)(138)$$
$$x = \frac{(80)(138)}{60}$$
$$x = \$184$$

13.
$$F_1 d_1 = F_2 d_2$$
$$(800 \text{ lb})(9 \text{ ft}) = (x)(3.6 \text{ ft})$$
$$x = \frac{(800 \text{ lb})(9 \text{ ft})}{3.6 \text{ ft}}$$
$$x = 2000 \text{ lb}$$

9.
$$\frac{x}{4.8 \text{ cm}} = \frac{10 \text{ km}}{1 \text{ cm}}$$
$$(1 \text{ cm})x = (4.8 \text{ cm})(10 \text{ km})$$
$$x = \frac{(4.8 \text{ cm})(10 \text{ km})}{1 \text{ cm}}$$
$$x = 48 \text{ km}$$

Chapter 8: Graphing Linear Equations

Section 8.1: Linear Equations with Two Variables

1. Solve for y.

$$x + y = 5$$
$$x + y - x = 5 - x$$
$$y = 5 - x$$

$(3, 2)$
$(8, -3)$
$(-2, 7)$

3. Solve for y.

$$6x + 2y = 10$$
$$6x + 2y - 6x = 10 - 6x$$
$$2y = 10 - 6x$$
$$\frac{2y}{2} = \frac{10 - 6x}{2}$$
$$y = 5 - 3x$$

$(2, -1)$
$(0, 5)$
$(-2, 11)$

5. Solve for y.

$$3x - 4y = 8$$
$$3x - 4y - 3x = 8 - 3x$$
$$-4y = 8 - 3x$$
$$\frac{-4y}{-4} = \frac{8 - 3x}{-4}$$
$$y = \frac{8 - 3x}{-4}$$
$$y = \frac{(8 - 3x)(-1}{(-4)(-1)}$$
$$y = \frac{3x - 8}{4}$$

$(0, -2)$
$\left(2, -\frac{1}{2}\right)$
$(-4, -5)$

7. Solve for y.

$$-2x + 5y = 10$$
$$-2x + 5y + 2x = 10 + 2x$$
$$5y = 2x + 10$$
$$\frac{5y}{5} = \frac{2x + 10}{5}$$
$$y = \frac{2x + 10}{5}$$

$(5, 4)$
$(0, 2)$
$\left(-3, \frac{4}{5}\right)$

9. Solve for y.

$$9x - 2y = 10$$
$$9x - 2y - 9x = 10 - 9x$$
$$-2y = 10 - 9x$$
$$\frac{-2y}{-2} = \frac{10 - 9x}{-2}$$
$$y = \frac{10 - 9x}{-2}$$
$$y = \frac{(10 - 9x)(-1)}{(-2)(-1)}$$
$$y = \frac{9x - 10}{2}$$

$(2, 4)$
$(0, -5)$
$(-4, -23)$

11. $(2, 10)$; $(0, 4)$; $(-3, -5)$.

13. Solve for y.

$$5x + y = 7$$
$$5x + y - 5x = 7 - 5x$$
$$y = 7 - 5x$$

$(2, -3)$
$(0, 7)$
$(-4, 27)$

15. Solve for y.

$$2x = y - 4$$
$$2x + 4 = y - 4 + 4$$
$$2x + 4 = y$$
$$y = 2x + 4$$

$(3, 10)$
$(0, 4)$
$(-1, 2)$

17. Solve for y.

$$5x - 2y = -8$$
$$5x - 2y - 5x = -8 - 5x$$
$$-2y = -5x - 8$$
$$\frac{-2y}{-2} = \frac{-5x - 8}{-2}$$
$$y = \frac{-5x - 8}{-2}$$
$$y = \frac{(-5x - 8)(-1)}{-2(-1)}$$
$$y = \frac{5x + 8}{2}$$

$(4, 14)$
$(0, 4)$
$(-2, -1)$

19. Solve for y.

$$9x - 2y = 5$$
$$9x - 2y - 9x = 5 - 2x$$
$$-2y = -9x + 5 \qquad (1,2)$$
$$\frac{-2y}{-2} = \frac{-9x + 5}{-2} \qquad \left(0, -\frac{5}{2}\right)$$
$$y = \frac{-9x + 5}{-2} \qquad (-3, -16)$$
$$y = \frac{(-9x + 5)(-1)}{-2(-1)}$$
$$y = \frac{9x - 5}{2}$$

21. $(2,3)$; $(0,3)$; $(-4,3)$.

23. $(5,4)$; $(5,0)$; $(5,-2)$.

25.

$$2x + 3y = 6$$
$$2x + 3y - 2x = 6 - 2x$$
$$3y = -2x + 6$$
$$\frac{3y}{3} = \frac{-2x + 6}{3}$$
$$y = \frac{-2x + 6}{3}$$

27.

$$x + 2y = 7$$
$$x + 2y - x = 7 - x$$
$$2y = -x + 7$$
$$\frac{2y}{2} = \frac{-x + 7}{2}$$
$$y = \frac{-x + 7}{2}$$

29.

$$x - 2y = 6$$
$$x - 2y - x = 6 - x$$
$$-2y = -x + 6$$
$$\frac{-2y}{-2} = \frac{-x + 6}{-2}$$
$$y = \frac{-x + 6}{-2}$$
$$y = \frac{(-x + 6)(-1)}{-2(-1)}$$
$$y = \frac{x - 6}{2}$$

31.

$$2x - 3y = 9$$
$$2x - 3y - 2x = 9 - 2x$$
$$-3y = -2x + 9$$
$$\frac{-3y}{-3} = \frac{-2x + 9}{-3}$$
$$y = \frac{-2x + 9}{-3}$$
$$y = \frac{(-2x + 9)(-1)}{-3(-1)}$$
$$y = \frac{2x - 9}{3}$$

33.

$$-2x + 3y = 6$$
$$-2x + 3y + 2x = 6 + 2x$$
$$3y = 2x + 6$$
$$\frac{3y}{3} = \frac{2x + 6}{3}$$
$$y = \frac{2x + 6}{3}$$

35.

$$-2x - 3y = -15$$
$$-2x - 3y + 2x = -15 + 2x$$
$$-3y = 2x - 8$$
$$\frac{-3y}{-3} = \frac{2x - 15}{-3}$$
$$y = \frac{2x - 15}{-3}$$
$$y = \frac{(2x - 15)(-1)}{-3(-1)}$$
$$y = \frac{-2x + 15}{3}$$

37. $A(-2,2)$

39. $C(5,-1)$

41. $E(-4,-5)$

43. $G(0,5)$

45. $I(4,2)$

47.–65.

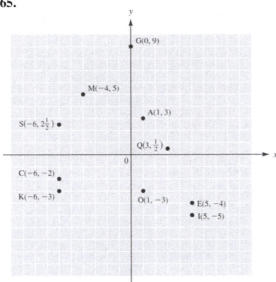

Section 8.2: Graphing Linear Equations

1.

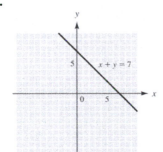

3.

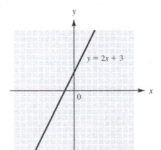

5.

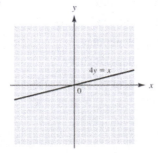

7.

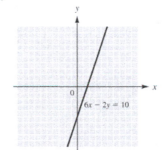

9.

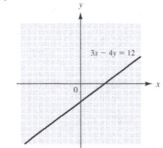

11.

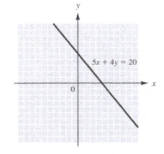

13.

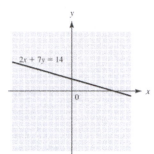

15.

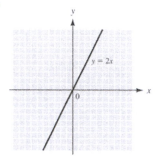

17.

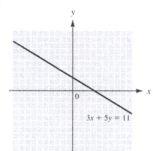

19.

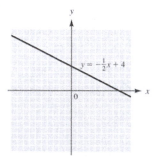

21.

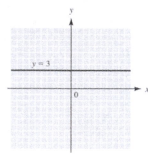

23.

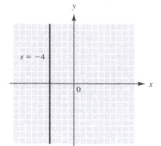

25.

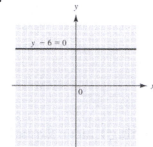

27.

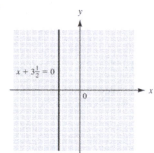

29.

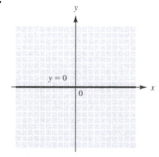

31. independent t ; dependent s

33. independent V ; dependent R

35. independent t ; dependent i

37. independent t ; dependent v

39. independent t ; dependent s

41.

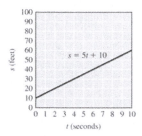

45.

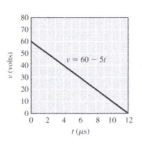

43.

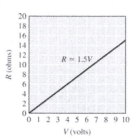

Section 8.3: The Slope of a Line

1. $m = \dfrac{6-1}{2-(-3)} = 1$

3. $m = \dfrac{1-(-5)}{-2-(-3)} = 6$

5. $m = \dfrac{0-5}{4-0} = -\dfrac{5}{4}$

7. $m = \dfrac{4-4}{5-(-2)} = \dfrac{0}{7} = 0$

9. $m = \dfrac{1-(-4)}{6-6} = \dfrac{5}{0}$; (undefined)

11. $m = \dfrac{-2-(-8)}{4-(-6)} = \dfrac{3}{5}$

13. $m = \dfrac{-6-4}{2-(-3)} = -2$

15. The points $(-1,-5)$ and $(2,2)$ are on the line,

so $m = \dfrac{2-(-1)}{2-(-5)} = \dfrac{3}{7}$.

17. The points $(-3,-5)$ and $(4,-5)$ are on the

line, so $m = \dfrac{-5-(-5)}{4-(-3)} = \dfrac{0}{7} = 0$.

19. $m = 6$

21. $m = -5$

23.
$$3x + 5y = 6$$
$$5y = -3x + 6$$
$$y = -\dfrac{3}{5}x + \dfrac{5}{6}$$
$$m = -\dfrac{3}{5}$$

25.
$$-2x + 8y = 3$$
$$8y = 2x + 3$$
$$y = \dfrac{1}{4}x + \dfrac{3}{8}$$
$$m = \dfrac{1}{4}$$

27.
$$5x - 2y = 16$$
$$-2y = -5x + 16$$
$$y = \dfrac{5}{2}x - 8$$
$$m = \dfrac{5}{2}$$

29. The line is vertical, so the slope is undefined.

31. $m_1 = 4$; $m_2 = 4$; parallel

33. $m_1 = \dfrac{3}{4}$; $m_2 = -\dfrac{4}{3}$; perpendicular

35.

$$x + 3y = 9$$
$$3y = -x + 9$$
$$y = -\frac{1}{3}x + 3$$
$$m_1 = -\frac{1}{3}$$

$$3x - y = 14$$
$$-y = -3x + 14$$
$$y = 3x - 14$$
$$m_2 = \frac{3}{1}$$

The lines are perpendicular.

37.

$$x - 4y = 12$$
$$-4y = -x + 12$$
$$y = \frac{1}{4}x - 3$$
$$m_1 = \frac{1}{4}$$

$$x + 4y = 12$$
$$4y = -x + 12$$
$$y = -\frac{1}{4}x + 3$$
$$m_2 = -\frac{1}{4}$$

The lines are neither parallel nor perpendicular.

39.

$$y - 5x = 12$$
$$y = 5x + 12$$
$$m_1 = 5$$

$$5x - y = -6$$
$$-y = -5x - 6$$
$$y = 5x + 6$$
$$m_2 = 5$$

The lines are parallel.

Section 8.4: The Equation of a Line

1.

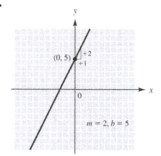

7.

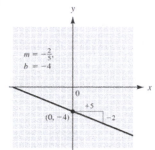

3.

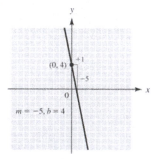

9.

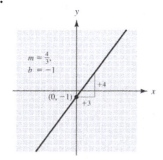

5.

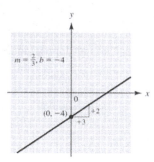

11.

$$2x + y = 6$$
$$y = -2x + 6$$
$$m = -2; \ b = 6$$

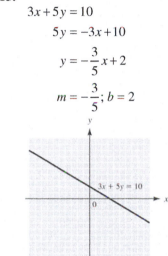

13.

$$3x + 5y = 10$$
$$5y = -3x + 10$$
$$y = -\frac{3}{5}x + 2$$
$$m = -\frac{3}{5}; \ b = 2$$

15.

$$3x - y = 7$$
$$-y = -3x + 7$$
$$y = 3x - 7$$
$$m = 3; \ b = -7$$

17.

$$3x - 2y = 12$$
$$-2y = -3x + 12$$
$$y = \frac{3}{2}x - 6$$
$$m = \frac{3}{2}; \ b = -6$$

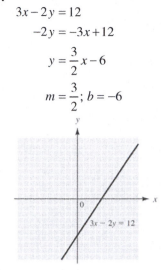

19.

$$2x - 6y = 0$$
$$-6y = -2x$$
$$y = \frac{1}{3}x + 0$$
$$m = \frac{1}{3}; \ b = 0$$

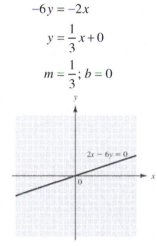

21.

$$y = 2x + 5 \text{ or}$$
$$-2x + y = 5$$
$$2x - y = -5$$

23.

$$y = -5x + 4 \text{ or}$$
$$5x + y = 4$$

25.

$$y = \frac{2}{3}x - 4 \text{ or}$$
$$3y = 2x - 12$$
$$-2x + 3y = -12$$
$$2x - 3y = 12$$

27.

$$y = -\frac{6}{5}x + 3 \text{ or}$$
$$5y = -6x + 15$$
$$6x + 5y = 15$$

29.

$$y = -\frac{3}{5}x \text{ or}$$
$$5y = -3x$$
$$3x + 5y = 0$$

31.

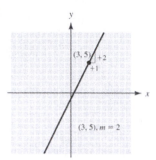

39.

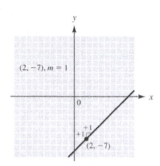

33.

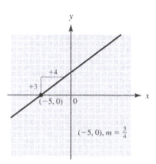

41.

$$y - y_1 = m(x - x_1)$$
$$y - 5 = 2(x - 3)$$
$$y - 5 = 2x - 6$$
$$y = 2x - 1 \text{ or}$$
$$-2x + y = -1$$
$$2x - y = 1$$

43.

$$y - y_1 = m(x - x_1)$$
$$y - 0 = \frac{3}{4}\left[x - (-5)\right]$$
$$y = \frac{3}{4}(x + 5)$$
$$y = \frac{3}{4}x + \frac{15}{4} \text{ or}$$
$$4y = 3x + 15$$
$$-3x + 4y = 15$$
$$3x - 4y = -15$$

35.

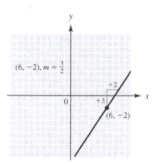

37.

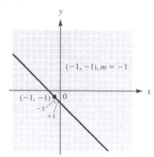

45.

$$y - y_1 = m(x - x_1)$$
$$y - (-2) = \frac{3}{2}(x - 6)$$
$$y + 2 = \frac{3}{2}x - 9$$
$$y = \frac{3}{2}x - 11 \text{ or}$$
$$2y = 3x - 22$$
$$-3x + 2y = -11$$
$$3x - 2y = 11$$

47.

$$y - y_1 = m(x - x_1)$$

$$y - (-1) = -\frac{10}{3}\left[x - (-3)\right]$$

$$y + 1 = -\frac{10}{3}(x + 3)$$

$$y + 1 = -\frac{10}{3}x - 10$$

$$y = -\frac{10}{3}x - 11 \text{ or}$$

$$3y = -10x - 33$$

$$10x + 3y = -33$$

49.

$$y - y_1 = m(x - x_1)$$

$$y - (-10) = -1(x - 12)$$

$$y + 10 = -x + 12$$

$$y = -x + 2 \text{ or}$$

$$x + y = 2$$

51. $m = \dfrac{1 - 3}{5 - 2} = -\dfrac{2}{3}$

$$y - y_1 = m(x - x_1)$$

$$y - 3 = -\frac{2}{3}(x - 2)$$

$$y - \frac{9}{3} = -\frac{2}{3}x + \frac{4}{3}$$

$$y = -\frac{2}{3}x + \frac{13}{3} \text{ or}$$

$$3y = -2x + 13$$

$$2x + 3y = 13$$

or

$$y - y_1 = m(x - x_1)$$

$$y - 1 = -\frac{2}{3}(x - 5)$$

$$y - \frac{3}{3} = -\frac{2}{3}x + \frac{10}{3}$$

$$y = -\frac{2}{3}x + \frac{13}{3} \text{ or}$$

$$3y = -2x + 13$$

$$2x + 3y = 13$$

53. $m = \dfrac{-2 - 4}{2 - (-1)} = -2$

$$y - y_1 = m(x - x_1)$$

$$y - (-2) = -2(x - 2)$$

$$y + 2 = -2x + 4$$

$$y = -2x + 2 \text{ or}$$

$$2x + y = 2$$

or

$$y - y_1 = m(x - x_1)$$

$$y - 4 = -2\left[(x - (-1))\right]$$

$$y - 4 = -2(x + 1)$$

$$y - 4 = -2x - 2$$

$$y = -2x + 2 \text{ or}$$

$$2x + y = 2$$

55. $m = \dfrac{1-0}{0-(-3)} = \dfrac{1}{3}$

$y - y_1 = m(x - x_1)$ $y - y_1 = m(x - x_1)$

$y - 1 = \dfrac{1}{3}(x - 0)$ $y - 0 = \dfrac{1}{3}\left[\left(x - (-3)\right)\right]$

$y - 1 = \dfrac{1}{3}x$ or $y = \dfrac{1}{3}(x + 3)$

$y = \dfrac{1}{3}x + 1$ or $y = \dfrac{1}{3}x + 1$ or

$3y = x + 3$ $3y = x + 3$

$x - 3y = -3$ $x - 3y = -3$

57. $m = \dfrac{4-0}{7-(-1)} = \dfrac{1}{2}$

$y - y_1 = m(x - x_1)$ $y - y_1 = m(x - x_1)$

$y - 4 = \dfrac{1}{2}(x - 7)$ $y - 0 = \dfrac{1}{2}\left[x - (-1)\right]$

$y - \dfrac{8}{2} = \dfrac{1}{2}x - \dfrac{7}{2}$ or $y - 0 = \dfrac{1}{2}(x + 2)$

$y = \dfrac{1}{2}x + \dfrac{1}{2}$ or $y = \dfrac{1}{2}x + \dfrac{1}{2}$ or

$2y = x + 1$ $2y = x + 1$

$x - 2y = -1$ $x - 2y = -1$

59. $m = \dfrac{3-5}{3-1} = -1$

$y - y_1 = m(x - x_1)$ $y - y_1 = m(x - x_1)$

$y - 3 = -1(x - 3)$ $y - 5 = -1(x - 1)$

$y - 3 = -x + 3$ or $y - 5 = -x + 1$

$y = -x + 6$ or $y = -x + 6$ or

$x + y = 6$ $x + y = 6$

Chapter 8 Review

1. Solve for y.

$x + 2y = 8$ $\left(3, \dfrac{5}{2}\right)$

$2y = -x + 8$ $(0, 4)$

$y = -\dfrac{1}{2}x + 4$ $(-4, 6)$

2. Solve for y.

$2x - 3y = 12$ $(3, -2)$

$-3y = -2x + 12$ $(0, -4)$

$y = \dfrac{2}{3}x - 4$ $(-3, -6)$

3.

$6x + y = 15$

$y = -6x + 15$

4.

$$3x - 5y = -10$$
$$-5y = -3x - 10$$
$$y = \frac{3}{5}x + 2$$

5. $A(3, 5)$

6. $B(-2, -6)$

7. $C(2, -1)$

8. $D(4, 0)$

9.–12.

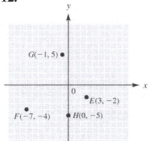

13.

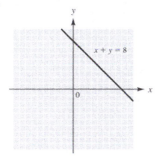

14.

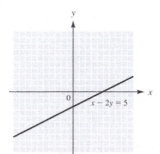

15.

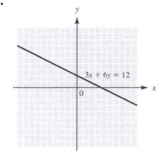

16.

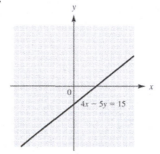

17.

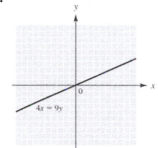

18.

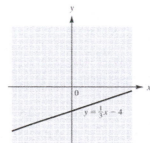

19.

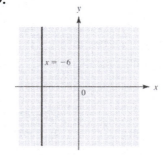

20.

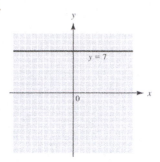

21. $m = \dfrac{5-(-4)}{10-3} = \dfrac{9}{7}$

22. $m = \dfrac{6-0}{2-(-4)} = 1$

23. $m = 4$

24.
$$2x + 5y = 8$$
$$5y = -2x + 8$$
$$y = -\dfrac{2}{5}x + \dfrac{8}{5}$$
$$m = -\dfrac{2}{5}$$

25.
$$5x - 9y = -2$$
$$-9y = -5x - 2$$
$$y = \dfrac{5}{9}x + \dfrac{2}{9}$$
$$m = \dfrac{5}{9}$$

26.

$$y = 3x - 5 \qquad\qquad y = -\dfrac{1}{3}x - 5$$
$$m_1 = 3 \qquad\qquad m_2 = -\dfrac{1}{3}$$

The lines are perpendicular.

27.

$$3x - 4y = 12 \qquad\qquad 8x - 6y = 15$$
$$-4y = -3x + 12 \qquad\qquad -6y = -8x + 15$$
$$y = \dfrac{3}{4}x - 3 \qquad\qquad y = \dfrac{4}{3}x - \dfrac{5}{2}$$
$$m_1 = \dfrac{3}{4} \qquad\qquad m_2 = \dfrac{4}{3}$$

The lines are neither parallel nor perpendicular.

28.

$$2x + 5y = 8 \qquad\qquad 4x + 10y = 25$$
$$5y = -2x + 8 \qquad\qquad 10y = -4x + 25$$
$$y = -\dfrac{2}{5}x + \dfrac{8}{5} \qquad\qquad y = -\dfrac{2}{5}x + \dfrac{5}{2}$$
$$m_1 = -\dfrac{2}{5} \qquad\qquad m_2 = -\dfrac{2}{5}$$

The lines are parallel.

29.

$$x = 4 \qquad\qquad\qquad y = -6$$
$$m_1 \text{ (undefined)} \qquad\qquad m_2 = 0$$

The lines are perpendicular.

30.

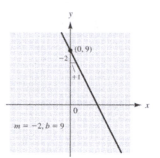

31.

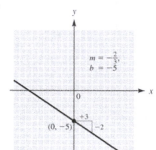

32.

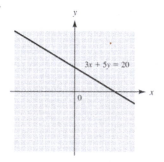

33.

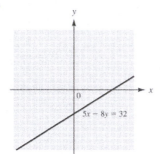

34.
$$y = -\frac{1}{2}x + 3 \text{ or}$$
$$2y = -x + 6$$
$$x + 2y = 6$$

35.
$$y = \frac{8}{3}x + 0$$
$$y = \frac{8}{3}x \text{ or}$$
$$3y = 8x$$
$$8x - 3y = 0$$

36.
$$y = 0x + 0$$
$$y = 0$$

37.

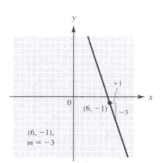

38.

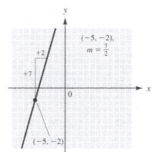

39.
$$y - y_1 = m(x - x_1)$$
$$y - 8 = -1\big[x - (-2)\big]$$
$$y - 8 = -1(x + 2)$$
$$y - 8 = -x - 2$$
$$y = -x + 6 \text{ or}$$
$$x + y = 6$$

40.
$$y - y_1 = m(x - x_1)$$
$$y - (-5) = -\frac{1}{4}(x - 0)$$
$$y + \frac{20}{4} = -\frac{1}{4}x$$
$$y = -\frac{1}{4}x - 5 \text{ or}$$
$$4y = -x - 20$$
$$x + 4y = -20$$

41. $m = \dfrac{5 - (-3)}{10 - 2} = 1$

$$y - y_1 = m(x - x_1) \qquad\qquad y - y_1 = m(x - x_1)$$
$$y - (-3) = 1(x - 2) \qquad\qquad y - 5 = 1(x - 10)$$
$$y + 3 = x - 2 \qquad \text{or} \qquad y - 5 = x - 10$$
$$y = x - 5 \text{ or} \qquad\qquad y = x - 5 \text{ or}$$
$$x - y = 5 \qquad\qquad\qquad x - y = 5$$

42. $m = \dfrac{0-(-5)}{12-2} = \dfrac{1}{2}$

$$y - y_1 = m(x - x_1)$$

$$y - 0 = \frac{1}{2}(x - 12)$$

$$y = \frac{1}{2}x - 6 \text{ or}$$ or

$$2y = x - 12$$

$$x - 2y = 12$$

$$y - y_1 = m(x - x_1)$$

$$y - (-5) = \frac{1}{2}(x - 2)$$

$$y + 5 = \frac{1}{2}x - 1$$

$$y = \frac{1}{2}x - 6 \text{ or}$$

$$2y = x - 12$$

$$x - 2y = 12$$

Chapter 8 Test

1. Solve for y.

$$3x - 4y = 24$$

$$-4y = -3x + 24$$

$$y = \frac{3}{4}x - 6$$

$$y = \frac{3}{4}(-4) - 6 = -9$$

$$(-4, -9)$$

3.

$$3x - 4y = 24$$

$$-4y = -3x + 24$$

$$y = \frac{3}{4}x - 6$$

$$y = \frac{3}{4}(4) - 6 = -3$$

$$(4, 3)$$

5. $(4, -2)$

7.

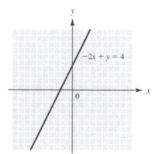

9. $m = \dfrac{6-4}{5-(-2)} = \dfrac{2}{7}$

11. Solve for y.

$$2x - 5y = 10$$

$$-5y = -2x + 10$$

$$y = \frac{2}{5}x - 2$$

$$m = \frac{2}{5}$$

The lines are parallel.

13.

$$y = \frac{1}{2}x - 3 \text{ or}$$

$$2y = x - 6$$

$$x - 2y = 6$$

15.

Cumulative Review Chapters 1-8

1.
$$2(6-5)+3 = 2(1)+3$$
$$= 2+5$$
$$= 5$$

2.
$$+6-(-9) = 6+9$$
$$= 15$$

3. a.
$$a = 3\frac{1}{2}\ \text{in.}+4\frac{1}{8}\ \text{in.}+3\frac{1}{2}\ \text{in.}-6\frac{15}{16}\ \text{in.}$$
$$= \frac{7}{2}\ \text{in.}+\frac{33}{8}\ \text{in.}+\frac{7}{2}\ \text{in.}-\frac{111}{16}\ \text{in.}$$
$$= \frac{56}{16}\ \text{in.}+\frac{66}{16}\ \text{in.}+\frac{56}{16}\ \text{in.}-\frac{111}{16}\ \text{in.}$$
$$= \frac{67}{16}\ \text{in.} = 4\frac{3}{16}\ \text{in.}$$

b.
$$b = 5\frac{1}{4}\ \text{in.}-2\frac{3}{8}\ \text{in.}$$
$$= \frac{21}{4}\ \text{in.}-\frac{19}{8}\ \text{in.}$$
$$= \frac{42}{8}\ \text{in.}-\frac{19}{8}\ \text{in.}$$
$$= \frac{23}{8}\ \text{in.} = 2\frac{7}{8}\ \text{in.}$$

4. $250\ \text{cm} \times \dfrac{1\ \text{in.}}{2.54\ \text{cm}} = 98.4\ \text{in.}$

5.
$$(6.2\times10^{-3})(1.8\times10^{5})$$
$$= (6.2)(1.8)\times(10^{-3})(10^{5})$$
$$= 1.116\times10^{3}$$

6. $61\ \text{mm} \times \dfrac{1\ \text{m}}{1000\ \text{mm}} = 0.061\ \text{mm}$

7. 5

8. 7.82 mm

9. 350 m^3

10.
$$(4x-5)-(6-3x) = 4x-5-6+3x$$
$$= (4+3)x+(-5-6)$$
$$= 7x-11$$

11.
$$(-5xy^2)(8x^3y^2)$$
$$= (-5)(8)(x)(x^3)(y^2)(y^2)$$
$$= -40x^4y^4$$

12.
$$2x(4x-3y) = 2x(4x)-2x(3y)$$
$$= 8x^2-6xy$$

13.
$$3(x-2)+4(3-2x) = 9$$
$$3x-6+12-8x = 9$$
$$-5x+6 = 9$$
$$-5x+6-6 = 9-6$$
$$-5x = 3$$
$$\frac{-5x}{-5} = \frac{3}{-5}$$
$$x = -\frac{3}{5}$$

14.
$$\frac{2x}{3}+\frac{1}{5} = \frac{x}{4}-\frac{2}{3}$$
$$60\left(\frac{2x}{3}+\frac{1}{5}\right) = \left(\frac{x}{4}-\frac{2}{3}\right)60$$
$$60\left(\frac{2x}{3}\right)+60\left(\frac{1}{5}\right) = \left(\frac{x}{4}\right)60-\left(\frac{2}{3}\right)60$$
$$20(2x)+12 = (x)15-(2)20$$
$$40x+12-15x = 15x-40-15x$$
$$25x+12 = -40$$
$$25x+12-12 = -40-12$$
$$25x = -52$$
$$\frac{25x}{25} = \frac{-52}{25}$$
$$x = -\frac{52}{25}$$

15.
$$s = \frac{2V + t}{3}$$

$$3(s) = \left(\frac{2V + t}{3}\right)3$$

$$3s = 2V + t$$

$$3s - t = 2V + t - t$$

$$3s - t = 2V$$

$$\frac{3s - t}{2} = \frac{2V}{2}$$

$$\frac{3s - t}{2} = V$$

$$\text{So } V = \frac{3s - t}{2}$$

16. $5 \text{ to } 65 = \frac{5}{65} = \frac{1}{13}$

17. $3 \text{ yd} \times \frac{36 \text{ in.}}{1 \text{ yd}} = 108 \text{ in.}$

$32 \text{ in.} : 3 \text{ yd} = 32 \text{ in.} : 108 \text{ in.}$

$$= \frac{32 \text{ in}}{108 \text{ in.}}$$

$$= \frac{8}{27}$$

18.
$$\frac{5}{13} = \frac{x}{156}$$

$$13x = 780$$

$$x = 60$$

19.
$$\frac{29.1}{73.8} = \frac{x}{104}$$

$$73.8x = (29.1)(104)$$

$$x = \frac{(29.1)(104)}{73.8}$$

$$x = 41.0$$

20.
$$\frac{286}{x} = \frac{11.8}{59.7}$$

$$11.8x = (286)(59.7)$$

$$x = \frac{(286)(59.7)}{11.8}$$

$$x = 1450$$

21.
$$\frac{x}{18 \text{ ft}^2} = \frac{\$28.50}{5 \text{ ft}^2}$$

$$\left(5 \text{ ft}^2\right)x = \left(18 \text{ ft}^2\right)(\$28.50)$$

$$x = \frac{\left(18 \text{ ft}^2\right)(\$28.50)}{5 \text{ ft}^2}$$

$$x = \$102.60$$

22.
$$\frac{x}{4\frac{1}{4} \text{ in.}} = \frac{40 \text{ mi}}{1 \text{ in.}}$$

$$(1 \text{ in.})x = \left(4\frac{1}{4} \text{ in.}\right)(40 \text{ mi})$$

$$x = \frac{\left(4\frac{1}{4} \text{ in.}\right)(40 \text{ mi})}{1 \text{ in.}}$$

$$x = 170 \text{ mi}$$

23.
$$\left(n_A\right)\left(\text{rpm}_A\right) = \left(n_B\right)\left(\text{rpm}_B\right)$$

$$(x)(64 \text{ rpm}) = (16 \text{ teeth})(40 \text{ rpm})$$

$$x = \frac{(16 \text{ teeth})(40 \text{ rpm})}{64 \text{ rpm}}$$

$$x = 10 \text{ teeth}$$

24. Solve for y.

$$2x + 3y = 12 \qquad\qquad (3, 2)$$

$$3y = -2x + 12 \qquad\quad (0, 4)$$

$$y = -\frac{2}{3}x + 4 \qquad\quad (-3, 6)$$

25.
$$4x + 2y = 7$$

$$2y = -4x + 7$$

$$y = -2x + \frac{7}{2}$$

26.

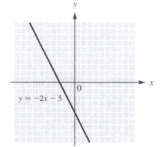

$$y = -2x - 5$$

27.

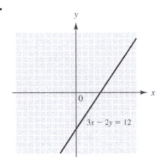

28. $m = \dfrac{6-3}{2-(-1)} = 1$

29.

$$y - y_1 = m(x - x_1)$$
$$y - (-4) = \frac{1}{2}(x - 2)$$
$$y + 4 = \frac{1}{2}x - 1$$
$$y = \frac{1}{2}x - 5 \text{ or}$$
$$2y = x - 10$$
$$x - 2y = 10$$

30.

$$2x - 3y = 6 \qquad\qquad 3x + 5y = 7$$
$$-3y = -2x + 6 \qquad\qquad 5y = -3x + 7$$
$$y = \frac{2}{3}x - 2 \qquad\qquad y = -\frac{3}{5}x - \frac{2}{5}$$
$$m_1 = \frac{2}{3} \qquad\qquad m_2 = -\frac{3}{5}$$

The lines are neither parallel nor perpendicular.

Chapter 9: Systems of Linear Equations

Section 9.1: Solving Pairs of Linear Equations by Graphing

1.

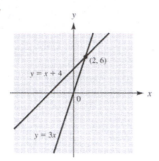

3.

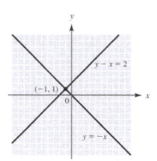

5.

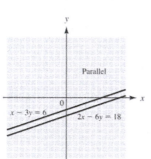

7.

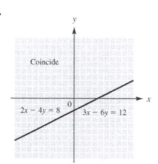

9.

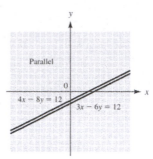

11.

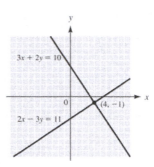

13.

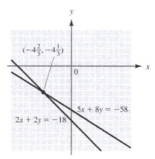

15.

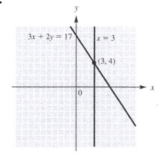

140

17.

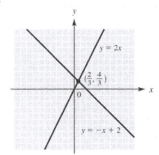

19.

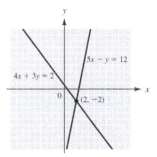

21.

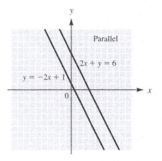

23.

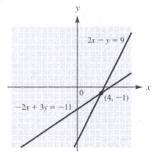

25.

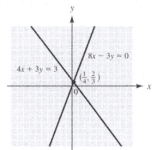

27.

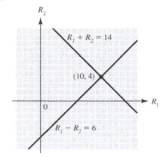

29.

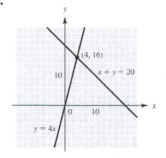

Section 9.2: Solving Pairs of Linear Equations by Addition

1.

$$3x + y = 7$$
$$\underline{x - y = 1}$$
$$4x \quad = 8$$
$$x \quad = 2$$
$$3x + y = 7$$
$$3(2) + y = 7$$
$$6 + y = 7$$
$$y = 1$$

The solution is $(2, 1)$.

3.

$$2x + 5y = 18$$
$$\underline{4x - 5y = 6}$$
$$6x \quad = 24$$
$$x \quad = 4$$
$$2x + 5y = 18$$
$$2(4) + 5y = 18$$
$$8 + 5y = 18$$
$$5y = 10$$
$$y = 2$$

The solution is $(4, 2)$.

5.

$$-2x + 5y = 39$$
$$\underline{2x - 3y = -25}$$
$$2y = 14$$
$$y = 7$$
$$2x - 3(7) = -25$$
$$2x - 21 = -25$$
$$2x = -4$$
$$x = -2$$

The solution is $(-2, 7)$.

7.

$$x + 3y = 6$$
$$x - y = 2$$

$$x + 3y = 6$$
$$3(x - y) = (2)3$$

$$x + 3y = 6$$
$$\underline{3x - 3y = 6}$$
$$4x \quad = 12$$
$$x \quad = 3$$

$$x - y = 2$$
$$-y = -1$$
$$y = 1$$

The solution is $(3, 1)$.

9.

$$2x + 5y = 15$$
$$7x + 5y = -10$$

$$(-1)(2x + 5y) = (15)(-1)$$
$$7x + 5y = -10$$

$$-2x - 5y = -15$$
$$\underline{7x + 5y = -10}$$
$$5x \quad = -25$$
$$x \quad = -5$$

$$2x + 5y = 15$$
$$2(-5) + 5y = 15$$
$$-10 + 5y = 15$$
$$5y = 35$$
$$y = 5$$

The solution is $(-5, 5)$.

11.

$$5x + 6y = 31$$
$$2x + 6y = 16$$

$$5x + 6y = 31$$
$$(-1)(2x + 6y) = (16)(-1)$$

$$5x + 6y = 31$$
$$\underline{-2x - 6y = -16}$$
$$3x \quad\ = 15$$
$$x \quad\ = 5$$

$$2x + 6y = 16$$
$$2(5) + 6y = 16$$
$$10 + 6y = 16$$
$$6y = 6$$
$$y = 1$$

The solution is $(5,1)$.

13.

$$4x - 5y = 14$$
$$2x + 3y = -4$$

$$4x - 5y = 14$$
$$(-2)(2x + 3y) = (-4)(-2)$$

$$4x - 5y = 14$$
$$\underline{-4x - 6y = 8}$$
$$-11y = 22$$
$$y = -2$$

$$2x + 3y = -4$$
$$2x + 3(-2) = -4$$
$$2x - 6 = -4$$
$$2x = 2$$
$$x = 1$$

The solution is $(1, -2)$.

15.

$$3x - 2y = -11$$
$$7x - 10y = -47$$

$$(-5)(3x - 2y) = (-11)(-5)$$
$$7x - 10y = -47$$

$$-15x + 10y = 55$$
$$\underline{7x - 10y = -47}$$
$$-8x \quad\ = 8$$
$$x \quad\ = -1$$

$$3x - 2y = -11$$
$$3(-1) - 2y = -11$$
$$-3 - 2y = -11$$
$$-2y = -8$$
$$y = 4$$

The solution is $(-1, 4)$.

17.

$$x + 2y = -3$$
$$2x + y = 9$$

$$(-2)(x + 2y) = (-3)(-2)$$
$$2x + y = 9$$

$$-2x - 4y = 6$$
$$\underline{2x + y = 9}$$
$$-3y = 15$$
$$y = -5$$

$$x + 2y = -3$$
$$x + 2(-5) = -3$$
$$x - 10 = -3$$
$$x = 7$$

The solution is $(7, -5)$.

19.

$$3x + 5y = 7$$
$$2x - 7y = 15$$

$$7(3x + 5y) = (7)7$$
$$5(2x - 7y) = (15)5$$

$$21x + 35y = 49$$
$$\underline{10x - 35y = 75}$$
$$31x \qquad = 124$$
$$x \qquad = 4$$

$$3x + 5y = 7$$
$$3(4) + 5y = 7$$
$$12 + 5y = 7$$
$$5y = -5$$
$$y = -1$$

The solution is $(4, -1)$.

21.

$$8x - 7y = -51$$
$$12x + 13y = 41$$

$$(-3)(8x - 7y) = (-51)(-3)$$
$$2(12x + 13y) = (41)2$$

$$-24x + 21y = 153$$
$$\underline{24x + 26y = 82}$$
$$47y = 235$$
$$y = 5$$

$$8x - 7y = -51$$
$$8x - 7(5) = -51$$
$$8x - 35 = -51$$
$$8x = -16$$
$$x = -2$$

The solution is $(-2, 5)$.

23.

$$5x - 12y = -5$$
$$9x - 16y = -2$$

$$(-4)(5x - 12y) = (-5)(-4)$$
$$3(9x - 16y) = (-2)3$$

$$-20x + 48y = 20$$
$$\underline{27x - 48y = -6}$$
$$7x \qquad = 14$$
$$x \qquad = 2$$

$$5x - 12y = -5$$
$$5(2) - 12y = -5$$
$$10 - 12y = -5$$
$$-12y = -15$$
$$y = \frac{5}{4} = 1\frac{1}{4}$$

The solution is $\left(2, 1\frac{1}{4}\right)$.

25.

$$2x + 3y = 8$$
$$x + y = 2$$

$$2x + 3y = 8$$
$$(-2)(x + y) = (2)(-2)$$

$$2x + 3y = 8$$
$$\underline{-2x - 2y = -4}$$
$$y = 4$$

$$x + y = 2$$
$$x + (4) = 2$$
$$x = -2$$

The solution is $(-2, 4)$.

27.

$$3x - 5y = 7$$
$$9x - 15y = 21$$

$$(-3)(3x - 5y) = (7)(-3)$$
$$9x - 15y = 21$$

$$-9x + 15y = -21$$
$$\underline{9x - 15y = 21}$$
$$0 = 0$$

The lines coincide.

29.

$$2x + 5y = -1$$
$$3x - 2y = 8$$

$$2(2x + 5y) = (-1)2$$
$$5(3x - 2y) = (8)5$$

$$4x + 10y = -2$$
$$\underline{15x - 10y = 40}$$
$$19x \quad\quad = 38$$
$$x \quad\quad = 2$$

$$2x + 5y = -1$$
$$2(2) + 5y = -1$$
$$4 + 5y = -1$$
$$5y = -5$$
$$y = -1$$

The solution is $(2, -1)$.

31.

$$16x - 36y = 70$$
$$4x - 9y = 17$$

$$16x - 36y = 70$$
$$(-4)(4x - 9y) = (17)(-4)$$

$$16x - 36y = 70$$
$$\underline{-16x + 36y = -68}$$
$$0 = 2$$

The lines are parallel.

33.

$$4x + 3y = 17$$
$$2x - y = -4$$

$$4x + 3y = 17$$
$$3(2x - y) = (-4)3$$

$$4x + 3y = 17$$
$$\underline{6x - 3y = -12}$$
$$10x \quad\quad = 5$$
$$x \quad\quad = \frac{1}{2}$$

$$2x - y = -4$$
$$2\left(\frac{1}{2}\right) - y = -4$$
$$1 - y = -4$$
$$-y = -5$$
$$y = 5$$

The solution is $\left(\frac{1}{2}, 5\right)$.

35.

$$2x - 5y = 8$$
$$4x - 10y = 16$$

$$(-2)(2x - 5y) = (8)(-2)$$
$$4x - 10y = 16$$

$$-4x + 10y = -16$$
$$\underline{4x - 10y = 16}$$
$$0 = 0$$

The lines coincide.

37.

$$5x - 8y = 10$$
$$-10x + 16y = 8$$

$$2(5x - 8y) = (10)2$$
$$-10x + 16y = 8$$

$$10x - 16y = 20$$
$$\underline{-10x + 16y = 8}$$
$$0 = 28$$

The lines are parallel.

39.

$$8x - 5y = 426$$
$$7x - 2y = 444$$

$$(-2)(8x - 5y) = (426)(-2)$$
$$5(7x - 2y) = (444)5$$

$$-16x + 10y = -852$$
$$\underline{35x - 10y = 2220}$$
$$19x \qquad = 1368$$
$$x \qquad = 72$$

$$7x - 2y = 444$$
$$7(72) - 2y = 444$$
$$504 - 2y = 444$$
$$-2y = -60$$
$$y = 30$$

The solution is $(72, 30)$.

41.

$$16x + 5y = 6$$
$$7x + \frac{5}{8}y = 2$$

$$16x + 5y = 6$$
$$(-8)\left(7x + \frac{5}{8}y\right) = (2)(-8)$$

$$16x + 5y = 6$$
$$\underline{-56x - 5y = -16}$$
$$-40x \qquad = -10$$
$$x = \frac{1}{4}$$

$$16x + 5y = 6$$
$$16\left(\frac{1}{4}\right) + 5y = 6$$
$$4 + 5y = 6$$
$$5y = 2$$
$$y = \frac{2}{5}$$

The solution is $\left(\frac{1}{4}, \frac{2}{5}\right)$.

43.

$$7x + 8y = 47$$
$$5x - 3y = 51$$

$$3(7x + 8y) = (47)3$$
$$8(5x - 3y) = (51)8$$

$$21x + 24y = 141$$
$$\underline{40x - 24y = 408}$$
$$61x \qquad = 549$$
$$x \qquad = 9$$

$$7x + 8y = 47$$
$$7(9) + 8y = 47$$
$$63 + 8y = 47$$
$$8y = -16$$
$$y = -2$$

The solution is $(9, -2)$.

Section 9.3: Solving Pairs of Linear Equations by Substitution

1.

$$2x + y = 12$$
$$y = 3x$$

$$2x + y = 12$$
$$2x + (3x) = 12$$
$$5x = 12$$
$$x = \frac{12}{5} = 2\frac{2}{5}$$

$$y = 3x$$
$$y = 3\left(\frac{12}{5}\right)$$
$$y = \frac{36}{5} = 7\frac{1}{5}$$

The solution is $\left(2\frac{2}{5}, 7\frac{1}{5}\right)$ or $(2.4, 7.2)$.

3.

$$5x - 2y = 46$$
$$x = 5y$$

$$5x - 2y = 46$$
$$5(5y) - 2y = 46$$
$$25y - 2y = 46$$
$$23y = 46$$
$$y = 2$$

$$x = 5y$$
$$x = 5(2)$$
$$x = 10$$

The solution is $(10, 2)$.

5.

$$3x + 2y = 30$$
$$x = y$$

$$3x + 2y = 30$$
$$3(y) + 2y = 30$$
$$3y + 2y = 30$$
$$5y = 60$$
$$y = 6$$

$$x = y$$
$$x = (6)$$
$$x = 6$$

The solution is $(6, 6)$.

7.

$$5x - y = 18$$
$$y = \frac{1}{2}x$$

$$5x - y = 18$$
$$5x - \left(\frac{1}{2}x\right) = 18$$
$$\frac{10x}{2} - \frac{1}{2}x = 18$$
$$\frac{9}{2}x = 18$$
$$x = 4$$

$$y = \frac{1}{2}x$$
$$y = \frac{1}{2}(4)$$
$$y = 2$$

The solution is $(4, 2)$.

9.

$$x - 6y = 3$$
$$3y = x$$

$$x - 6y = 3$$
$$(3y) - 6y = 3$$
$$-3y = 3$$
$$y = -1$$

$$x = 3y$$
$$x = 3(-1)$$
$$x = -3$$

The solution is $(-3, -1)$.

11.

$$3x + y = 7$$
$$4x - y = 0$$

$$4x - y = 0$$
$$y = 4x$$

$$3x + y = 7$$
$$3x + (4x) = 7$$
$$7x = 7$$
$$x = 1$$

$$y = 4x$$
$$y = 4(1)$$
$$y = 4$$

The solution is $(1, 4)$.

13.

$$4x + 3y = -2$$
$$x + y = 0$$

$$x + y = 0$$
$$y = -x$$

$$4x + 3y = -2$$
$$4x + 3(-x) = -2$$
$$4x - 3x = -2$$
$$x = -2$$

$$y = -x$$
$$y = -(-2)$$
$$y = 2$$

The solution is $(-2, 2)$.

15.

$$6x - 8y = 115$$
$$x = -\frac{y}{5}$$

$$6x - 8y = 115$$
$$6\left(-\frac{y}{5}\right) - 8y = 115$$
$$-\frac{6y}{5} - \frac{40y}{5} = 115$$
$$-\frac{46y}{5} = 115$$
$$y = -\frac{25}{2} = -12\frac{1}{2}$$

$$x = -\frac{y}{5}$$
$$x = -\frac{\left(-\frac{25}{2}\right)}{5}$$
$$x = \frac{5}{2} = 2\frac{1}{2}$$

The solution is $\left(2\frac{1}{2}, -12\frac{1}{5}\right)$.

17.

$$3x + 8y = 27$$
$$y = 2x + 1$$

$$3x + 8y = 27$$
$$3x + 8(2x + 1) = 27$$
$$3x + 16x + 8 = 27$$
$$19x + 8 = 27$$
$$19x = 19$$
$$x = 1$$

$$y = 2x + 1$$
$$y = 2(1) + 1$$
$$y = 3$$

The solution is $(1, 3)$.

19.

$$8y - 2x = -34$$
$$x = 1 - 4y$$

$$8y - 2x = -34$$
$$8y - 2(1 - 4y) = -34$$
$$8y - 2 + 8y = -34$$
$$16y - 2 = -34$$
$$16y = -32$$
$$y = -2$$

$$x = 1 - 4y$$
$$x = 1 - 4(-2)$$
$$x = 9$$

The solution is $(9, -2)$.

21.

$$3x + 4y = 25$$
$$x - 5y = -17$$

$$x - 5y = -17$$
$$x = 5y - 17$$

$$3x + 4y = 25$$
$$3(5y - 17) + 4y = 25$$
$$15y - 51 + 4y = 25$$
$$19y - 51 = 25$$
$$19y = 76$$
$$y = 4$$

$$x = 5y - 17$$
$$x = 5(4) - 17$$
$$x = 3$$

The solution is $(3, 4)$.

23.

$$4x + y = 30$$
$$-2x + 5y = 18$$

$$4x + y = 30$$
$$y = -4x + 30$$

$$-2x + 5y = 18$$
$$-2x + 5(-4x + 30) = 18$$
$$-2x - 20x + 150 = 18$$
$$-22x + 150 = 18$$
$$-22x = -132$$
$$x = 6$$

$$y = -4x + 30$$
$$y = -4(6) + 30$$
$$y = 6$$

The solution is $(6, 6)$.

25.

$$3x + 4y = 22$$
$$-5x + 2y = -2$$

$$-5x + 2y = -2$$
$$y = \frac{5}{2}x - 1$$

$$3x + 4y = 22$$
$$3x + 4\left(\frac{5}{2}x - 1\right) = 22$$
$$3x + 10x - 4 = 22$$
$$13x = 26$$
$$x = 2$$

$$y = \frac{5}{2}x - 1$$
$$y = \frac{5}{2}(2) - 1$$
$$y = 4$$

The solution is $(2, 4)$.

Section 9.4: Applications Involving Pairs of Linear Equations

1.

let x = length of longer piece

y = length of shorter piece

$$x + y = 96$$
$$x = y + 12$$

$$x + y = 96$$
$$(y + 12) + y = 96$$
$$2y + 12 = 96$$
$$2y = 84$$
$$y = 42$$

$$x = y + 12$$
$$x = (42) + 12$$
$$x = 54$$

The longer piece is 54 cm and the shorter piece is 42 m.

3.

let x = operation time of 180 gal/h pump

y = operation time of 250 gal/h pump

$$x + y = 6$$
$$180x + 250y = 1325$$

$$x + y = 6$$
$$y = 6 - x$$

$$180x + 250y = 1325$$
$$180x + 250(6 - x) = 1325$$
$$180x + 1500 - 250x = 1325$$
$$1500 - 70x = 1325$$
$$-70x = -125$$
$$x = 2.5$$

$$y = 6 - x$$
$$y = 6 - (2, 5)$$
$$y = 3.5$$

The 180 gal/h pump ran for 2.5 hours and the 250 gal/h pump ran for 3.5 hours.

5.

let x = time of \$32/h welder

y = time of \$41/h welder

$x + y = 48$

$32x + 41y = 1734$

$x + y = 48$

$y = 48 - x$

$32x + 41y = 1734$

$32x + 41(48 - x) = 1734$

$32x + 1968 - 41x = 1734$

$1968 - 9x = 1734$

$-9x = -234$

$x = 26$

$y = 48 - x$

$y = 48 - (26)$

$y = 22$

The \$32/h welder worked for 26 h, the \$41/h welder worked for 22 h.

7.

let x = amount of 5% feed

y = amount of 15% feed

$x + y = 100$

$0.05x + 0.15y = 0.12(100) = 12$

$x + y = 100$

$y = 100 - x$

$0.05x + 0.15y = 12$

$0.05x + 0.15(100 - x) = 12$

$0.05x + 15 - 0.15x = 12$

$15 - 0.10x = 12$

$-0.10x = -3$

$x = 30$

$y = 100 - x$

$y = 100 - (30)$

$y = 70$

30 lb of 5% feed and 70 lb of 15% feed are required.

9.

let x = bushels of corn

y = bushels of soybeans

$x + y = 3150$

$6x + 15y = 22,950$

$x + y = 3150$

$y = 3150 - x$

$6x + 15y = 22,950$

$6x + 15(3150 - x) = 22,950$

$6x + 47,250 - 15x = 22,950$

$-9x = -24,300$

$-9x = -321,300$

$x = 2700$

$y = 3150 - x$

$y = 3150 - (2700)$

$y = 450$

2700 bu of corn and 450 bu of soybeans were sold.

11.

let x = amount of 6% pesticide solution

y = amount of 12% pesticide solution

$x + y = 300$

$0.06x + 0.12y = 0.08(300) = 24$

$x + y = 300$

$y = 300 - x$

$0.06x + 0.12y = 24$

$0.06x + 0.12(300 - x) = 24$

$0.06x + 36 - 0.12x = 24$

$36 - 0.06x = 24$

$-0.06x = -12$

$x = 200$

$y = 300 - x$

$y = 300 - (200)$

$y = 100$

200 gal of 6% pesticide solution and 100 gal of 12% pesticide solution are required.

13.

let x = number of 3 V batteries

y = number of 4.5 V batteries

$$x + y = 9$$
$$3x + 4.5y = 33$$

$$x + y = 9$$
$$y = 9 - x$$

$$3x + 4.5y = 33$$
$$3x + 4.5(9 - x) = 33$$
$$3x + 40.5 - 4.5x = 33$$
$$40.5 - 1.5x = 33$$
$$-1.5x = -7.5$$
$$x = 5$$

$$y = 9 - x$$
$$y = 9 - (5)$$
$$y = 4$$

Five 3 V batteries and four 4.5 V batteries were used.

15.

let x = amount of 8% solution

y = amount of 12% solution

$$x + y = 140$$
$$0.08x + 0.12y = 0.09(140) = 12.6$$

$$x + y = 140$$
$$y = 140 - x$$

$$0.08x + 0.12y = 12.6$$
$$0.08x + 0.12(140 - x) = 12.6$$
$$0.08x + 16.8 - 0.12x = 12.6$$
$$16.8 - 0.04x = 12.6$$
$$-0.04x = -4.2$$
$$x = 105$$

$$y = 140 - x$$
$$y = 140 - (105)$$
$$y = 35$$

105 mL of 95% electrolyte solution and 35 mL of 80% electrolyte solution are required.

17.

let x = time at 850 rpm

y = time at 1250 rpm

$$x + y = 14$$
$$850x + 1250y = 15,500$$

$$x + y = 14$$
$$y = 14 - x$$

$$850x + 1250y = 15,500$$
$$850x + 1250(14 - x) = 15,500$$
$$850x + 17,500 - 1250x = 15,500$$
$$17,500 - 400x = 15,500$$
$$-400x = -2000$$
$$x = 5$$

$$y = 14 - x$$
$$y = 14 - (5)$$
$$y = 9$$

The engine ran at 850 rpm for 5 min and at 1250 rpm for 9 min.

19.

let x = time at first setting

y = time at second setting

$$5\text{ h} \times \frac{60 \text{ min}}{1 \text{ h}} = 300 \text{ min}$$

$$x + y = 300$$

$$\frac{x}{12} + \frac{y}{15} = 22$$

$$x + y = 300$$

$$y = 300 - x$$

$$\frac{x}{12} + \frac{y}{15} = 22$$

$$\frac{x}{12} + \frac{300 - x}{15} = 22$$

$$60\left(\frac{x}{12} + \frac{300 - x}{15}\right) = (22)60$$

$$5x + 4(300 - x) = 1320$$

$$x + 1200 = 1320$$

$$x = 120$$

$$y = 300 - x$$

$$y = 300 - (120)$$

$$y = 180$$

The engine ran at the first setting for

$120 \text{ min} \times \dfrac{1 \text{ hr}}{60 \text{ min}} = 2 \text{ hr}$ and at the second

setting for $180 \text{ min} \times \dfrac{1 \text{ hr}}{60 \text{ min}} = 3 \text{ hr}$.

21.

let x = amount of 3% pesticide solution

y = amount of 8% pesticide solution

$$x + y = 200$$

$$0.08x + 0.03y = 0.04(200) = 8$$

$$x + y = 200$$

$$y = 200 - x$$

$$0.08x + 0.03y = 8$$

$$0.08x + 0.03(200 - x) = 8$$

$$0.08x + 6 - 0.03x = 8$$

$$0.05x + 6 = 8$$

$$0.05x = 2$$

$$x = 40$$

$$y = 200 - x$$

$$y = 200 - (40)$$

$$y = 160$$

40 L of 8% pesticide solution and 160 L of 3% pesticide solution are required.

23.

let x = current for one compressor

y = current for one air-handling unit

$$3x + 5y = 26.4$$

$$2x + 3y = 17.2$$

$$2(3x + 5y) = (26.4)2$$

$$(-3)(2x + 3y) = (17.2)(-3)$$

$$6x + 10y = 52.8$$

$$\underline{-6x - 9y = -51.6}$$

$$y = 1.2$$

$$2x + 3y = 17.2$$

$$2x + 3(1.2) = 17.2$$

$$2x + 3.6 = 17.2$$

$$2x = 13.6$$

$$x = 6.8$$

a. One compressor requires 6.8 A.

b. One air-handling unit requires 1.2 A.

25.

let x = initial flow time

y = final flow time

$$x + y = 8$$
$$140x + 100y = 1000$$

$$x + y = 8$$
$$y = 8 - x$$

$$140x + 100y = 1000$$
$$140x + 100(8 - x) = 1000$$
$$140x + 800 - 100x = 1000$$
$$40x + 800 = 1000$$
$$40x = 200$$
$$x = 5$$

$$y = 8 - x$$
$$y = 8 - (5)$$
$$y = 3$$

The solution flowed at 140 mL/h for 5 hours and 100 mL/h for 3 hours.

27.

let x = number of 2-mL vials

y = number of 5-mL vials

$$x + y = 42$$
$$2x + 5y = 117$$

$$x + y = 42$$
$$y = 42 - x$$

$$2x + 5y = 117$$
$$2x + 5(42 - x) = 117$$
$$2x + 210 - 5x = 117$$
$$210 - 3x = 117$$
$$-3x = -93$$
$$x = 31$$

$$y = 42 - x$$
$$y = 42 - (31)$$
$$y = 11$$

31 2-mL vials and 11 5-mL vials were used.

29.

let x = number of one-bedroom apts

y = number of two-bedroom apts

$$x + y = 13$$
$$575x + 650y = 8150$$

$$x + y = 13$$
$$y = 13 - x$$

$$575x + 650y = 8150$$
$$575x + 650(13 - x) = 8150$$
$$575x + 8450 - 650x = 8150$$
$$8450 - 75x = 8150$$
$$-75x = -300$$
$$x = 4$$

$$y = 13 - x$$
$$y = 13 - (4)$$
$$y = 9$$

She has 4 one-bedroom and 9 two-bedroom apartments.

31.

let x = larger resistance

y = smaller resistance

$$x + y = 550$$
$$x = 4.5y$$

$$x + y = 550$$
$$(4.5y) + y = 550$$
$$5.5y = 550$$
$$y = 100$$

$$x = 4.5y$$
$$x = 4.5(100)$$
$$x = 450$$

The larger resistance is $450\,\Omega$ and the smaller resistance is $100\,\Omega$.

33.

let x = length of longer piece

y = length of shorter piece

$$x + y = 120$$
$$x = 3y$$

$$x + y = 120$$
$$(3y) + y = 120$$
$$4y = 120$$
$$y = 30$$

$$x = 3y$$
$$x = 3(90)$$
$$x = 90$$

The longer piece is 90 cm and the shorter piece is 30 m.

35.

let x = length of rectangle

y = width of rectangle

$$2x + 2y = 240$$
$$x = 2y$$

$$2x + 2y = 240$$
$$2(2y) + 2y = 240$$
$$4y + 2y = 240$$
$$6y = 240$$
$$y = 40$$

$$x = 2y$$
$$x = 2(40)$$
$$x = 80$$

The length is 80 cm and the width is 40 ft.

37.

let x = current in equal currents

y = current in third current

$$2x + y = 210$$
$$y = 5x$$

$$2x + y = 210$$
$$2x + (5x) = 210$$
$$7x = 210$$
$$x = 30$$

$$y = 5x$$
$$y = 5(30)$$
$$y = 150$$

The third current is 150mA.

39.

let x = length of plot

y = width of plot

$$2x + 2y = 2800$$
$$x = y + 40$$

$$2x + 2y = 2800$$
$$2(y + 40) + 2y = 2800$$
$$2y + 80 + 2y = 2800$$
$$4y + 80 = 2800$$
$$4y + 80 = 2720$$
$$4y = 2760$$
$$y = 680$$

$$x = y + 40$$
$$x = (680) + 40$$
$$x = 720$$

The length is 720 ft and the width is 680 ft.

41. a.

let x = original length of room

y = original width of room

$2x + 2y = 40$ (Original room)

$2(x - 6) + 2(2y) = 40$ (Modified room)

$2x + 2y = 40$

$2x - 12 + 4y = 40$

$2x + 2y = 40$

$2x + 4y = 52$

$(-1)(2x + 2y) = (40)(-1)$

$2x + 4y = 52$

$-2x - 2y = -40$

$2x + 4y = 52$

$2y = 12$

$y = 6$

$2x + 2y = 40$

$2x + 2(6) = 40$

$2x + 12 = 40$

$2x = 28$

$x = 14$

The original room has a length of 14 ft and a width of 6 ft.

b.

Area of original room

$14 \text{ ft} \times 6 \text{ ft} = 84 \text{ ft}^2$

Area of modified room

$(14 \text{ ft} - 6 \text{ ft}) \times (2 \times 6 \text{ ft}) = 84 \text{ ft}^2 =$

$= 8 \text{ ft} \times 12 \text{ ft} = 96 \text{ ft}^2$

The new room would have more area.

43.

let x = length of original building

y = width of original building

$2(x + 5) + 2(y + 5) = 230$

$x = 2.5y$

$2x + 10 + 2y + 10 = 230$

$x = 2.5y$

$2x + 2y = 210$

$x = 2.5y$

$2x + 2y = 210$

$2(2.5y) + 2y = 210$

$5y + 2y = 210$

$7y = 210$

$y = 30$

$x = 2.5y$

$x = 2.5(30)$

$x = 75$

The original building had a length of 75 ft and a width of 30 ft.

45.

let x = length of longer piece

y = length of shorter piece

$8 \text{ ft} \times \dfrac{12 \text{ in.}}{1 \text{ ft}} = 144 \text{ in.}$

$x + y = 144$

$x - y = 8$

$2x \quad = 152$

$x \quad = 76$

$x + y = 144$

$(76) + y = 144$

$y = 68$

The longer board is 76 in. long and the shorter board is 68 in. long.

47.

let x = pounds of soybean meal

y = pounds of corn

$$x + y = 100$$

$$y = \frac{8.5}{0.10} = 85$$

$$x + y = 100$$

$$x + (85) = 100$$

$$x = 15$$

85 lb of corn and 15 lb of soybean meal are required.

49.

let x = amount in municipal bonds

y = amount in corporate bonds

$$x + y = 22,500$$

$$0.035x + 0.085y = 1662.50$$

$$x + y = 22,500$$

$$y = 22,500 - x$$

$$0.035x + 0.085y = 1662.50$$

$$0.035x + 0.085(22,500 - x) = 1662.50$$

$$0.035x + 1912.5 - 0.085x = 1662.50$$

$$1912.5 - 0.05x = 1662.5$$

$$-0.05x = -250$$

$$x = 5000$$

$$y = 22,500 - x$$

$$y = 22,500 - (5000)$$

$$y = 17,500$$

Jim bought $5000 worth of municipal bonds and $17,500 worth of corporate bounds.

51.

let x = number of 6-guest tables

y = number of 8-guest tables

$$x + y = 21$$

$$6x + 8y = 148$$

$$x + y = 21$$

$$y = 21 - x$$

$$6x + 8y = 148$$

$$6x + 8(21 - x) = 148$$

$$6x + 168 - 8x = 148$$

$$168 - 2x = 148$$

$$-2x = -20$$

$$x = 10$$

$$y = 21 - x$$

$$y = 21 - (10)$$

$$y = 11$$

10 tables for 6 and 11 tables for 8 were used.

Chapter 9 Review

1.

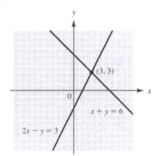

2.

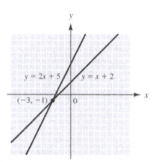

3.

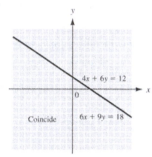

4.

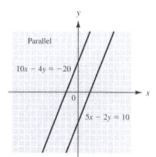

5.

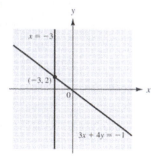

6.

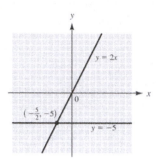

7.

$$x + y = 7$$
$$\underline{2x - y = 2}$$
$$3x \quad = 9$$
$$x \quad = 3$$

$$x + y = 7$$
$$(3) + y = 7$$
$$y = 4$$

The solution is $(3, 4)$.

8.

$$3x + 2y = 11$$
$$x + 2y = 5$$

$$x + 2y = 5$$
$$x = -2y + 5$$

$$3x + 2y = 11$$
$$3(-2y + 5) + 2y = 11$$
$$-6y + 15 + 2y = 11$$
$$-4y + 15 = 11$$
$$-4y = -4$$
$$y = 1$$

$$x = -2y + 5$$
$$x = -2(1) + 5$$
$$x = 3$$

The solution is $(3, 1)$.

9.

$$3x - 5y = -3$$
$$2x - 3y = -1$$

$$(-2)(3x - 5y) = (-3)(-2)$$
$$3(2x - 3y) = (-1)3$$

$$-6x + 10y = 6$$
$$\underline{6x - 9y = -3}$$
$$y = -3$$
$$y = 3$$

$$2x - 3y = -1$$
$$2x - 3(3) = -1$$
$$2x - 9 = -1$$
$$2x = 8$$
$$x = 4$$

The solution is $(4, 3)$.

10.

$$2x - 3y = 1$$
$$4x - 6y = 5$$

$$(-2)(2x - 3y) = (1)(-2)$$
$$4x - 6y = 5$$

$$-4x + 6y = -2$$
$$\underline{4x - 6y = 5}$$
$$0 = 3$$

No solution, the lines are parallel.

11.

$$3x + 5y = 8$$
$$6x - 4y = 44$$

$$(-2)(3x + 5y) = (8)(-2)$$
$$6x - 4y = 44$$

$$-6x - 10y = -16$$
$$\underline{6x - 4y = 44}$$
$$-14y = 28$$
$$y = -2$$

$$3x + 5y = 8$$
$$3x + 5(-2) = 8$$
$$3x - 10 = 8$$
$$3x = 18$$
$$x = 6$$

The solution is $(6, -2)$.

12.

$$5x + 7y = 22$$
$$4x + 8y = 20$$

$$4(5x + 7y) = (22)4$$
$$(-5)(4x + 8y) = (20)(-5)$$

$$20x + 28y = 88$$
$$\underline{-20x - 40y = -100}$$
$$-12y = -12$$
$$y = 1$$

$$5x + 7y = 22$$
$$5x + 7(1) = 22$$
$$5x + 7 = 22$$
$$5x = 15$$
$$x = 3$$

The solution is $(3, 1)$.

13.

$$x + 2y = 3$$
$$3x + 6y = 9$$

$$x + 2y = 3$$
$$x = -2y + 3$$

$$3x + 6y = 9$$
$$3(-2y + 3) + 6y = 9$$
$$-6y + 9 + 6y = 9$$
$$9 = 9$$

There are many solutions, the lines coincide.

14.

$$3x + 5y = 52$$
$$y = 2x$$

$$3x + 5y = 52$$
$$3x + 5(2x) = 52$$
$$3x + 10x = 52$$
$$13x = 52$$
$$x = 4$$

$$y = 2x$$
$$y = 2(4)$$
$$y = 8$$

The solution is $(4, 8)$.

15.

$$5y - 4x = -6$$
$$x = \frac{1}{2}y$$

$$5y - 4x = -6$$
$$5y - 4\left(\frac{1}{2}y\right) = -6$$
$$5y - 2y = -6$$
$$3y = -6$$
$$y = -2$$

$$x = \frac{1}{2}y$$
$$x = \frac{1}{2}(-2)$$
$$x = -1$$

The solution is $(-1, -2)$.

16.

$$3x - 7y = -69$$
$$y = 4x + 5$$

$$3x - 7y = -69$$
$$3x - 7(4x + 5) = -69$$
$$3x - 28x - 35 = -69$$
$$-25x - 35 = -69$$
$$-25x = -34$$
$$x = \frac{34}{25} = 1\frac{9}{25}$$

$$y = 4x + 5$$
$$y = 4\left(\frac{34}{25}\right) + 5$$
$$y = \frac{136}{25} + \frac{125}{25}$$
$$x = \frac{261}{25} = 10\frac{11}{25}$$

The solution is $\left(1\frac{9}{25}, 10\frac{11}{25}\right)$.

17.

let x = number of 20-amp switches

y = number of 15-amp switches

$$20x + 8y = 162$$
$$60x + 40y = 670$$

$$(-3)(20x + 8y) = (162)(-3)$$
$$60x + 40y = 670$$

$$-60x - 24y = -486$$
$$\underline{60x + 40y = 670}$$
$$16y = 184$$
$$y = 11.5$$

$$20x + 8y = 162$$
$$20x + 8(11.5) = 162$$
$$20x + 92 = 162$$
$$20x = 70$$
$$x = 3.5$$

20-amp switches cost $3.50 each and 15-amp switches cost $11.50 each.

18.

let x = length of rectangle

y = width of rectangle

$$x + y = 190$$
$$x = y + 75$$

$$(y + 75) + y = 190$$
$$2y + 75 = 190$$
$$2y = 115$$
$$y = 57.5$$

$$x = y + 75$$
$$x = (57.5) + 75$$
$$x = 132.5$$

The length is 132.5 ft and the width is 57.5 ft.

19.

let x = larger inductor

y = smaller inductor

$$x + y = 90$$
$$x = 3.5y$$

$$(3.5y) + y = 90$$
$$4.5y = 90$$
$$y = 20$$

$$x = 3.5y$$
$$x = 3.5(20)$$
$$x = 70$$

The larger inductor is 70 mH and the smaller inductor is 20 mH.

20.

let x = longer length

y = shorter length

$$x + y = 90$$
$$\underline{x - y = 20}$$
$$2x \quad = 110$$
$$x \quad = 55$$

$$x + y = 90$$
$$(55) + y = 90$$
$$y = 35$$

The longer length is 55 m and the shorter length is 35 m.

Chapter 9 Test

1.

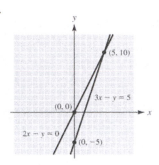

3.

$$2x + 7y = -1$$
$$x + 2y = 1$$

$$2x + 7y = -1$$
$$(-2)(x + 2y) = (1)(-2)$$

$$2x + 7y = -1$$
$$\underline{-2x - 4y = -2}$$
$$3y = -3$$
$$y = -1$$

$$x + 2y = 1$$
$$x + 2(-1) = 1$$
$$x - 2 = 1$$
$$x = 3$$

The solution is $(3, -1)$.

5.

$$y = -3x$$
$$2x + 3y = 13$$

$$2x + 3y = 13$$
$$2x + 3(-3x) = 13$$
$$2x - 9x = 13$$
$$-7x = 13$$
$$x = -\frac{13}{7} = -1\frac{6}{7}$$

$$y = -3x$$
$$y = -3\left(-\frac{13}{7}\right)$$
$$y = \frac{39}{7} = 5\frac{4}{7}$$

The solution is $\left(-1\frac{6}{7}, 5\frac{4}{7}\right)$.

7.

$$4x - 5y = 10$$
$$-8x + 10y = 6$$

$$2(4x - 5y) = (10)2$$
$$-8x + 10y = 6$$

$$8x - 10y = 20$$
$$\underline{-8x + 10y = 6}$$
$$0 = 26$$

There is no solution, the lines are parallel.

9.

$$x - 3y = -8$$
$$2x + y = 5$$

$$x - 3y = -8$$
$$x = 3y - 8$$

$$2x + y = 5$$
$$2(3y - 8) + y = 5$$
$$6y - 16 + y = 5$$
$$7y - 16 = 5$$
$$7y = 21$$
$$y = 3$$

$$x = 3y - 8$$
$$x = 3(3) - 8$$
$$x = 1$$

The solution is $(1, 3)$.

11.

let x = larger resistance
    y = smaller resistance

$$x + y = 550$$
$$\underline{x - y = 250}$$
$$2x \quad\; = 800$$
$$x \quad\;\; = 400$$

$$x + y = 550$$
$$(400) + y = 550$$
$$y = 150$$

The larger resistance is $400\,\Omega$ and the smaller resistance is $150\,\Omega$.

Chapter 10: Factoring Algebraic Expressions

Section 10.1: Finding Monomial Factors

1. $4(a+1)$

3. $b(x+y)$

5. $5(3b-4)$

7. $x(x-7)$

9. $a(a-4)$

11. $4n(n-2)$

13. $5x(2x+5)$

15. $3r(r-2)$

17. $4x^2(x^2+2x+3)$

19. $9a(a-x^2)$

21. $10(x+y-z)$

23. $3(y-2)$

25. $7xy(2-xy)$

27. $m(12x^2-7)$

29. $12a(5x-1)$

31. $13mn(4mn-1)$

33. $2(26m^2-7m+1)$

35. $18y^2(2-y+3y^2)$

37. $3m(2m^3-4m+1)$

39. $-2x^2y^3(2+3y+5y^2)$

41. $3abc(abc+9a^2b^2c-27)$

43. $4xz^2(x^2z^2-2xy^2z+3y)$

10.2: Finding the Product of Two Binomials Mentally

1.
$$(x+5)(x+2) = x^2+(2x+5x)+10$$
$$= x^2+7x+10$$

3.
$$(2x+3)(3x+4) = 6x^2+(8x+9x)+12$$
$$= 6x^2+17x+12$$

5.
$$(x-5)(x-6) = x^2+(-6x-5x)+30$$
$$= x^2-11x+30$$

7.
$$(x-12)(x-2) = x^2+(-2x-12x)+24$$
$$= x^2-14x+24$$

9.
$$(x+8)(2x+3) = 2x^2+(3x+16x)+24$$
$$= 2x^2+19x+24$$

11.
$$(x+6)(x-2) = x^2+(-2x+6x)-12$$
$$= x^2+4x-12$$

13.
$$(x-9)(x-10) = x^2+(-10x-9x)+90$$
$$= x^2-19x+90$$

15.
$$(x-12)(x+6) = x^2+(6x-12x)-72$$
$$= x^2-6x-72$$

17.
$$(2x-7)(4x+5) = 8x^2+(10x-28x)-35$$
$$= 8x^2-18x-35$$

19.
$$(2x+5)(4x-7)$$
$$= 8x^2+(-14x+20x)-35$$
$$= 8x^2+6x-35$$

21.
$$(7x+3)(2x+5) = 14x^2+(35x+6x)+15$$
$$= 14x^2+41x+15$$

23.
$$(x-9)(3x+8) = 3x^2+(8x-27x)-72$$
$$= 3x^2-19x-72$$

25.
$$(6x+5)(x+7) = 6x^2 + (42x+5x) + 35$$
$$= 6x^2 + 47x + 35$$

27.
$$(13x-4)(13x-4)$$
$$= 169x^2 + (-52x-52x) + 16$$
$$= 169x^2 - 104x + 16$$

29.
$$(10x+7)(12x-3)$$
$$= 120x^2 + (-30x+84x) - 21$$
$$= 120x^2 + 54x - 21$$

31.
$$(10x-7)(10x-3)$$
$$= 100x^2 + (-30x-70x) + 21$$
$$= 100x^2 - 100x + 21$$

33.
$$(2x-3)(2x-5) = 4x^2 + (-10x-6x) + 15$$
$$= 4x^2 - 16x + 15$$

35.
$$(2x-3)(2x+5) = 4x^2 + (10x-6x) - 15$$
$$= 4x^2 + 4x - 15$$

37.
$$(3x-8)(2x+7) = 6x^2 + (21x-16) - 56$$
$$= 6x^2 + 5x - 56$$

39.
$$(3x+8)(2x+7) = 6x^2 + (21x+16) - 56$$
$$= 6x^2 + 37x + 56$$

41.
$$(8x-5)(2x+3) = 16x^2 + (24x-10x) - 15$$
$$= 16x^2 + 14x - 15$$

43.
$$(y-7)(2y+3) = 2y^2 + (3y-14y) - 21$$
$$= 2y^2 - 11y - 21$$

45.
$$(3n-6y)(2n+5y)$$
$$= 6n^2 + (15ny-12ny) - 30y^2$$
$$= 6n^2 + 3ny - 30y^2$$

47.
$$(4x-y)(2x+7y)$$
$$= 8x^2 + (28xy-2xy) - 7y^2$$
$$= 8x^2 + 26xy - 7y^2$$

49.
$$\left(\frac{1}{2}x-8\right)\left(\frac{1}{4}x-6\right)$$
$$= \frac{1}{8}x^2 + (-3x-2x) + 48$$
$$= \frac{1}{8}x^2 - 5x + 48$$

Section 10.3: Finding Binomial Factors

1. $x^2 + 6x + 8 = (x+2)(x+4)$

3. $y^2 + 9y + 20 = (y+4)(y+5)$

5.
$$3r^2 + 30r + 75 = 3\left[r^2 + 10r + 25\right]$$
$$= 3(r+5)(r+5)$$
$$= 3(r+5)^2$$

7. $b^2 + 11b + 30 = (b+5)(b+6)$

9. $x^2 + 17x + 72 = (x+8)(x+9)$

11.
$$5a^2 + 35a + 60 = 5\left[a^2 + 7a + 12\right]$$
$$= 5(a+4)(a+3)$$

13. $x^2 - 7x + 12 = (x-4)(x-3)$

15.
$$2a^2 - 18a + 28 = 2\left[a^2 - 9a + 14\right]$$
$$= 2(a-7)(a-2)$$

17.
$$3x^2 - 30x + 63 = 3\left[x^2 - 10x + 21\right]$$
$$= 3(x-7)(x-3)$$

19. $w^2 - 13w + 42 = (w-6)(w-7)$

21. $x^2 - 19x + 90 = (x-9)(x-10)$

23. $t^2 - 12t + 20 = (t - 10)(t - 2)$

25. $x^2 + 2x - 8 = (x + 4)(x - 2)$

27. $y^2 + y - 20 = (y + 5)(y - 4)$

29. $a^2 + 5a - 24 = (a + 8)(a - 3)$

31. $c^2 - 15c - 54 = (c - 18)(c + 3)$

33.
$$3x^2 - 3x - 36 = 3\left[x^2 - x - 12\right]$$
$$= 3(x - 4)(x + 3)$$

35. $c^2 + 3c - 18 = (c + 6)(c - 3)$

37. $y^2 + 17y + 42 = (y + 14)(y + 3)$

39. $r^2 - 2r - 35 = (r - 7)(r + 5)$

41. $m^2 - 22m + 40 = (m - 20)(m - 2)$

43. $x^2 - 9x - 90 = (x - 15)(x + 6)$

45. $a^2 + 27a + 92 = (a + 23)(a + 4)$

47.
$$2a^2 - 12a - 110 = 2\left[a^2 - 6a - 55\right]$$
$$= 2(a - 11)(a + 5)$$

49. $a^2 + 29a + 100 = (a + 25)(a + 4)$

51. $y^2 - 14y - 95 = (y - 19)(y + 5)$

53. $y^2 - 18y + 32 = (y - 16)(y - 2)$

55.
$$7x^2 + 7x - 14 = 7\left[x^2 + x - 2\right]$$
$$= 7(x + 2)(x - 1)$$

57. $6x^2 + 12x - 6 = 6\left(x^2 + 2x - 1\right)$

59. $y^2 - 12y + 35 = (y - 7)(y - 5)$

61. $a^2 + 2a - 63 = (a + 9)(a - 7)$

63. $x^2 + 18x + 56 = (x + 4)(x + 14)$

65.
$$2y^2 - 36y + 90 = 2\left[y^2 - 18y + 45\right]$$
$$= 2(y - 15)(y - 3)$$

67.
$$3xy^2 - 18xy + 27x = 3x\left[y^2 - 6y + 9\right]$$
$$= 3x(y - 3)(y - 3)$$
$$= 3x(y - 3)^2$$

69.
$$x^2 + 30x + 225 = (x + 15)(x + 15)$$
$$= (x + 15)^2$$

71. $x^2 - 26x + 153 = (x - 9)(x - 17)$

73. $x^2 + 28x + 192 = (x + 12)(x + 16)$

75. $x^2 + 14x - 176 = (x + 22)(x - 8)$

77.
$$2a^2b + 4ab - 48b = 2b\left[a^2 + 2b - 24\right]$$
$$= 2b(a + 6)(a - 4)$$

79. $y^2 - y - 72 = (y - 9)(y + 8)$

Section 10.4: Special Products

1. $(x + 3)(x - 3) = x^2 - 9$

3. $(a + 5)(a - 5) = a^2 - 25$

5. $(2b + 11)(2b - 11) = 4b^2 - 121$

7. $(100 + 3)(100 - 3) = 10,000 - 9 = 9991$

9. $(3y^2 + 14)(3y^2 - 14) = 9y^4 - 196$

11. $(r - 12)^2 = r^2 - 24r + 144$

13. $(4y + 5)(4y - 5) = 16y^2 - 25$

15. $(xy - 4)^2 = x^2y^2 - 8xy + 16$

17. $(ab + d)^2 = a^2b^2 + 2abd + d^2$

19. $(z - 11)^2 = z^2 - 22z + 121$

21. $(st - 7)^2 = s^2t^2 - 14st + 49$

23. $(x + y^2)(x - y^2) = x^2 - y^4$

25. $(x + 5)^2 = x^2 + 10x + 25$

27. $(x+7)(x-7) = x^2 - 49$

29. $(x-3)^2 = x^2 - 6x + 9$

31. $(ab+2)(ab-2) = a^2b^2 - 4$

33. $(x^2+2)(x^2-2) = x^4 - 4$

35. $(r-15)^2 = r^2 - 30r + 225$

37. $(y^3-5)^2 = y^6 - 10y^3 + 25$

39. $(10-x)(10+x) = 100 - x^2$

Section 10.5: Finding Factors of Special Products

1. $a^2 + 8a + 16 = (a+4)^2$

3. $b^2 - c^2 = (b+c)(b-c)$

5. $x^2 - 4x + 4 = (x-2)^2$

7. $4 - x^2 = (2+x)(2-x)$

9. $y^2 - 36 = (y+6)(y-6)$

11.
$$5a^2 + 10a + 5 = 5\left[a^2 + 2a + 1\right]$$
$$= 5(a+1)^2$$

13. $1 - 81y^2 = (1+9y)(1-9y)$

15. $49 - a^4 = (7+a^2)(7-a^2)$

17. $49x^2 - 64y^2 = (7x+8y)(7x-8y)$

19. $1 - x^2y^2 = (1+xy)(1-xy)$

21. $4x^2 - 12x + 9 = (2x-3)^2$

23. $R^2 - r^2 = (R+r)(R-r)$

25. $49x^2 - 25 = (7x+5)(7x-5)$

27. $y^2 - 10y + 25 = (y-5)^2$

29. $b^2 - 9 = (b+3)(b-3)$

31. $m^2 + 22m + 121 = (m+11)^2$

33. $4m^2 - 9 = (2m+3)(2m-3)$

35.
$$4x^2 + 24x + 36 = 4\left[x^2 + 6x + 9\right]$$
$$= 4(x+3)^2$$

37.
$$27x^2 - 3 = 3\left[9x^2 - 1\right]$$
$$= 3(3x+1)(3x-1)$$

39.
$$am^2 - 14am + 49a = a\left[m^2 - 14m + 49\right]$$
$$= a(m-7)^2$$

Section 10.6: Factoring General Trinomials

1. $5x^2 - 28x - 12 = (5x+2)(x-6)$

3. $10x^2 - 29x + 21 = (2x-3)(5x-7)$

5. $12x^2 - 28x + 15 = (6x-5)(2x-3)$

7. $8x^2 + 26x - 45 = (2x+9)(4x-5)$

9. $16x^2 - 11x - 5 = (16x+5)(x-1)$

11.
$$12x^2 - 16x - 16 = 4\left[3x^2 - 4x - 4\right]$$
$$= 4(3x+2)(x-2)$$

13. $15y^2 - y - 6 = (5y+3)(3y-2)$

15. $8m^2 - 10m - 3 = (4m+1)(2m-3)$

17. $35a^2 - 2a - 1 = (7a+1)(5a-1)$

19. $16y^2 - 8y + 1 = (4y-1)^2$

21. $3x^2 + 20x - 63 = (3x-7)(x+9)$

23. $12b^2 + 5b - 2 = (4b-1)(3b+2)$

25. $15y^2 - 14y - 8 = (5y+2)(3y-4)$

27. $90 + 17c - 3c^2 = (10+3c)(9-c)$

29. $6x^2 - 13x + 5 = (3x-5)(2x-1)$

31. $2y^4 + 9y^2 - 35 = (2y^2-5)(y^2+7)$

33. $4b^2 + 52b + 169 = (2b+13)^2$

35. $14x^2 - 51x + 40 = (7x - 8)(2x - 5)$

37.

$$28x^3 + 140x^2 + 175x$$
$$= 7x\left[4x^2 + 20x + 25\right]$$
$$= 7x(2x + 5)^2$$

39.

$$10ab^2 - 15ab - 175a$$
$$= 5a\left[2b^2 - 3b - 35\right]$$
$$= 5a(2b + 7)(b - 5)$$

Chapter 10 Review

1. $(c + d)(c - d) = c^2 - d^2$

2. $(x - 6)(x + 6) = x^2 - 36$

3.

$$(y + 7)(y - 4) = y^2 + (7y - 4) - 28$$
$$= y^2 + 3y - 28$$

4.

$$(2x + 5)(2x - 9)$$
$$= 4x^2 + (-18x + 10x) - 45$$
$$= 4x^2 - 8x - 45$$

5.

$$(x + 8)(x - 3) = x^2 + (-3x + 8x) + 24$$
$$= x^2 + 5x - 24$$

6.

$$(x - 4)(x - 9) = x^2 + (-9x - 4x) + 36$$
$$= x^2 - 13x + 36$$

7. $(x - 3)^2 = x^2 - 6x + 9$

8. $(2x - 6)^2 = 4x^2 - 24x + 36$

9. $\left(1 - 5x^2\right)^2 = 1 - 10x^2 + 25x^4$

10. $6a + 6 = 6(a + 1)$

11. $5x - 15 = 5(x - 3)$

12. $xy + 2xz = x(y + 2z)$

13.

$$y^4 + 17y^3 - 18y^2 = y^2\left[y^2 + 17y - 18\right]$$
$$= y^2(y + 18)(y - 1)$$

14. $y^2 - 6y - 7 = (y - 7)(y + 1)$

15. $z^2 + 18z + 81 = (z + 9)^2$

16. $x^2 + 10x + 16 = (x + 8)(x + 2)$

17. $4a^2 + 4x^2 = 4\left(a^2 + x^2\right)$

18. $x^2 - 17x + 72 = (x - 9)(x - 8)$

19. $x^2 - 18x + 81 = (x - 9)^2$

20. $x^2 + 19x + 60 = (x + 15)(x + 4)$

21. $y^2 - 2y + 1 = (y - 1)^2$

22. $x^2 - 3x - 28 = (x - 7)(x + 4)$

23. $x^2 - 4x - 96 = (x - 12)(x + 8)$

24. $x^2 + x - 110 = (x + 11)(x - 10)$

25. $x^2 - 49 = (x + 7)(x - 7)$

26. $16y^2 - 9x^2 = (4y + 3x)(4y - 3x)$

27. $x^2 - 144 = (x + 12)(x - 12)$

28. $25x^2 - 81y^2 = (5x + 9y)(5x - 9y)$

29.

$$4x^2 - 24x - 364 = 4\left[x^2 - 6x - 91\right]$$
$$= 4(x - 13)(x + 7)$$

30.

$$5x^2 - 5x - 780 = 5\left[x^2 - x - 156\right]$$
$$= 5(x + 12)(x - 13)$$

31. $2x^2 + 11x + 14 = (2x + 7)(x + 2)$

32. $12x^2 - 19x + 4 = (4x - 1)(3x - 4)$

33. $30x^2 + 7x - 15 = (6x + 5)(5x - 3)$

34. $12x^2 + 143x - 12 = (12x - 1)(x + 12)$

35.

$$4x^2 - 6x + 2 = 2\left[2x^2 - 3x + 1\right]$$
$$= 2(2x - 1)(x - 1)$$

36. $36x^2 - 49y^2 = (6x + 7y)(6x - 7y)$

37.
$$28x^2 + 82x + 30 = 2\left[14x^2 + 41x + 15\right]$$
$$= 2(7x + 3)(2x + 5)$$

38.
$$30x^2 - 27x - 21 = 3\left[10x^2 - 9x - 7\right]$$
$$= 3(5x - 7)(2x + 1)$$

39.
$$4x^3 - 4x = 4x\left[x^2 - 1\right]$$
$$= 4x(x + 1)(x - 1)$$

40.
$$25y^2 - 100 = 25\left[y^2 - 4\right]$$
$$= 25(y + 2)(y - 2)$$

Chapter 10 Test

1.
$$(x + 8)(x - 3) = x^2 + (8x - 3x) - 24$$
$$= x^2 + 5x - 24$$

3. $(2x - 8)(2x + 8) = 4x^2 - 64$

5.
$$(4x - 7)(2x + 3) = 8x^2 + (12x - 14x) - 21$$
$$= 8x^2 - 2x - 21$$

7. $x^2 + 4x + 3 = (x + 3)(x + 1)$

9. $6x^2 - 7x - 90 = (3x + 10)(2x - 9)$

11. $x^2 + 7x - 18 = (x + 9)(x - 2)$

13. $6x^2 + 13x + 6 = (3x + 2)(2x + 3)$

15. $3x^2 - 11x - 4 = (3x + 1)(x - 4)$

17. $5x^2 + 7x - 6 = (5x - 3)(x + 2)$

19. $9x^2 - 121 = (3x + 11)(3x - 11)$

Cumulative Review Chapters 1-10

1.
$$2 + 6^2 - 24 \div 3(4) = 2 + 36 - 8(4)$$
$$= 2 + 36 - 32$$
$$= 6$$

2. a. 746.8

b. 750

3.
$$-\frac{2}{3} \div \frac{1}{5} + \frac{2}{3} = -\frac{2}{3} \times \frac{5}{1} + \frac{2}{3}$$
$$= -\frac{10}{3} + \frac{2}{3}$$
$$= -\frac{8}{3} = -2\frac{2}{3}$$

4. a. 3.18×10^{-4}

b. 318×10^{-6}

5. $625 \text{ g} \times \dfrac{1 \text{ kg}}{1000 \text{ g}} = 0.625 \text{ kg}$

6.
$$6 \text{ m}^2 \times \left(\frac{1.094 \text{ yd}}{1 \text{ m}}\right)^2 \times \left(\frac{3 \text{ ft}}{1 \text{ yd}}\right)^2$$
$$= 6406 \text{ ft}^2$$

7. 70 V

8. 95 cm^3

9.
$$3(x - 2) - 4(2 - 3x) = 3x - 6 - 8 + 12x$$
$$= 15x - 14$$

10.
$$(6a - 3b + 2c) - (-2a - 3b + c)$$
$$= (6a - 3b + 2c) + (2a + 3b - c)$$
$$= 8a + c$$

11.
$$\frac{x}{3} - 4 = \frac{2x}{5}$$
$$15\left(\frac{x}{3} - 4\right) = \left(\frac{2x}{5}\right)15$$
$$5x - 60 = 6x$$
$$-60 = x$$
$$x = -60$$

12.

Let x = width of rectangle

$x + 5$ = length of rectangle

$P = 2l + 2w$

$58 = 2(w + 5) + 2w$

$58 = 2w + 10 + 2w$

$58 = 4w + 10$

$48 = 4w$

$w = 12$

The rectangle has a length of $12 + 5 = 17$ m and a width of 12 m.

13.

$\dfrac{15.7}{8.2} = \dfrac{x}{10}$

$8.2x = (15.7)(10)$

$x = \dfrac{(15.7)(10)}{8.2}$

$x = 19.1$

14.

$(d_A)(\text{rpm}_A) = (d_B)(\text{rpm}_B)$

$(x)(225 \text{ rpm}) = (18 \text{ in.})(125 \text{ rpm})$

$x = \dfrac{(18 \text{ in.})(125 \text{ rpm})}{225 \text{ rpm}}$

$= 10 \text{ in.}$

15.

$2x + 3y = 12$

$2x + 3y - 2x = 12 - 2x$

$3y = 12 - 2x$

$\dfrac{3y}{3} = \dfrac{12 - 2x}{3}$

$y = -\dfrac{2}{3}x + 4$

$(3, 2), (0, 4), (-3, 6)$

16.

$3x - y = 5$

$3x - y - 3x = 5 - 3x$

$-y = 5 - 3x$

$\dfrac{-y}{-1} = \dfrac{5 - 3x}{-1}$

$y = 3x - 5$

17.

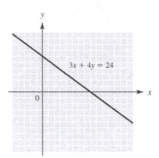

18.

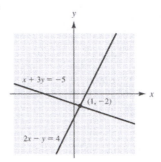

19.

$\dfrac{1}{2}x - \dfrac{2}{3}y = 1$

$6x - 8y = 12$

$6\left(\dfrac{1}{2}x - \dfrac{2}{3}y\right) = (1)6$

$6x - 8y = 12$

$(-2)(3x - 4y) = (6)(-2)$

$6x - 8y = 12$

$-6x + 8y = -12$

$\underline{6x - 8y = 12}$

$0 = 0$

There are many solutions, the lines coincide.

20.

$$y = 3x - 5$$
$$x + 3y = 8$$

$$x + 3y = 8$$
$$x + 3(3x - 5) = 8$$
$$x + 9x - 15 = 8$$
$$10x - 15 = 8$$
$$10x = 23$$
$$x = 2.3$$

$$y = 3x - 5$$
$$y = 3(2.3) - 5$$
$$y = 1.9$$

The solution is $(2.3, 1.9)$.

21.

$$x - y = 6$$
$$\underline{3x + y = 2}$$
$$4x \quad = 8$$
$$x \quad = 2$$

$$3x + y = 2$$
$$3(2) + y = 2$$
$$6 + y = 2$$
$$y = -4$$

The solution is $(2, -4)$.

22.

$$3x - 5y = 7$$
$$-6x + 10y = 5$$

$$2(3x - 5y) = (7)2$$
$$-6x + 10y = 5$$

$$6x - 10y = 14$$
$$\underline{-6x + 10y = 5}$$
$$0 = 19$$

No solution, the lines are parallel.

23.

$$135x + 40y = 29$$
$$60x - 45y = 38$$

$$9(135x + 40y) = (29)9$$
$$8(60x - 45y) = (38)8$$

$$1215x + 360y = 261$$
$$\underline{480x - 360y = 304}$$
$$1695x \qquad = 565$$

$$x \qquad = \frac{1}{3}$$

$$135x + 40y = 29$$
$$135\left(\frac{1}{3}\right) + 40y = 29$$
$$45 + 40y = 29$$
$$40y = -16$$
$$y = -\frac{2}{5}$$

The solution is $\left(\frac{1}{3}, -\frac{2}{5}\right)$.

24.

let $x =$ days \$53.95 car rented
$y =$ days \$42.95 car rented
$$x + y = 16$$
$$53.95x + 42.95y = 753.20$$

$$x + y = 16$$
$$y = 16 - x$$

$$53.95x + 42.95y = 753.20$$
$$53.95x + 42.95(16 - x) = 753.20$$
$$53.95x + 687.2 - 42.95x = 753.20$$
$$11x + 687.2 = 753.2$$
$$11x = 66$$
$$x = 6$$

$$y = 16 - x$$
$$y = 16 - (6)$$
$$y = 10$$

The \$53.95/day car was rented for 6 days and the \$42.95/day car was rented for 10 days.

25.

$$(2x-5)(3x+8)$$
$$=6x^2+(16x-15x)-40$$
$$=6x^2+x-40$$

26. $(5x-7y)^2=25x^2-70xy+49y^2$

27.

$$(3x-5)(5x-7)$$
$$=15x^2+(-21x-25x)+35$$
$$=15x^2-46x+35$$

28.

$$7x^3-63x=7x\left[x^2-9\right]$$
$$=7x(x+3)(x-3)$$

29. $4x^3+12x^2=4x^2(x+3)$

30. $2x^2-7x-4=(2x+1)(x-4)$

Chapter 11: Quadratic Equations

Section 11.1: Solving Quadratic Equations by Factoring

1.
$$x^2 + x = 12$$
$$x^2 + x - 12 = 0$$
$$(x+4)(x-3) = 0$$
$$x+4 = 0 \quad \text{or} \quad x-3 = 0$$
$$x = -4 \quad \text{or} \quad x = 3$$

3.
$$x^2 + x - 20 = 0$$
$$(x+5)(x-4) = 0$$
$$x+5 = 0 \quad \text{or} \quad x-4 = 0$$
$$x = -5 \quad \text{or} \quad x = 4$$

5.
$$x^2 - 2 = x$$
$$x^2 - x - 2 = 0$$
$$(x+1)(x-2) = 0$$
$$x+1 = 0 \quad \text{or} \quad x-2 = 0$$
$$x = -1 \quad \text{or} \quad x = 2$$

7.
$$x^2 - 1 = 0$$
$$(x+1)(x-1) = 0$$
$$x+1 = 0 \quad \text{or} \quad x-1 = 0$$
$$x = -1 \quad \text{or} \quad x = 1$$

9.
$$x^2 - 49 = 0$$
$$(x+7)(x-7) = 0$$
$$x+7 = 0 \quad \text{or} \quad x-7 = 0$$
$$x = -7 \quad \text{or} \quad x = 7$$

11.
$$w^2 + 5w + 6 = 0$$
$$(w+3)(w+2) = 0$$
$$w+3 = 0 \quad \text{or} \quad w+2 = 0$$
$$w = -3 \quad \text{or} \quad w = -2$$

13.
$$y^2 - 4y = 21$$
$$y^2 - 4y - 21 = 0$$
$$(y+3)(y-7) = 0$$
$$y+3 = 0 \quad \text{or} \quad y-7 = 0$$
$$y = -3 \quad \text{or} \quad y = 7$$

15.
$$n^2 - 6n - 40 = 0$$
$$(n+4)(n-10) = 0$$
$$n+4 = 0 \quad \text{or} \quad n-10 = 0$$
$$n = -4 \quad \text{or} \quad n = 10$$

17.
$$9m = m^2$$
$$m^2 - 9m = 0$$
$$m(m-9) = 0$$
$$m = 0 \quad \text{or} \quad m-9 = 0$$
$$m = 0 \quad \text{or} \quad m = 9$$

19.
$$x^2 = 108 + 3x$$
$$x^2 - 3x - 108 = 0$$
$$(x+12)(x-9) = 0$$
$$x+12 = 0 \quad \text{or} \quad x-9 = 0$$
$$x = -12 \quad \text{or} \quad x = 9$$

21.
$$c^2 + 6c = 16$$
$$c^2 + 6c - 16 = 0$$
$$(c+8)(c-2) = 0$$
$$c+8 = 0 \quad \text{or} \quad c-2 = 0$$
$$c = -8 \quad \text{or} \quad c = 2$$

23.
$$10x^2 + 29x + 10 = 0$$
$$(2x+5)(5x+2) = 0$$
$$2x+5 = 0 \quad \text{or} \quad 5x+2 = 0$$
$$2x = -5 \quad \text{or} \quad 5x = -2$$
$$x = -\frac{5}{2} \quad \text{or} \quad x = -\frac{2}{5}$$

25.
$$4x^2 = 25$$
$$4x^2 - 25 = 0$$
$$(2x+5)(2x-5) = 0$$
$$2x+5 = 0 \quad \text{or} \quad 2x+5 = 0$$
$$2x = -5 \quad \text{or} \quad 2x = 5$$
$$x = -\frac{5}{2} \quad \text{or} \quad x = \frac{5}{2}$$

27.
$$9x^2 + 16 = 24x$$
$$9x^2 - 24x + 16 = 0$$
$$(3x - 4)(3x - 4) = 0$$
$$3x - 4 = 0 \quad \text{or} \quad 3x - 4 = 0$$
$$3x = 4$$
$$x = \frac{4}{3}$$

29.
$$3x^2 + 9x = 0$$
$$3x(x + 3) = 0$$
$$3x = 0 \quad \text{or} \quad x + 3 = 0$$
$$x = 0 \quad \text{or} \quad x = -3$$

33.

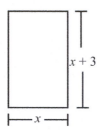

Lengths must be nonnegative so 5 is the only solution. The length is $5 + 3 = 8$ in. and the width is 5 in.

31.
$$A = \frac{1}{2}bh$$
$$66 = \frac{1}{2}(x + 1)x$$
$$2[66] = 2\left[\frac{1}{2}(x + 1)x\right]$$
$$132 = (x + 1)x$$
$$132 = x^2 + x$$
$$0 = x^2 + x - 132$$
$$0 = (x + 12)(x - 11)$$
$$x + 12 = 0 \quad \text{or} \quad x - 11 = 0$$
$$x = -12 \quad \text{or} \quad x = 11$$

Lengths must be nonnegative so 11 is the only solution. The length is $11 + 1 = 12$ ft and the width is 11 ft.

$$A = lw$$
$$40 = x(x + 3)$$
$$40 = x^2 + 3x$$
$$0 = x^2 + 3x - 40$$
$$0 = (x + 8)(x - 5)$$
$$x + 8 = 0 \text{ or } x - 5 = 0$$
$$x = -8 \text{ or } x = 5$$

11.2: The Quadratic Formula

1. $a = 1,\ b = -7,\ c = 4$

3. $a = 3,\ b = 4,\ c = 9$

5. $a = -3,\ b = 4,\ c = 7$

7. $a = 3,\ b = 0,\ c = -14$

9.
$$x = \frac{-b \pm \sqrt{b^2 - 4ac}}{2a}$$
$$a = 1,\ b = 1,\ c = -6$$
$$\text{So } x = \frac{-(1) \pm \sqrt{(1)^2 - 4(1)(-6)}}{2(1)}$$
$$= \frac{-1 \pm \sqrt{1 + 24}}{2}$$
$$= \frac{-1 \pm \sqrt{25}}{2}$$
$$= \frac{-1 \pm 5}{2}$$
$$= -3 \ \text{or} \ 2$$

11.
$$x = \frac{-b \pm \sqrt{b^2 - 4ac}}{2a}$$
$$a = 1,\ b = 8,\ c = -9$$
$$\text{So } x = \frac{-(8) \pm \sqrt{(8)^2 - 4(1)(-9)}}{2(1)}$$
$$= \frac{-8 \pm \sqrt{64 + 36}}{2}$$
$$= \frac{-8 \pm \sqrt{100}}{2}$$
$$= \frac{-8 \pm 10}{2}$$
$$= -9 \ \text{or} \ 1$$

13.

$$x = \frac{-b \pm \sqrt{b^2 - 4ac}}{2a}$$

$$a = 5, \ b = 2, \ c = 0$$

$$\text{So } x = \frac{-(2) \pm \sqrt{(2)^2 - 4(5)(0)}}{2(5)}$$

$$= \frac{-2 \pm \sqrt{4 - 0}}{2(5)}$$

$$= \frac{-2 \pm \sqrt{4}}{10}$$

$$= \frac{-2 \pm 2}{10}$$

$$= -\frac{2}{5} \ \text{ or } \ 0$$

15.

$$x = \frac{-b \pm \sqrt{b^2 - 4ac}}{2a}$$

$$a = 48, \ b = -32, \ c = -35$$

$$\text{So } x = \frac{-(-32) \pm \sqrt{(-32)^2 - 4(48)(-35)}}{2(48)}$$

$$= \frac{32 \pm \sqrt{1024 + 6720}}{96}$$

$$= \frac{32 \pm \sqrt{7744}}{96}$$

$$= \frac{32 \pm 88}{96}$$

$$= -\frac{7}{12} \ \text{ or } \ \frac{5}{4}$$

17.

$$x = \frac{-b \pm \sqrt{b^2 - 4ac}}{2a}$$

$$a = 2, \ b = 1, \ c = -5$$

$$\text{So } x = \frac{-(1) \pm \sqrt{(1)^2 - 4(2)(-5)}}{2(2)}$$

$$= \frac{-1 \pm \sqrt{1 + 40}}{4}$$

$$= \frac{-1 \pm \sqrt{41}}{4}$$

$$= \frac{-1 \pm 6.40}{4}$$

$$= -1.85 \ \text{ or } \ 1.35$$

19.

$$x = \frac{-b \pm \sqrt{b^2 - 4ac}}{2a}$$

$$a = 3, \ b = -5, \ c = 0$$

$$\text{So } x = \frac{-(-5) \pm \sqrt{(-5)^2 - 4(3)(0)}}{2(3)}$$

$$= \frac{5 \pm \sqrt{25 - 0}}{6}$$

$$= \frac{5 \pm \sqrt{25}}{6}$$

$$= \frac{5 \pm 5}{6}$$

$$= 0 \ \text{ or } \ \frac{5}{3}$$

21.

$$x = \frac{-b \pm \sqrt{b^2 - 4ac}}{2a}$$

$$a = -2, \ b = 1, \ c = 3$$

$$\text{So } x = \frac{-(1) \pm \sqrt{(1)^2 - 4(-2)(3)}}{2(-2)}$$

$$= \frac{-1 \pm \sqrt{1 + 24}}{-4}$$

$$= \frac{-1 \pm \sqrt{25}}{-4}$$

$$= \frac{-1 \pm 5}{-4}$$

$$= \frac{3}{2} \ \text{ or } \ -1$$

23.

$$x = \frac{-b \pm \sqrt{b^2 - 4ac}}{2a}$$

$$a = 6, \ b = 9, \ c = 1$$

$$\text{So } x = \frac{-(9) \pm \sqrt{(9)^2 - 4(6)(1)}}{2(6)}$$

$$= \frac{-9 \pm \sqrt{81 - 24}}{12}$$

$$= \frac{-9 \pm \sqrt{57}}{12}$$

$$= \frac{-9 \pm 7.55}{12}$$

$$= -1.38 \ \text{ or } \ -0.121$$

25. Rewrite the equation as $4x^2 + 5x + 1 = 0$.

$$x = \frac{-b \pm \sqrt{b^2 - 4ac}}{2a}$$

$$a = 4, \ b = 5, \ c = 1$$

$$\text{So } x = \frac{-(5) \pm \sqrt{(5)^2 - 4(4)(1)}}{2(4)}$$

$$= \frac{-5 \pm \sqrt{25 - 16}}{8}$$

$$= \frac{-5 \pm \sqrt{9}}{8}$$

$$= \frac{-5 \pm 3}{8}$$

$$= -1 \ \text{or} \ -\frac{1}{4}$$

27. Rewrite the equation as $3x^2 - 17 = 0$.

$$x = \frac{-b \pm \sqrt{b^2 - 4ac}}{2a}$$

$$a = 3, \ b = 0, \ c = -17$$

$$\text{So } x = \frac{-(0) \pm \sqrt{(0)^2 - 4(3)(-17)}}{2(3)}$$

$$= \frac{0 \pm \sqrt{0 + 204}}{6}$$

$$= \frac{\pm \sqrt{204}}{6}$$

$$= \frac{\pm 14.3}{6}$$

$$= -2.38 \ \text{or} \ 2.38$$

29. Rewrite the equation as $x^2 - 15x - 7 = 0$.

$$x = \frac{-b \pm \sqrt{b^2 - 4ac}}{2a}$$

$$a = 1, \ b = -15, \ c = -7$$

$$\text{So } x = \frac{-(-15) \pm \sqrt{(-15)^2 - 4(1)(-7)}}{2(1)}$$

$$= \frac{15 \pm \sqrt{225 + 28}}{2}$$

$$= \frac{15 \pm \sqrt{253}}{2}$$

$$= \frac{15 \pm 15.9}{2}$$

$$= 0.45 \ \text{or} \ 15.5$$

31. Rewrite the equation as $3x^2 - 5x - 31 = 0$.

$$x = \frac{-b \pm \sqrt{b^2 - 4ac}}{2a}$$

$$a = 3, \ b = -5, \ c = -31$$

$$\text{So } x = \frac{-(-5) \pm \sqrt{(-5)^2 - 4(3)(-31)}}{2(3)}$$

$$= \frac{5 \pm \sqrt{25 + 372}}{6}$$

$$= \frac{5 \pm \sqrt{397}}{6}$$

$$= \frac{5 \pm 19.9}{6}$$

$$= -2.48 \ \text{or} \ 4.15$$

33. Rewrite the equation as
$$23.8x^2 + 52.3x - 11.8 = 0.$$

$$x = \frac{-b \pm \sqrt{b^2 - 4ac}}{2a}$$

$$a = 23.8, \ b = 52.3, \ c = -11.8$$

$$\text{So } x = \frac{-(52.3) \pm \sqrt{(52.3)^2 - 4(23.8)(-11.8)}}{2(23.8)}$$

$$= \frac{-52.3 \pm \sqrt{3858.65}}{47.6}$$

$$= \frac{-52.3 \pm 62.1}{47.6}$$

$$= -2.40 \ \text{or} \ 0.206$$

Section 11.3: Applications Involving Quadratic Equations

1. a.

$$V = t^2 - 12t + 40$$
$$8 = t^2 - 12t + 40$$
$$0 = t^2 - 12t + 32$$
$$0 = (t-4)(t-8)$$
$$t-4 = 0 \quad \text{or} \quad t-8 = 0$$
$$t = 4 \quad \text{or} \quad t = 8$$

The voltage is 8 V at 4 seconds and 8 seconds.

b.

$$V = t^2 - 12t + 40$$
$$25 = t^2 - 12t + 40$$
$$0 = t^2 - 12t + 15$$
$$t = \frac{-b \pm \sqrt{b^2 - 4ac}}{2a}$$
$$a = 1, \ b = -12, \ c = 15$$

So $\ t = \dfrac{-(-12) \pm \sqrt{(-12)^2 - 4(1)(15)}}{2(1)}$

$$= \frac{12 \pm \sqrt{84}}{2}$$
$$= \frac{12 \pm 9.17}{2}$$
$$= \frac{12 - 9.17}{2} \quad \text{or} \quad \frac{12 + 9.17}{2}$$
$$= 1.42 \quad \text{or} \quad 10.6$$

The voltage is 25 V at 1.42 seconds and 10.6

c.

$$V = t^2 - 12t + 40$$
$$104 = t^2 - 12t + 40$$
$$0 = t^2 - 12t - 64$$
$$0 = (t+4)(t-16)$$
$$t+4 = 0 \ \text{or} \ t-16 = 0$$
$$t = -4 \ \text{or} \ t = 16$$

Since time must be nonnegative, the voltage is 104 V only at 16 seconds.

3.

$$A = lw$$
$$21 = (x+4)x$$
$$21 = x^2 + 4x$$
$$0 = x^2 + 4x - 21$$
$$0 = (x+7)(x-3)$$
$$x+7 = 0 \quad \text{or} \quad x-3 = 0$$
$$x = -7 \ \text{or} \ x = 3$$

Lengths must be nonnegative so only $x = 3$ works, so the length is $3 + 4 = 7$ ft and the width is 3 ft.

5.

$$A = lw$$
$$175 = (x+30)x$$
$$175 = x^2 + 30x$$
$$0 = x^2 + 30x - 175$$
$$0 = (x+35)(x-5)$$
$$x+35 = 0 \quad \text{or} \quad x-5 = 0$$
$$x = -35 \ \text{or} \ x = 5$$

Lengths must be nonnegative so only $x = 5$ works, so the length is $5 + 8 = 13$ ft and the width is 5 ft.

7. The perimeter is 160 m, therefore one length plus one width is 80 m.

$$A = lw$$
$$1200 = (80 - x)x$$
$$1200 = -x^2 + 80x$$
$$x^2 - 80x + 1200 = 0$$
$$(x-20)(x-60) = 0$$
$$x-20 = 0 \quad \text{or} \quad x-60 = 0$$
$$x = 20 \ \text{or} \ x = 60$$

Using $x = 20$, the length is $80 - 20 = 40$ m and the width is 20 m. Using $x = 60$, the length is $80 - 60 = 20$ m and the width is 60 m. So, the rectangle is 20 m by 60 m.

9.

$$A = lw$$

$$3 \times 7.5 + 18 = (3 + x)(7.5 + x)$$

$$40.5 = x^2 + 10.5x + 22.5$$

$$0 = x^2 + 10.5x - 18 \quad \text{Multiply both sides by 2.}$$

$$0 = 2x^2 + 21x - 36$$

$$0 = (x + 12)(2x - 3)$$

$$x + 12 = 0 \quad \text{or} \quad 2x - 3 = 0$$

$$x = -12 \quad \text{or} \quad x = 1.5$$

Lengths must be nonnegative so only $x = 1.5$ works, so the door is $3 + 1.5 = 4.5$ ft by $7.5 + 1.5 = 9$ ft.

11. a.

$$A = lw$$

$$1500 = (60 - 2x)(40 - 2x)$$

$$1500 = 4x^2 - 200x + 2400$$

$$0 = 4x^2 - 200x + 900 \quad \text{Divide both sides by 4.}$$

$$0 = x^2 - 50x + 225$$

$$0 = (x - 5)(x - 45)$$

$$x - 5 = 0 \quad \text{or} \quad x - 45 = 0$$

$$x = 5 \quad \text{or} \quad x = 45$$

$x = 45$ is not physically possible, so only $x = 5$ works, so each cut-out square is 5 cm by 5 cm.
b. The length of the base of the box is $60 - 2 \times 5 = 50$ cm, the width is $40 - 2 \times 5 = 30$ cm, and the height is 5 cm, so the volume is $50 \text{ cm} \times 30 \text{ cm} \times \text{cm} 5 = 7500 \text{ cm}^3$.

13.

$$A = lw$$

$$20 \times 16 + 160 = (2x + 20)(2x + 16)$$

$$480 = 4x^2 + 72x + 320$$

$$0 = 4x^2 + 72x - 160 \quad \text{Divide both sides by 4.}$$

$$0 = x^2 + 18x - 40$$

$$0 = (x + 20)(x - 2)$$

$$x + 20 = 0 \quad \text{or} \quad x - 2 = 0$$

$$x = -20 \quad \text{or} \quad x = 2$$

Lengths must be nonnegative so only $x = 2$ works, so the border is 2 ft wide.

15. a. The perimeter is 300 ft, therefore one length plus one width is 150 ft. The area is $(10 \text{ ft})(140 \text{ ft}) = 1400 \text{ ft}^2$
b. The perimeter is 300 ft, therefore one length plus one width is 150 ft. The area is $(20 \text{ ft})(130 \text{ ft}) = 2600 \text{ ft}^2$
c. The perimeter is 300 ft, therefore one length plus one width is 150 ft. $A = l(150 - l)$

15. (continued)

d.

Length (ft)	30	40	50	60	70	80
Area (ft^2)	3600	4400	5000	5400	5600	5600
Length (ft)	90	100	110	120	130	140
Area (ft^2)	5400	5000	4400	3600	2600	1400

e. The maximum area in the table is 5600 ft^2 when the length is 70 ft or 80ft.

f.

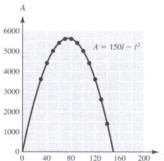

g. When the length and width are both equal to 17 ft, the maximum area is $\boxed{5625 \text{ ft}^2}$.

17.

$$A = lw$$
$$9165 = (x + 76)(x)$$
$$9165 = x^2 + 76x$$
$$0 = x^2 + 76x - 9165$$
$$0 = (x + 141)(x - 65)$$
$$x + 141 = 0 \quad \text{or} \quad x - 65 = 0$$
$$x = -141 \quad \text{or} \quad x = 65$$

Lengths must be nonnegative so only $x = 65$ works, so the plot is $65 + 76 = 141$ ft by 65 ft.

Section 11.4: Graphs of Quadratic Equations

1.

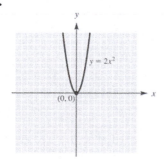

3.

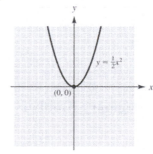

5.

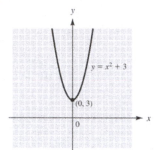

7.

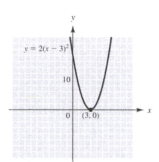

9.

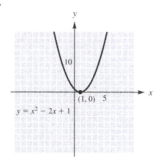

11.

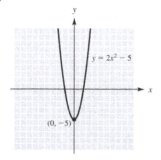

13.

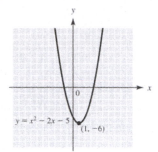

15.

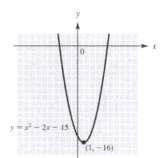

17.

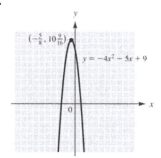

19.

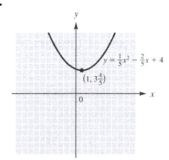

Section 11.5: Imaginary Numbers

1. $7j$

3. $3.74j$

5. $1.41j$

7. $7.48j$

9. $13j$

11. $5.20j$

13. $j^3 = \left(j^2\right)(j) = (-1)(j) = -j$

15. $j^{13} = \left(j^{12}\right)\left(j^1\right) = (1)(j) = j$

17. $j^{19} = \left(j^{16}\right)\left(j^3\right) = (1)(-j) = -j$

19. $j^{24} = 1$

21. $j^{38} = \left(j^{36}\right)\left(j^2\right) = (1)(-1) = -1$

23. $\dfrac{1}{j} = \dfrac{1 \times j}{j \times j} = \dfrac{j}{j^2} = \dfrac{j}{-1} = -j$

25.

$$a = 1, \ b = 3, \ c = -10$$

$$b^2 - 4ac = (3)^2 - 4(1)(-10) = 49$$

Since 49 is a positive perfect square, both roots are rational.

27.

$$a = 5, \ b = 4, \ c = 1$$

$$b^2 - 4ac = (4)^2 - 4(5)(1) = -4$$

Since -4 is negative, both roots are imaginary.

29.

$$2x^2 - 3x - 1 = 0$$

$$a = 2, \ b = -3, \ c = -1$$

$$b^2 - 4ac = (-3)^2 - 4(1)(-1) = 13$$

Since 13 is positive and not a perfect square, both roots are irrational.

31.

$$2x^2 - x + 6 = 0$$

$$a = 2, \ b = -1, \ c = 6$$

$$b^2 - 4ac = (-1)^2 - 4(2)(6) = -47$$

Since -47 is negative, both roots are imaginary.

33.

$$a = 1, \ b = 0, \ c = 25$$

$$b^2 - 4ac = (0)^2 - 4(1)(25) = -100$$

Since -100 is negative, both roots are imaginary.

35.

$$x = \frac{-b \pm \sqrt{b^2 - 4ac}}{2a}$$

$$a = 1, \ b = -6, \ c = 10$$

$$\text{So } x = \frac{-(-6) \pm \sqrt{(-6)^2 - 4(1)(10)}}{2(1)}$$

$$= \frac{6 \pm \sqrt{-4}}{2}$$

$$= \frac{6 \pm 2j}{2}$$

$$= 3 - j \ \text{ or } \ 3 + j$$

37.

$$x = \frac{-b \pm \sqrt{b^2 - 4ac}}{2a}$$

$$a = 1, \ b = -14, \ c = 53$$

$$\text{So } x = \frac{-(-14) \pm \sqrt{(-14)^2 - 4(1)(53)}}{2(1)}$$

$$= \frac{14 \pm \sqrt{-16}}{2(1)}$$

$$= \frac{14 \pm 4j}{2}$$

$$= 7 - 2j \ \text{ or } \ 7 + 22j$$

39.

$$x = \frac{-b \pm \sqrt{b^2 - 4ac}}{2a}$$

$$a = 1, \ b = 8, \ c = 41$$

$$\text{So } x = \frac{-(8) \pm \sqrt{(8)^2 - 4(1)(41)}}{2(1)}$$

$$= \frac{-8 \pm \sqrt{-100}}{2}$$

$$= \frac{-8 \pm 10j}{2}$$

$$= -4 - 5j \ \text{ or } \ -4 + 5j$$

41.

$$x = \frac{-b \pm \sqrt{b^2 - 4ac}}{2a}$$

$$a = 6, \ b = 5, \ c = 8$$

$$\text{So } x = \frac{-(5) \pm \sqrt{(5)^2 - 4(6)(8)}}{2(6)}$$

$$= \frac{-5 \pm \sqrt{-167}}{12}$$

$$= \frac{-5 \pm 12.9j}{12}$$

$$= -0.417 - 1.08j \ \text{ or } \ -0.417 + 1.08j$$

43. Rewrite the equation as $3x^2 - 6x + 7 = 0$.

$$x = \frac{-b \pm \sqrt{b^2 - 4ac}}{2a}$$

$a = 3,\ b = -6,\ c = 7$

So $x = \dfrac{-(-6) \pm \sqrt{(-6)^2 - 4(3)(7)}}{2(3)}$

$$= \frac{6 \pm \sqrt{-48}}{6}$$

$$= \frac{6 \pm 6.93j}{6}$$

$$= 1 - 1.16j \quad \text{or} \quad 1 + 1.16j$$

45.

$$x = \frac{-b \pm \sqrt{b^2 - 4ac}}{2a}$$

$a = 5,\ b = 8,\ c = 4$

So $x = \dfrac{-(8) \pm \sqrt{(8)^2 - 4(5)(4)}}{2(5)}$

$$= \frac{-8 \pm \sqrt{-16}}{10}$$

$$= \frac{-8 \pm 4j}{10}$$

$$= -0.8 - 0.4j \quad \text{or} \quad -0.8 - 0.4j$$

47. Rewrite the equation as $5x^2 + 14x - 3 = 0$.

$$x = \frac{-b \pm \sqrt{b^2 - 4ac}}{2a}$$

$a = 5,\ b = 14,\ c = -3$

So $x = \dfrac{-(14) \pm \sqrt{(14)^2 - 4(5)(-3)}}{2(5)}$

$$= \frac{-14 \pm \sqrt{256}}{10}$$

$$= \frac{-14 \pm 16}{10}$$

$$= -3 \quad \text{or} \quad \frac{1}{5}$$

49.

$$x = \frac{-b \pm \sqrt{b^2 - 4ac}}{2a}$$

$a = 1,\ b = 1,\ c = 1$

So $x = \dfrac{-(1) \pm \sqrt{(1)^2 - 4(1)(1)}}{2(1)}$

$$= \frac{-1 \pm \sqrt{-3}}{2}$$

$$= \frac{-1 \pm 1.73j}{2}$$

$$= -0.5 - 0.865j \quad \text{or} \quad -0.5 + 0.865j$$

Chapter 11 Review

1. Either $a = 0$ or $b = 0$.

2.

$$3x = 0 \text{ or } x - 2 = 0$$
$$x = 0 \text{ or } x = 2$$

3.

$$x^2 - 4 = 0$$
$$(x + 2)(x - 2) = 0$$
$$x + 2 = 0 \text{ or } x - 2 = 0$$
$$x = -2 \text{ or } x = 2$$

4.

$$x^2 - x = 6$$
$$x^2 - x - 6 = 0$$
$$(x + 2)(x - 3) = 0$$
$$x + 2 = 0 \text{ or } x - 3 = 0$$
$$x = -2 \text{ or } x = 3$$

5.

$$5x^2 - 6x = 0$$
$$x(5x - 6) = 0$$
$$x = 0 \text{ or } 5x - 6 = 0$$
$$x = 0 \text{ or } x = \frac{6}{5}$$

6.

$$x^2 - 3x - 28 = 0$$
$$(x + 4)(x - 7) = 0$$
$$x + 4 = 0 \text{ or } x - 7 = 0$$
$$x = -4 \text{ or } x = 7$$

7.

$$x^2 - 14x = -45$$
$$x^2 - 14x + 45 = 0$$
$$(x-5)(x-9) = 0$$
$$x - 5 = 0 \text{ or } x - 9 = 0$$
$$x = 5 \text{ or } x = 9$$

8.

$$x^2 - 18 - 3x = 0$$
$$x^2 - 3x - 18 = 0$$
$$(x+3)(x-6) = 0$$
$$x + 3 = 0 \text{ or } x - 6 = 0$$
$$x = -3 \text{ or } x = 6$$

9.

$$3x^2 + 20x + 32 = 0$$
$$(x+4)(3x+8) = 0$$
$$x + 4 = 0 \text{ or } 3x + 8 = 0$$
$$x = -4 \text{ or } x = -\frac{8}{3}$$

10.

$$x = \frac{-b \pm \sqrt{b^2 - 4ac}}{2a}$$
$$a = 3, \ b = -16, \ c = -12$$
$$\text{So } x = \frac{-(-16) \pm \sqrt{(-16)^2 - 4(3)(-12)}}{2(3)}$$
$$= \frac{16 \pm \sqrt{400}}{6}$$
$$= \frac{16 \pm 20}{6}$$
$$= -\frac{2}{3} \text{ or } 6$$

11.

$$x = \frac{-b \pm \sqrt{b^2 - 4ac}}{2a}$$
$$a = 1, \ b = 7, \ c = -5$$
$$\text{So } x = \frac{-(7) \pm \sqrt{(7)^2 - 4(1)(-5)}}{2(1)}$$
$$= \frac{-7 \pm \sqrt{69}}{2}$$
$$= \frac{-7 \pm \sqrt{69}}{2}$$
$$= \frac{-7 - 8.31}{2} \text{ or } \frac{-7 + 8.31}{2}$$
$$= -7.66 \text{ or } 0.655$$

12. Rewrite the equation as $2x^2 + x - 15 = 0$.

$$x = \frac{-b \pm \sqrt{b^2 - 4ac}}{2a}$$
$$a = 2, \ b = 1, \ c = -15$$
$$\text{So } x = \frac{-(1) \pm \sqrt{(1)^2 - 4(2)(-15)}}{2(2)}$$
$$= \frac{-1 \pm \sqrt{121}}{4}$$
$$= \frac{-1 \pm 11}{4}$$
$$= -3 \text{ or } \frac{5}{2}$$

13. Rewrite the equation as $x^2 - 4x - 2 = 0$.

$$x = \frac{-b \pm \sqrt{b^2 - 4ac}}{2a}$$
$$a = 1, \ b = -4, \ c = -2$$
$$\text{So } x = \frac{-(-4) \pm \sqrt{(-4)^2 - 4(1)(-2)}}{2(1)}$$
$$= \frac{4 \pm \sqrt{24}}{2}$$
$$= \frac{4 \pm 4.90}{2}$$
$$= -0.45 \text{ or } 4.45$$

14. Rewrite the equation as $3x^2 - 4x - 5 = 0$.

$$x = \frac{-b \pm \sqrt{b^2 - 4ac}}{2a}$$
$$a = 3, \ b = -4, \ c = -5$$
$$\text{So } x = \frac{-(-4) \pm \sqrt{(-4)^2 - 4(3)(-5)}}{2(3)}$$
$$= \frac{4 \pm \sqrt{76}}{6}$$
$$= \frac{4 \pm 8.72}{6}$$
$$= -0.787 \text{ or } 2.12$$

15.

$$A = lw$$
$$36 = (x+5)x$$
$$36 = x^2 + 5x$$
$$0 = x^2 + 5x - 36$$
$$0 = (x+9)(x-4)$$
$$x+9 = 0 \quad \text{or} \quad x-4 = 0$$
$$x = -9 \quad \text{or} \quad x = 4$$

Lengths must be nonnegative so only $x = 4$ works, so the length is $4 + 5 = 9$ ft and the width is 4ft.

16. a.

$$i = t^2 - 12t + 36$$
$$4 = t^2 - 12t + 36$$
$$0 = t^2 - 12t + 32$$
$$0 = (t-4)(t-8)$$
$$t - 4 = 0 \text{ or } t - 8 = 0$$
$$t = 4 \text{ or } t = 8$$

The current is 4 A at 4 and 8 seconds.

b.

$$i = t^2 - 12t + 36$$
$$0 = t^2 - 12t + 36$$
$$0 = (t-6)(t-6)$$
$$t - 6 = 0$$
$$t = 6$$

The current is 0 A at 6 seconds.

c.

$$i = t^2 - 12t + 36$$
$$10 = t^2 - 12t + 36$$
$$0 = t^2 - 12t + 26$$
$$t = \frac{-b \pm \sqrt{b^2 - 4ac}}{2a}$$
$$a = 1, \ b = -12, \ c = 26$$

So $t = \dfrac{-(-12) \pm \sqrt{(-12)^2 - 4(1)(26)}}{2(1)}$

$$= \frac{12 \pm \sqrt{40}}{2}$$
$$= \frac{12 \pm 6.32}{2}$$
$$= 2.84 \text{ or } 916$$

The current is 10 A at 2.84 and 9.16 seconds.

17.

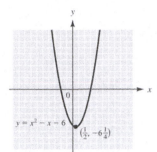

18.

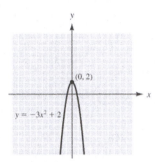

19. $6j$

20. $8.55j$

21. $j^{12} = 1$

22. $j^{27} = (j^{24})(j^3) = (1)(-j) = -j$

23.

$$a = 9, \ b = 30, \ c = 25$$
$$b^2 - 4ac = (30)^2 - 4(9)(25) = 0$$

Since the value is 0, there is only one rational root.

24.

$$a = 3, \ b = -2, \ c = 4$$
$$b^2 - 4ac = (-2)^2 - 4(3)(4) = -44$$

Since –44 is negative, both roots are imaginary.

25.

$$x = \frac{-b \pm \sqrt{b^2 - 4ac}}{2a}$$

$$a = 1, \ b = -4, \ c = 5$$

So $x = \dfrac{-(-4) \pm \sqrt{(-4)^2 - 4(1)(5)}}{2(1)}$

$$= \frac{4 \pm \sqrt{-4}}{2}$$

$$= \frac{4 \pm 2j}{2}$$

$$= 2 - j \ \text{ or } \ 2 + j$$

26.

$$x = \frac{-b \pm \sqrt{b^2 - 4ac}}{2a}$$

$$a = 5, \ b = -6, \ c = 4$$

So $x = \dfrac{-(-6) \pm \sqrt{(-6)^2 - 4(5)(4)}}{2(5)}$

$$= \frac{6 \pm \sqrt{-44}}{10}$$

$$= \frac{6 \pm 6.63j}{10}$$

$$= 0.6 - 0.663j \ \text{ or } \ 0.6 + 0.663j$$

28.

$$A = lw$$

$$672 = (26 + 2x)(15 + 2x)$$

$$672 = 4x^2 + 82x + 390$$

$$0 = 4x^2 + 82x - 282 \ \text{ Divide both sides by 2}$$

$$0 = 2x^2 + 41x - 141$$

$$0 = (2x + 47)(x - 3)$$

$$2x + 47 = 0 \ \text{ or } \ x - 3 = 0$$

$$x = -\frac{47}{2} \ \text{ or } \ x = 3$$

Lengths must be nonnegative so only $x = 3$ works, so the strip must be 3 in. wide.

27.

$$A = lw$$

$$21.25 = (3x + 1)x$$

$$21.25 = 3x^2 + x \ \text{ Multiply both sides by 4}$$

$$85 = 12x^2 + 4x$$

$$0 = 12x^2 + 4x - 85$$

$$0 = (6x + 17)(2x - 5)$$

$$6x + 17 = 0 \ \text{ or } \ 2x - 5 = 0$$

$$x = -\frac{17}{6} \ \text{ or } \ x = 2.5$$

Lengths must be nonnegative so only $x = 2.5$ works, so the length is $3 \times 2.5 + 1 = 8.5$ ft and the width is 2.5ft.

Chapter 11 Test

1.

$$x^2 = 64$$

$$x^2 - 64 = 0$$

$$(x + 8)(x - 8) = 0$$

$$x + 8 = 0 \text{ or } x - 8 = 0$$

$$x = -8 \text{ or } x = 8$$

3.

$$x^2 + 9x - 36 = 0$$

$$(x + 12)(x - 3) = 0$$

$$x + 12 = 0 \text{ or } x - 3 = 0$$

$$x = -12 \text{ or } x = 3$$

5.

$$x = \frac{-b \pm \sqrt{b^2 - 4ac}}{2a}$$

$$a = 5,\ b = 6,\ c = -10$$

So $x = \dfrac{-(6) \pm \sqrt{(6)^2 - 4(5)(-10)}}{2(5)}$

$$= \frac{-6 \pm \sqrt{236}}{10}$$

$$= \frac{-6 \pm 15.4}{10}$$

$$= \frac{-6 - 15.4}{10} \quad \text{or} \quad \frac{-6 + 15.4}{10}$$

$$= -2.14 \quad \text{or} \quad 0.94$$

7.

$$21x^2 - 29x - 10 = 0$$

$$(7x + 2)(3x - 5) = 0$$

$$7x + 2 = 0 \text{ or } 3x - 5 = 0$$

$$x = -\frac{2}{7} \text{ or } x = \frac{5}{3}$$

9.

$3x^2 - 39x + 90 = 0$ Divide both sides by 3.

$$x^2 - 13x + 30 = 0$$

$$(x - 3)(x - 10) = 0$$

$$x - 3 = 0 \text{ or } x - 10 = 0$$

$$x = 3 \text{ or } x = 10$$

11.

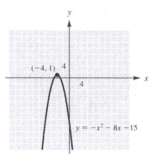

13. $4j$

15. $j^9 = \left(j^8\right)\left(j^1\right) = (1)(j) = j$

17.

$$a = 3,\ b = -1,\ c = 4$$

$$b^2 - 4ac = (-1)^2 - 4(3)(4) = -47$$

Since -47 is negative, both roots are imaginary.

Chapter 12: Geometry

Section 12.1: Angles and Polygons

1. acute angle

3. right angle

5. acute angle

7. obtuse angle

9. $\angle 1$ is a right angle. Lines l and m are perpendicular.

11. a. $\angle 1$ and $\angle 2$, $\angle 2$ and $\angle 4$, $\angle 3$ and $\angle 4$, $\angle 1$ and $\angle 3$, $\angle 5$ and $\angle 6$, $\angle 6$ and $\angle 8$, $\angle 7$ and $\angle 8$, $\angle 5$ and $\angle 7$.

    b. $\angle 1$ and $\angle 4$, $\angle 2$ and $\angle 3$, $\angle 5$ and $\angle 8$, $\angle 6$ and $\angle 7$.

13.
$$\angle 2 = \angle 1 = 57°$$
$$\angle 3 = \angle 1 = 57°$$
$$\angle 4 = 180° - 57° = 123°$$

15. $\angle COB = 180° - 119° = 61°$

17. $\angle 3 = 90° - 38° = 52°$

19. a. Yes; $\angle 2 = \angle 3 = 38°$, so $\angle 1 + \angle 2 = 90°$.

    b. Yes; $\angle 4 = \angle 1 = 52°$, so $\angle 1 + \angle 2 = 90°$.

21.
$$\angle 1 = \angle 3$$
$$\angle 1 + \angle 3 = 180°$$
$$\angle 1 + (\angle 1) = 180°$$
$$2\angle 1 = 180°$$
$$\angle 1 = 90°$$
$$\angle 3 = \angle 1 = 90°$$
$$\angle 2 = 180° - 90° = 90°$$
$$\angle 4 = \angle 2 = 90°$$

23. $\angle 1$ and $\angle 3$ are supplementary
$$\angle 1 = 180° - \angle 3$$
$$4x + 55 = 180 - (10x - 85)$$
$$4x + 55 = 265 - 10x$$
$$14x = 210$$
$$x = 15°$$

25. Supplementary angles: $x = 180° - 32° = 148°$

27. Alternate-exterior angles are equal: $z = 152°$

29. triangle

31. hexagon

33. quadrilateral

35. heptagon

Section 12.2: Quadrilaterals

1.
$$P = 2(a + b)$$
$$P = 2(15.0 \text{ cm} + 15.0 \text{ cm})$$
$$= 2(30.0 \text{ cm})$$
$$= 60.0 \text{ cm}$$
$$A = bh$$
$$A = (15.0 \text{ cm})(15.0 \text{ cm})$$
$$= 225 \text{ cm}^2$$

3.
$$P = 2(a + b)$$
$$P = 2(8.0 \text{ m} + 8.0 \text{ m})$$
$$= 32.0 \text{ m}$$
$$A = bh$$
$$A = (8.0 \text{ m})(6.0 \text{ m})$$
$$= 48 \text{ m}^2$$

187

5.

$P = a + b + c + d$

$P = 10.0 \text{ m} + 24.5 \text{ m} + 8.0 \text{ m} + 11.0 \text{ m}$

$\quad = 53.5 \text{ m}$

$A = \left(\dfrac{a+b}{2}\right)h$

$A = \left(\dfrac{10.0 \text{ m} + 24.5 \text{ m}}{2}\right)(6.0 \text{ m})$

$\quad = 1\bar{0}0 \text{ m}^2$

7.

$P = 2(a + b)$

$P = 2(20.8 \text{ in.} + 23.9 \text{ in.})$

$\quad = 89.4 \text{ in.}$

$A = bh$

$A = (23.9 \text{ in.})(17.2 \text{ in.})$

$\quad = 411 \text{ in}^2$

9.

$P = 2(a + b)$

$P = 2(9.2 \text{ cm} + 9.2 \text{ cm})$

$\quad = 36.8 \text{ cm}$

$A = bh$

$A = (9.2 \text{ cm})(9.2 \text{ cm})$

$\quad = 85 \text{ cm}^2$

11.

$A = bh$

$h = \dfrac{A}{b}$

$h = \dfrac{24\bar{0} \text{ cm}^2}{10.0 \text{ cm}}$

$\quad = 24.0 \text{ cm}$

13.

$A = bh$

$h = \dfrac{A}{b}$

$h = \dfrac{486 \text{ ft}^2}{36.2 \text{ ft}}$

$\quad = 13.4 \text{ ft}$

15.

$A = \left(\dfrac{a+b}{2}\right)h$

$A = \left(\dfrac{21.0 \text{ in.} + 23.0 \text{ in.}}{2}\right)(16.0 \text{ in.})$

$\quad = 352 \text{ in}^2$

17. a.

$P = a + b + c + d$

$P = 20.0 \text{ mi} + 14.0 \text{ mi} + 12.0 \text{ mi} + 13.42 \text{ mi}$

$\quad = 59.4 \text{ m}$

b.

$A = \left(\dfrac{a+b}{2}\right)h$

$A = \left(\dfrac{20.0 \text{ mi} + 14.0 \text{ mi}}{2}\right)(11.6 \text{ mi})$

$\quad = 197 \text{ m}^2$

19.

$A = bh$

$h = \dfrac{A}{b}$

$h = \dfrac{128 \text{ in}^2}{8.0 \text{ in.}}$

$\quad = 16 \text{ in.}$

$P = 2(8.0 \text{ in.} + 16 \text{ in.})$

$\quad = 48 \text{ in}$

21.

$A = bh$

$A = (24 \text{ ft})(36 \text{ ft})$

$\quad = 860 \text{ ft}^2$

$\dfrac{864 \text{ ft}^2}{96 \text{ ft}^2} = 9 \text{ beds}$

23.

$36 \text{ in.} \times \dfrac{1 \text{ ft}}{12 \text{ in.}} = 3 \text{ ft}$

$A = lw$

$l = \dfrac{A}{w}$

$l = \dfrac{108 \text{ ft}^2}{3 \text{ ft}}$

$\quad = 36 \text{ ft}$

25.

$$72 \text{ in.} \times \frac{1 \text{ ft}}{12 \text{ in.}} = 6 \text{ ft}$$

$$\frac{36 \text{ ft}}{6 \text{ ft}} = 6 \text{ pieces}$$

29. a.

$$A = bh$$

$$A = (16.0 \text{ in.})(13.0 \text{ in.})$$

$$= 208 \text{ in}^2$$

b.

Parallelogram

$$A = bh$$

$$A = (42.0 \text{ in.})(19.0 \text{ in.})$$

$$= 798 \text{ in}^2$$

Remaining metal

33.

$$A = 2 \times bh$$

$$A = 2(28 \text{ ft})(12 \text{ ft})$$

$$= 672 \text{ ft}^2$$

$$672 \text{ ft}^2 \times \frac{1 \text{ square}}{100 \text{ ft}^2} = 7 \text{ squares}$$

35.

Sides of Building

$$A = 2 \times bh$$

$$A = 2(9.0 \text{ ft})(48.0 \text{ ft})$$

$$= 864 \text{ ft}^2$$

Front and Back

$$A = 2 \times \left(2 \times \left(\frac{a+b}{2} \right) h \right)$$

$$A = 4 \times \left(\frac{9 \text{ ft} + 14 \text{ ft}}{2} \right) (12 \text{ ft})$$

$$= 552 \text{ ft}^2$$

House

$$864 \text{ ft}^2 + 552 \text{ ft}^2 - 235 \text{ ft}^2 = 1091 \text{ ft}^2$$

$$1091 \text{ ft}^2 \times \$0.85/\text{ft}^2 = \$927.35$$

37.

$$A = bh$$

$$A = (29.0 \text{ ft})(8.7 \text{ ft})$$

$$= 250 \text{ ft}^2$$

27.

$$A = bh$$

$$A = (2.9 \text{ in.})(9.75 \text{ in.})$$

$$= 28 \text{ in}^2$$

$$798 \text{ in}^2 - 208 \text{ in}^2 = 59\overline{0} \text{ in}^2$$

31.

$$A = bh$$

$$A = (25.0 \text{ mi})(25.0 \text{ mi})$$

$$= 625 \text{ mi}^2$$

$$A = 625 \text{ mi}^2 \times \frac{640 \text{ acres}}{1 \text{ mi}^2}$$

$$= 40\overline{0},000 \text{ acres}$$

39.

$$P = 2(a+b)$$

$$P = 2(192.7 \text{ ft} + 91.6 \text{ ft})$$

$$= 568.6 \text{ ft}$$

$$\text{Cost} = 568.6 \text{ ft} \times \frac{\$9.50}{1 \text{ ft}}$$

$$\$5401.70$$

41.

100-ft side

$$A = bh$$

$$A = (100 \text{ ft})(4 \text{ ft})$$

$$= 400 \text{ ft}^2$$

80-ft side

$$A = bh$$

$$A = (76 \text{ ft})(4 \text{ ft})$$

$$= 304 \text{ ft}^2$$

$$400 \text{ ft}^2 + 304 \text{ ft}^2 = 704 \text{ ft}^2 \text{ will be}$$
unavailable.

43.

$$A = \left(\frac{a+b}{2} \right) h$$

$$A = \left(\frac{25.0 \text{ ft} + 15.0 \text{ ft}}{2} \right) (10.0 \text{ ft})$$

$$= 20\overline{0} \text{ ft}^2$$

45.

$$A = bh$$

$$h = \frac{A}{b}$$

$$h = \frac{27{,}800 \text{ ft}^2}{265 \text{ ft}}$$

$$= 105 \text{ ft}$$

Section 12.3: Triangles

1.

$$c = \sqrt{a^2 + a^2}$$

$$c = \sqrt{(6.00 \text{ m})^2 + (8.00 \text{ m})^2}$$

$$= 10.0 \text{ m}$$

3.

$$c = \sqrt{a^2 + a^2}$$

$$c = \sqrt{(24.0 \text{ m})^2 + (7.00 \text{ m})^2}$$

$$= 25.0 \text{ m}$$

5.

$$c = \sqrt{a^2 + a^2}$$

$$c = \sqrt{(15.0 \text{ m})^2 + (8.00 \text{ m})^2}$$

$$= 17.0 \text{ m}$$

7.

$$a = \sqrt{c^2 - b^2}$$

$$a = \sqrt{(18.5 \text{ cm})^2 - (12.6 \text{ cm})^2}$$

$$= 13.5 \text{ cm}$$

9.

$$b = \sqrt{c^2 - a^2}$$

$$b = \sqrt{(2460 \text{ km})^2 - (1980 \text{ km})^2}$$

$$= 1460 \text{ km}$$

11.

$$a = \sqrt{c^2 - b^2}$$

$$a = \sqrt{(360 \text{ ft})^2 - (95 \text{ft})^2}$$

$$= 350 \text{ ft}$$

13.

$$c = \sqrt{a^2 + a^2}$$

$$c = \sqrt{(45.0 \text{ in.})^2 + (39.5 \text{ in.})^2}$$

$$= 59.9 \text{ in.}$$

15.

$$c = \sqrt{a^2 + a^2}$$

$$c = \sqrt{(25.5 \text{ in.})^2 + (18.3 \text{ in.})^2}$$

$$= 31.4 \text{ in.}$$

$$\text{Total} = 31.4 \text{ in.} + 1.0 \text{ in.} + 1.0 \text{ in.}$$

$$= 33.4 \text{ in.}$$

17.

$$c = \sqrt{a^2 + a^2}$$

$$c = \sqrt{(2.00 \text{ in.})^2 + (2.00 \text{ in.})^2}$$

$$= 2.83 \text{ in.}$$

19.

$$a = \sqrt{c^2 - b^2}$$

$$a = \sqrt{(7 \text{ in.})^2 - (18 \text{ in.})^2}$$

$$= 16.6 \text{ in.}$$

21.

$$c = \sqrt{a^2 + b^2}$$

$$c = \sqrt{(85.2 \text{ V})^2 + (78.4 \text{ V})^2}$$

$$= 116 \text{ V}$$

23.

$$a = \sqrt{c^2 - b^2}$$

$$a = \sqrt{(32 \text{ A})^2 - (24 \text{ A})^2}$$

$$= 21 \text{ A}$$

25.

$$a = \sqrt{c^2 - b^2}$$

$$a = \sqrt{(165 \text{ } \Omega)^2 - (105 \text{ } \Omega)^2}$$

$$= 127 \text{ } \Omega$$

27.

$$a = \sqrt{c^2 - b^2}$$

$$a = \sqrt{(4.5\ \Omega)^2 - (3.7\ \Omega)^2}$$

$$= 2.6\ \Omega$$

29.

$$c = \sqrt{(17.8\ \text{m})^2 + (17.8\ \text{m})^2}$$

$$= 25.2\ \text{m}$$

$$P = a + b + c$$

$$P = (17.8\ \text{m}) + (17.8\ \text{m}) + (25.2\ \text{m})$$

$$= 60.8\ \text{m}$$

$$A = \frac{1}{2}bh$$

$$A = \frac{1}{2}(17.8\ \text{m})(17.8\ \text{m})$$

$$= 158\ \text{m}^2$$

31.

$$P = a + b + c$$

$$P = (6.00\ \text{cm}) + (6.00\ \text{cm}) + (6.00\ \text{cm})$$

$$= 18.0\text{cm}$$

$$h = \sqrt{(6.00\ \text{cm})^2 - (3.00\ \text{cm})^2}$$

$$= 5.20\ \text{cm}$$

$$A = \frac{1}{2}bh$$

$$A = \frac{1}{2}(6.00\ \text{cm})(5.20\ \text{cm})$$

$$= 15.6\ \text{cm}^2$$

33.

$$P = a + b + c$$

$$P = (26.1\ \text{m}) + (32.9\ \text{m}) + (49.7\ \text{m})$$

$$= 108.7\ \text{m}$$

$$A = \frac{1}{2}bh$$

$$A = \frac{1}{2}(49.7\ \text{m})(15.9\ \text{m})$$

$$= 395\ \text{m}^2$$

35.

$$P = a + b + c$$

$$P = (7.4\ \text{m}) + (17.3\ \text{m}) + (11.3\ \text{m})$$

$$= 36.0\ \text{m}$$

$$A = \frac{1}{2}bh$$

$$A = \frac{1}{2}(11.3\ \text{m})(5.29\ \text{m})$$

$$= 29.9\ \text{m}^2$$

37.

$$P = a + b + c$$

$$P = (12.0\ \text{cm}) + (16.0\ \text{cm}) + (20.0\ \text{cm})$$

$$= 48.0\ \text{cm}$$

$$s = \frac{1}{2}(a + b + c)$$

$$= \frac{1}{2}(48.0) = 24.0$$

$$A = \sqrt{s(s-a)(s-b)(s-c)}$$

$$A = \sqrt{24(24 - 12.0)(24 - 16.0)(24 - 20.0)}$$

$$= 96.0\ \text{cm}^2$$

39.

$$c = \sqrt{a^2 + a^2}$$

$$c = \sqrt{(6.0\ \text{ft})^2 + (9.0\ \text{ft})^2}$$

$$= 10.8\ \text{ft}$$

41.

$$c = \sqrt{a^2 + a^2}$$

$$c = \sqrt{(62.0\ \text{mi})^2 + (41.0\ \text{mi})^2}$$

$$= 74.3\ \text{mi}$$

43.

$$A = \frac{1}{2}bh$$

$$A = \frac{1}{2}(26.4\ \text{in.})(11.0\ \text{in.})$$

$$= 145\ \text{in}^2$$

45. $x = 180° - 58° - 41° = 81°$

47. $x = \dfrac{180°}{3} = 60°$

49.

$$A = \frac{1}{2}bh$$

$$A = \frac{1}{2}(20.0 \text{ ft})(36.0 \text{ ft})$$

$$= 360 \text{ ft}^2$$

$$\frac{360 \text{ ft}^2}{75 \text{ ft}^2} = 5 \text{ bags}$$

51. a. $\dfrac{4.50 \text{ ft}}{50.0 \text{ ft}} = 9\%$

b.

$$c = \sqrt{a^2 + a^2}$$

$$c = \sqrt{(50.0 \text{ ft})^2 + (4.50 \text{ ft})^2}$$

$$= 50.2 \text{ ft}$$

53.

Area A

$$s = \frac{1}{2}(a + b + c)$$

$$s = \frac{1}{2}(14\overline{0}0 + 14\overline{0}0 + 18\overline{0}0) = 23\overline{0}0$$

$$A = \sqrt{s(s-a)(s-b)(s-c)}$$

$$A = \sqrt{23\overline{0}0(23\overline{0}0 - 14\overline{0}0)(23\overline{0}0 - 14\overline{0}0)(23\overline{0}0 - 18\overline{0}0)}$$

$$= 695{,}000 \text{ ft}^2$$

Area B

$$A = lw$$

$$A = (10\overline{0}0)(18\overline{0}0)$$

$$= 1{,}8\overline{0}0{,}000 \text{ ft}^2$$

Area C

$$a = \sqrt{c^2 - b^2}$$

$$a = \sqrt{(12\overline{0}0 \text{ ft})^2 - (10\overline{0}0 \text{ ft})^2}$$

$$= 663 \text{ ft}$$

$$A = \frac{1}{2}bh$$

$$A = \frac{1}{2}(10\overline{0}0 \text{ ft})(663 \text{ ft})$$

$$= 333{,}000 \text{ ft}^2$$

Total area

$$3{,}100{,}000 \text{ ft}^2 \times \frac{1 \text{ acre}}{43{,}560 \text{ ft}^2}$$

$$= 71.1 \text{ acres}$$

Section 12.4: Similar Polygons

1. $AB = AD + DB = 6 + 6 = 12$.

$$\frac{DE}{BC} = \frac{AD}{DB}$$

$$\frac{DE}{10} = \frac{6}{12}$$

$$12(DE) = (10)(6)$$

$$DE = \frac{(10)(6)}{12} = 5$$

3. $\angle A = \angle D$ and $\angle B = \angle B$ (alternate-interior angles.) $\angle AOB = \angle COD = COD\angle$ (vertical angles.) Yes, all corresponding angles of $\triangle ABO$ and of $\triangle DCO$ are equal.

5.

$$\frac{YZ}{BC} = \frac{XY}{AB}$$

$$\frac{YZ}{8} = \frac{8}{12}$$

$$12(YZ) = (8)(8)$$

$$YZ = \frac{(8)(8)}{12} = \frac{16}{3} = 5\frac{1}{3}$$

7.

$$\frac{x}{6.0} = \frac{8\overline{0}}{4.0}$$

$$4.0x = (6.0)(8\overline{0})$$

$$x = \frac{(6.0)(8\overline{0})}{4.00} = 120 \text{ ft}$$

9. a.

$$AC = AD + DC = 8.2 \text{ m} + 12.3 \text{ m}$$

$$= 20.5 \text{ m}$$

$$\frac{DE}{AB} = \frac{CD}{AC}$$

$$\frac{DE}{20.0} = \frac{12.3}{20.5}$$

$$20.5(DE) = (20.0)(12.3)$$

$$DE = \frac{(20.0)(12.3)}{20.5} = 12.0 \text{ m}$$

b.

$$\frac{BC}{CE} = \frac{AC}{CD}$$

$$\frac{BC}{10.2} = \frac{20.5}{12.3}$$

$$12.3(BC) = (10.2)(20.5)$$

$$BC = \frac{(10.2)(20.5)}{12.3} = 17.0 \text{ m}$$

11.

$$\frac{x}{17.0} = \frac{5.00}{8.00}$$

$$8.00(x) = (17.0)(5.00)$$

$$x = \frac{(17.0)(5.00)}{8.00} = 10.6 \text{ in.}$$

13. Each side of the smaller landing pad is

$$\frac{300\bar{0} \text{ ft}}{6} = 500.0 \text{ ft}$$

$$\frac{x}{500.0} = \frac{330\bar{0}}{300\bar{0}}$$

$$300\bar{0}(x) = (500.0)(330\bar{0})$$

$$x = \frac{(500.0)(330\bar{0})}{300\bar{0}} = 55\bar{0} \text{ ft}$$

15.

$$\frac{x}{10.0} = \frac{12.0}{6.0}$$

$$6.0(x) = (10.0)(12.0)$$

$$x = \frac{(10.0)(126.0)}{6.0} = 20.0 \text{ in.}$$

17.

$$\frac{x}{48} = \frac{20}{16}$$

$$16(x) = (48)(20)$$

$$x = \frac{(48)(20)}{16} = 6\bar{0} \text{ in.}$$

19.

$$\frac{A}{1.00} = \frac{8.00}{6.00}$$

$$6.00A = (1.00)(8.00)$$

$$A = \frac{(1.00)(8.00)}{6.00}$$

$$= 1.33 \text{ ft}$$

$$\frac{B}{2.00} = \frac{8.00}{6.00}$$

$$6.00B = (2.00)(8.00)$$

$$B = \frac{(2.00)(8.00)}{6.00}$$

$$= 2.67 \text{ ft}$$

$$\frac{C}{3.00} = \frac{8.00}{6.00}$$

$$6.00C = (3.00)(8.00)$$

$$C = \frac{(3.00)(8.00)}{6.00}$$

$$= 4.00 \text{ ft}$$

$$\frac{D}{4.00} = \frac{8.00}{6.00}$$

$$6.00D = (4.00)(8.00)$$

$$D = \frac{(4.00)(8.00)}{6.00}$$

$$= 5.33 \text{ ft}$$

$$\frac{E}{5.00} = \frac{8.00}{6.00}$$

$$6.00E = (5.00)(8.00)$$

$$E = \frac{(5.00)(8.00)}{6.00}$$

$$= 6.67 \text{ ft}$$

Cross-piece

$$c = \sqrt{(6.00 \text{ ft})^2 + (8.00 \text{ ft})^2}$$

$$= 10.0 \text{ ft}$$

21. a.

$$1 \text{ in.} \times \frac{1}{4} = \frac{1}{4} \text{ in.}; \ 2 \text{ in.} \times \frac{1}{4} = \frac{1}{2} \text{ in.}$$

$$2.5 \text{ in.} \times \frac{1}{4} = \frac{5}{8} \text{ in.}; \ 3 \text{ in.} \times \frac{1}{4} = \frac{3}{4} \text{ in.}$$

$$6 \text{ in.} \times \frac{1}{4} = 1\frac{1}{2} \text{ in.}$$

b.

$$V_{big \ block} = \left(\frac{1}{2} \text{ in.}\right)\left(\frac{3}{4} \text{ in.}\right)\left(1\frac{1}{2} \text{ in.}\right)$$

$$= \frac{9}{16} \text{ in}^3$$

$$V_{small \ block} = \left(\frac{1}{4} \text{ in.}\right)\left(\frac{1}{4} \text{ in.}\right)\left(\frac{5}{8} \text{ in.}\right)$$

$$= \frac{5}{128} \text{ in}^3$$

$$V_{part} = 2\left(\frac{9}{16} \text{ in}^3\right) + \frac{5}{128} \text{ in}^3$$

$$= 1.16 \text{ in}^3$$

$$Weight = 1.16 \text{ in}^3 \times \frac{2.00 \text{ oz}}{1 \text{ in}^3}$$

$$= 2.33 \text{ oz}$$

23.

$$\frac{base_{new}}{height_{new}} = \frac{base_{old}}{height_{old}}$$

$$\frac{x}{36} = \frac{26}{24}$$

$$24(x) = (36)(26)$$

$$x = \frac{(36)(26)}{24} = 39 \text{ in.}$$

$$c = \sqrt{(39 \text{ in.})^2 + (36 \text{ in.})^2}$$

$$= 53 \text{ in.}$$

$$l_{total} = 36 \text{ in.} + 39 \text{ in.} + 53 \text{ in.}$$

$$= 128 \text{ in.}$$

$$l_{total} = 128 \text{ in.} \times \frac{1 \text{ ft}}{12 \text{ in.}} = 10.7 \text{ ft}$$

One 12-ft board must be purchased.

Section 12.5: Circles

1.

a.
$$C = 2\pi r$$
$$C = 2\pi (5.00 \text{ in.})$$
$$= 31.4 \text{ in.}$$

b.
$$A = \pi r^2$$
$$A = \pi (5.00 \text{ in.})^2$$
$$= 78.5 \text{ in}^2$$

3.

a.
$$r = \frac{1}{2}d = \frac{1}{2}(9.21 \text{ mm}) = 4.605 \text{ mm}$$
$$C = 2\pi r$$
$$C = 2\pi (4.605 \text{ mm})$$
$$= 28.9 \text{ mm}$$

b.
$$A = \pi r^2$$
$$A = \pi (4.605 \text{ mm})^2$$
$$= 66.6 \text{ mm}^2$$

5.

a.
$$C = 2\pi r$$
$$C = 2\pi (56.10 \text{ mi})$$
$$= 352 \text{ mi}$$

b.
$$A = \pi r^2$$
$$A = \pi (56.10 \text{ mi})^2$$
$$= 9890 \text{ mi}^2$$

7. $x = 360° - 97° - 92° = 171°$

9.

$$x = 360° - 111.1° - 143.9° - 29.8° - 31.8°$$
$$= 43.4°$$

11.

$$A = \pi r^2$$
$$\frac{A}{\pi} = r^2$$
$$\frac{28.2 \text{ cm}^2}{\pi} = r^2$$
$$\sqrt{\frac{28.2 \text{ cm}^2}{\pi}} = r$$
$$3.00 \text{ cm} = r$$

13.

$$C = 2\pi r$$
$$\frac{C}{2\pi} = r$$
$$\frac{62.9 \text{ m}}{2\pi} = r$$
$$10.0 \text{ m} = r$$

15. $\dfrac{1}{4} \times 360° = 90°$

17.

$$A = \pi r^2$$
$$A = \pi (24.0 \text{ in.})^2$$
$$= 1810 \text{ in}^2$$
$$A = 1810 \text{ in}^2 \times \left(\frac{1 \text{ ft}}{12 \text{ in.}}\right)^2$$
$$= 12.6 \text{ ft}^2$$

19.

$$r = \frac{1}{2}d = \frac{1}{2}(2.25 \text{ in.}) = 1.125 \text{ in.}$$
$$A = \pi r^2$$
$$A = \pi (1.125 \text{ in.})^2$$
$$= 3.98 \text{ in}^2$$
$$C = 2\pi r$$
$$C = 2\pi (1.125 \text{ in.})$$
$$= 7.07 \text{ in.}$$

21.

$$r = \frac{1}{2}r = \frac{1}{2}(16.0 \text{ in.}) = 8.00 \text{ in.}$$
$$C = 2\pi r$$
$$C = 2\pi (8.00 \text{ in.})$$
$$= 50.3 \text{ in.}$$

23.

$$C = 2\pi r$$
$$C = 2\pi (1.80 \text{ ft})$$
$$= 11.3 \text{ in.}$$
$$L = 236 \times 11.3 \text{ in.} = 2670 \text{ in.}$$

25.

$$C = 2\pi r$$
$$C = 2\pi \left(\frac{4.25 \text{ in.}}{2}\right) = 13.4 \text{ in.}$$

27.

$$A_{large} = \pi \left(\frac{2.50 \text{ in.}}{2}\right)^2$$
$$= 4.91 \text{ in}^2$$
$$A_{small} = \frac{4.91 \text{ in}^2}{4} = 1.23 \text{ in}^2$$
$$A_{small} = \pi r^2$$
$$\sqrt{\frac{A_{small}}{\pi}} = r$$
$$\sqrt{\frac{1.23 \text{ in}^2}{\pi}} = r$$
$$0.626 \text{ in.} = r$$
$$d = 2 \times 0.626 \text{ in.} = 1.25 \text{ in.}$$

29.

$$A_{rectangle} = (60.00 \text{ cm})(30.00 \text{ cm})$$
$$= 180\overline{0} \text{ cm}^2$$
$$A_{large} = \pi \left(\frac{20.00 \text{ cm}}{2}\right)^2$$
$$= 314.2 \text{ cm}^2$$
$$A_{small} = \pi \left(\frac{14.00 \text{ cm}}{2}\right)^2$$
$$= 153.9 \text{ cm}^2$$
$$A = 180\overline{0} \text{ cm}^2 - 314.2 \text{ cm}^2 - 153.9 \text{ cm}^2$$
$$= 1332 \text{ cm}^2$$

31. The lengths of the long vertical and horizontal straps equal the radius, which is 6 in.

$$C_{curve} = \frac{1}{4}(2\pi r) = \frac{\pi r}{2}$$

$$C = \frac{\pi\left(\dfrac{12.00 \text{ in.}}{2}\right)}{2}$$

$$= 9.425 \text{ in.}$$

$$L = 2(1.50 \text{ in.}) + 2(6.00 \text{ in.}) + 9.425 \text{ in.}$$

$$= 24.43 \text{ in.}$$

33. a. $\dfrac{5.00 \text{ ft}}{2} = 2.50 \text{ ft}$

b. $\sqrt{(2.50 \text{ ft})^2 + (2.50 \text{ ft})^2} = 3.54 \text{ ft}$

35. $\dfrac{360°}{5} = 72°$

37.

$$\angle ACB = \frac{134.8°}{2} = 67.4°$$

$$\angle ABC = 180° - 90° - 67.4°$$

$$= 22.6°$$

$$\angle 1 = 2(22.6°) = 45.2°$$

39.

$$CP = \sqrt{CQ^2 + PQ^2}$$

$$CP = \sqrt{(2000 \text{ mi})^2 + (20{,}500 \text{ mi})^2}$$

$$= 20{,}600 \text{ mi}$$

$$SP = 20{,}600 \text{ mi} - 2000 \text{ mi}$$

$$= 18{,}600 \text{ mi}$$

41.

$$\angle DEB = 180° - 52° = 128°$$

$$128° = \frac{1}{2}\left(85° + \overset{\frown}{CF}\right)$$

$$256 = 85 + \overset{\frown}{CF}$$

$$\overset{\frown}{CF} = 171°$$

43.

$$\angle CEF = 180° - 78° = 102°$$

$$\angle CFE = \frac{1}{2}(142°) = 71°$$

$$\angle DCF = 180° - 102° - 71° = 7°$$

$$\overset{\frown}{DF} = 2(7°) = 14°$$

45. a. $\dfrac{360°}{3} = 120°$

b. $\dfrac{1}{2}(120°) = 60°$

c. $120°$

47. a. $\dfrac{360°}{6} = 60°$

b. $\dfrac{1}{2}(60°) = 30°$

c. $60°$

49. a. 5.6 in^2

b. 4.13 in.

c. 4.13 in.

d. $5.6 \text{ in}^2 \times \dfrac{1.5 \text{ oz}}{1 \text{ in}^2} = 8.4 \text{ oz}$

51.

$$V = A_{circle} \times h \times \frac{7.48 \text{ gal}}{1 \text{ ft}}$$

$$V = \left(\pi(60.0 \text{ ft})^2\right) \times \left(\frac{1}{12} \text{ ft}\right) \times \frac{7.48 \text{ gal}}{1 \text{ ft}}$$

$$= 7050 \text{ gal}$$

53. Rectangular tables will be 156 in. by 90 in. and round tables will be 132 in. in diameter.

$$A_{rectangular} = (156 \text{ in.})(90 \text{ in.})$$

$$A_{round} = \pi\left(\frac{132 \text{ in.}}{2}\right)^2$$

$$A_{difference} = (156 \text{ in.})(90 \text{ in.}) - \pi\left(\frac{132 \text{ in.}}{2}\right)^2$$

$$= 355 \text{ in}^2 = 2.50 \text{ ft}^2$$

The round tables require 2.50 ft^2 less space.

55. a. $10 \times 12 \times \dfrac{1}{3}$ cup $= 40$ cups

b. Each sheet can hold $2 \times 3 = 6$ cookies, so $\dfrac{10 \times 12}{6} = 20$ sheets will be needed.

Section 12.6: Radian Measure

1. $\pi \text{ rad} \times \dfrac{180°}{\pi \text{ rad}} = 180°$

3. $21.0° \times \dfrac{\pi \text{ rad}}{180°} = \dfrac{7\pi}{60} \text{ rad} = 0.367 \text{ rad}$

5. $\dfrac{\pi}{3} \text{ rad} \times \dfrac{180°}{\pi \text{ rad}} = 60°$

7. $135.0° \times \dfrac{\pi \text{ rad}}{180°} = \dfrac{3\pi}{4} \text{ rad} = 2.36 \text{ rad}$

9. $\dfrac{2}{3} \times 2\pi = \dfrac{4\pi}{3} \text{ rad} = 4.19 \text{ rad}$

11. $\dfrac{2}{5} \times 2\pi = \dfrac{4\pi}{5} \text{ rad} = 2.51 \text{ rad}$

13.

$s = r\theta$

$s = (25.0 \text{ cm})\left(\dfrac{2\pi}{5}\right)$

$\quad = 31.4 \text{ cm}$

15. $45.0° \times \dfrac{\pi \text{ rad}}{180°} = \dfrac{\pi}{4} \text{ rad}$

$s = r\theta$

$s = (6.00 \text{ cm})\left(\dfrac{\pi}{4}\right)$

$\quad = 4.71 \text{ cm}$

17. $330.0° \times \dfrac{\pi \text{ rad}}{180°} = \dfrac{11\pi}{6} \text{ rad}$

$s = r\theta$

$s = (18.0 \text{ cm})\left(\dfrac{11\pi}{6}\right)$

$\quad = 104 \text{ cm}$

19.

$s = r\theta$

$\theta = \dfrac{s}{r}$

$\theta = \dfrac{112 \text{ cm}}{40.0 \text{ cm}}$

$\quad = 2.80 \text{ rad}$

21.

$s = r\theta$

$\theta = \dfrac{s}{r}$

$\theta = \dfrac{0.860 \text{ m}}{0.500 \text{ m}}$

$\quad = 1.72 \text{ rad}$

$1.72 \text{ rad} \times \dfrac{180°}{\pi \text{ rad}} = 98.5°$

23.

$s = r\theta$

$r = \dfrac{s}{\theta}$

$r = \dfrac{18.5 \text{ cm}}{\dfrac{2\pi}{3}}$

$\quad = 8.83 \text{ cm}$

25. $\dfrac{10.0 \text{ rad}}{1 \text{ s}} \times \dfrac{1 \text{ rev}}{2\pi \text{ rad}} = \dfrac{5}{\pi} \text{ rps} = 1.59 \text{ rps}$

27. $240.0° \times \dfrac{\pi \text{ rad}}{180°} = \dfrac{4\pi}{3} \text{ rad}$

$s = r\theta$

$s = (22.0 \text{ cm})\left(\dfrac{4\pi}{3}\right)$

$\quad = 92.2 \text{ cm}$

29.

$r = \dfrac{1}{2}(15.2 \text{ cm}) = 7.60 \text{ cm}$

$s = r\theta$

$s = (7.60 \text{ cm})(3.40)$

$\quad = 25.8 \text{ cm}$

31. $60.0° \times \dfrac{\pi \text{ rad}}{180°} = \dfrac{\pi}{3} \text{ rad}$

a.

$s = r\theta$

$s = (25.0 \text{ m})\left(\dfrac{\pi}{3}\right)$

$\quad = 26.2 \text{ m}$

b.

$A = \dfrac{1}{2}r^2\theta$

$A = \dfrac{1}{2}(25.0 \text{ m})^2\left(\dfrac{\pi}{3}\right)$

$\quad = 327 \text{ m}^2$

c.

$\qquad c = 25.0 \text{ (equilateral triangle)}$

$A_{triangle} = \dfrac{c\sqrt{4r^2 - c^2}}{4}$

$A_{triangle} = \dfrac{(25.0 \text{ m})\sqrt{4(25.0 \text{ m})^2 - (25.0 \text{ m})^2}}{4}$

$\qquad = 271 \text{ m}^2$

$A_{sector} = 327 \text{ m}^2 - 271 \text{ m}^2 = 56 \text{ m}^2$

33.

$r = \dfrac{1}{2}(12.0 \text{ in.}) = 6.00 \text{ in.}$

$\theta = \dfrac{2\pi \text{ rad}}{8} = \dfrac{\pi}{4} \text{ rad}$

$A = \dfrac{1}{2}r^2\theta$

$A = \dfrac{1}{2}(6.00 \text{ in.})^2\left(\dfrac{\pi}{4}\right)$

$\quad = 14.1 \text{ in}^2$

Section 12.7: Prisms

1.

a.

$A_{front} = (16.0 \text{ in.})(10.0 \text{ in.}) = 16\overline{0} \text{ in}^2$

$A_{back} = (15.0 \text{ in.})(10.0 \text{ in.}) = 15\overline{0} \text{ in}^2$

$A_{right} = (13.0 \text{ in.})(10.0 \text{ in.}) = \underline{13\overline{0} \text{ in}^2}$

$\quad A = \qquad\qquad\qquad 440 \text{ in}^2$

b.

$B = \dfrac{1}{2}(16.0 \text{ in.})(11.4 \text{ in.}) = 91.2 \text{ in}^2$

$A = 440 \text{ in}^2 + 2(91.2 \text{ in}^2) = 622 \text{ in}^2$

c.

$V = Bh$

$V = (91.2 \text{ in}^2)(10.0 \text{ in.})$

$V = 912 \text{ in}^3$

3.

$$B = \frac{1}{2}(3.0 \text{ in.})(4.0 \text{ in.}) = 6.0 \text{ in}^2$$

$$A_{lateral} = (3.0 \text{ in.})(6.0 \text{ in.})$$
$$+(4.0 \text{ in.})(6.0 \text{ in.})$$
$$+(5.0 \text{ in.})(6.0 \text{ in.})$$
$$= 72 \text{ in}^2$$

$$A_{surface} = 72 \text{ in}^2 + 2(6.0 \text{ in}^2) = 84 \text{ in}^2$$

$$V = Bh$$

$$V = (6.0 \text{ in}^2)(6.0 \text{ in.}) = 36 \text{ in}^3$$

5.

$$V = lwh$$

$$V = (3.0 \text{ ft})(4.0 \text{ ft})(5.0 \text{ ft})$$
$$= 6\bar{0} \text{ ft}^3$$

7. a. $A = 2(42.0 \text{ m})(17.0 \text{ m}) + 2(28.0 \text{ m})(17.0 \text{ m}) + 2\left[\frac{1}{2}2(28.0 \text{ m})(4.00 \text{ m})\right] = 2490 \text{ m}^2$

   b. $A = 2(42.0 \text{ m})(14.6 \text{ m}) = 1230 \text{ m}^2$

   c. $V = (42.0 \text{ m})(28.0 \text{ m})\left(16.0 \text{ cm} \times \dfrac{1 \text{ m}}{100 \text{ cm}}\right) = 188 \text{ m}^3$

   d. $A = 2490 \text{ m}^2 + 1230 \text{ m}^2 + (28.0 \text{ m})(4.00 \text{ m}) = 490\bar{0} \text{ m}^2$

9.

$$V_{cube} = (6.00 \text{ ft})^3 = 216 \text{ ft}^3$$

$$A_{triangle} = \frac{1}{2}(6.00 \text{ ft})(6.00 \text{ ft}) = 18.0 \text{ ft}^2$$

$$V = Bh$$

$$V = (18.0 \text{ ft}^2)(6.00 \text{ ft}) = 108 \text{ ft}^3$$

$$V_{bin} = 216 \text{ ft}^3 + 108 \text{ ft}^3 = 324 \text{ ft}^3$$

11.

$$42.0 \text{ lb} \times \frac{1 \text{ in}^3}{0.28 \text{ lb}} = 150 \text{ in}^3$$

$$V = Bh$$

$$\frac{V}{B} = h$$

$$\frac{150 \text{ in}^3}{5.0 \text{ in}^2} = h$$

$$3\bar{0} \text{ in.} = h$$

13.

$$l_{sheet} = 3(4.00 \text{ ft}) + 1 \text{ in.} = 12 \text{ ft } 1 \text{ in.}$$

$$h_{triangle} = \sqrt{(4.00 \text{ ft})^2 - (2.00 \text{ ft})^2}$$
$$= 3.46 \text{ ft}$$

$$h_{sheet} = 2(3.46 \text{ ft}) + 3.00 \text{ ft} = 9.92 \text{ ft}$$
$$= 9 \text{ ft } 11 \text{ in.}$$

15.

$$l_{sheet} = 4(28 \text{ in}) + 1 \text{ in} = 113 \text{ in}$$

$$h_{sheet} = 2(28 \text{ in}) + 32 \text{ in} + 2(1 \text{ in}) = 9\bar{0} \text{ in}$$

17.

$$V = (40.0 \text{ ft})(80.0 \text{ ft})(4.00 \text{ ft})$$
$$= 12{,}\bar{0}00 \text{ ft}^3$$

$$V = 12{,}\bar{0}00 \text{ ft}^3 \times \frac{62.2 \text{ lb}}{1 \text{ ft}^3} \times \frac{1 \text{ gal}}{8.34 \text{ lb}}$$
$$= 95{,}500 \text{ gal}$$

$$\frac{95{,}500 \text{ gal}}{12{,}000 \text{ gal}} = 8 \text{ scuppers}$$

19.

$$B = \left(\frac{3.00 \text{ ft} + 6.00 \text{ ft}}{2}\right)(40.0 \text{ ft})$$
$$= 180 \text{ ft}^2$$

$$V = Bh$$

$$V = (180 \text{ ft}^2)(20.0 \text{ ft})$$
$$= 360\bar{0} \text{ ft}^3$$

21. a.

$$V_{one\ layer} = (12\ \text{in.})(18\ \text{in.})(2\ \text{in.})\left(\frac{2}{3}\right) = 288\ \text{in}^3$$

$$V_{cakes} = 8 \times 288\ \text{in}^3 \times \frac{1\ \text{cup}}{14.4\ \text{in}^3} = 160\ \text{cups}$$

b. Note: The heights of the sides are increased to compensate for the thickness of the frosting. For one cake:

$$V = V_{top\ and\ middle} + V_{sides} + V_{front/back}$$

$$= (12\ \text{in.})(18\ \text{in.})(0.5\ \text{in.}+0.25\ \text{in.}) + 2(12\ \text{in.})(4.25\ \text{in.})(0.5\ \text{in.}) + 2(18\ \text{in.})(4.25\ \text{in.})(0.5\ \text{in.})$$

$$= 289\ \text{in}^3$$

For all four cakes:

$$V = 4 \times 289\ \text{in}^3 \times \frac{1\ \text{cup}}{14.4\ \text{in}^3} = 80\ \text{cups}$$

Section 12.8: Cylinders

1.

$$V = \pi r^2 h$$

$$V = \pi (12.0\ \text{mm})^2 (30.0\ \text{mm})$$

$$= 13,600\ \text{mm}^3$$

3.

$$V = \pi r^2 h$$

$$V = \pi (8.20\ \text{m})^2 (39.2\ \text{m})$$

$$= 8280\ \text{m}^3$$

$$V = 8280\ \text{m}^3 \times \frac{1000\ \text{L}}{1\ \text{m}^3} = 8,280,000\ \text{L}$$

5. $400,\overline{0}00\ \text{gal} \times \dfrac{1\ \text{ft}^3}{7.48\ \text{gal}} = 53,500\ \text{ft}^3$

$$V = \pi r^2 h$$

$$h = \frac{V}{\pi r^2}$$

$$h = \frac{53,500\ \text{ft}^3}{\pi (20.0\ \text{ft})^2}$$

$$h = 42.6\ \text{ft}$$

7.

$$V_{outer} = \pi r^2 h$$

$$V_{outer} = \pi (5.00\ \text{in.})^2 (2.00\ \text{in.})$$

$$= 157\ \text{in}^3$$

$$V_{inner} = \pi r^2 h$$

$$V_{inner} = \pi (4.00\ \text{in.})^2 (2.00\ \text{in.})$$

$$= 101\ \text{in}^3$$

$$V_{filter} = 157\ \text{in}^3 - 101\ \text{in}^3$$

$$= 56.0\ \text{in}^3$$

9.

$$25\ \text{ft}\ 9\ \text{in.} = 25.75\ \text{ft}$$

$$7\ \text{ft}\ 6\ \text{in.} = 7.5\ \text{ft}$$

$$V = \pi r^2 h$$

$$V = \pi \left(\frac{7.5\ \text{ft}}{2}\right)^2 (25.75\ \text{ft})$$

$$= 1140\ \text{ft}^3$$

11.

$$V = \pi r^2 h$$

$$r = \sqrt{\frac{V}{\pi h}}$$

$$r = \sqrt{\frac{25.3 \text{ in}^3}{\pi (10.0 \text{ in.})}}$$

$$= 0.897 \text{ in.}$$

$$d = 2(0.897 \text{ in.})$$

$$= 1.79 \text{ in.}$$

13.

$$12 \text{ ft} = 144 \text{ in.}$$

$$V = \pi r^2 h$$

$$V = \pi \left(\frac{0.5 \text{ in.}}{2}\right)^2 (144 \text{ in.})$$

$$= 28.3 \text{ in}^3$$

15. a.

$$A_{lateral} = 2\pi rh$$

$$A_{lateral} = 2\pi (8.21 \text{ mm})(39.7 \text{ mm})$$

$$= 2050 \text{ mm}^2$$

b.

$$A_{ends} = 2 \times \pi r^2$$

$$A_{ends} = 2 \times \pi (8.21 \text{ mm})^2$$

$$= 424 \text{ mm}^2$$

$$A_{surface} = 2050 \text{ mm}^2 + 424 \text{ mm}^2$$

$$= 2470 \text{ mm}^2$$

17.

$$A_{can} = 2\pi rh$$

$$A_{can} = 2\pi (2.18 \text{ cm})(7.38 \text{ cm})$$

$$= 101 \text{ cm}^2$$

$$A_{total} = 1000(101 \text{ cm}^2)$$

$$= 101,000 \text{ cm}^2$$

19.

$$V_{outer} = \pi r^2 h$$

$$V_{outer} = \pi \left(\frac{3.10 \text{ in.}}{2}\right)^2 (5.00 \text{ in.})$$

$$= 37.4 \text{ in}^3$$

$$V_{inner} = \pi r^2 h$$

$$V_{inner} = \pi \left(\frac{2.24 \text{ in.}}{2}\right)^2 (5.00 \text{ in.})$$

$$= 19.7 \text{ in}^3$$

$$V_{removed} = 37.4 \text{ in}^3 - 19.7 \text{ in}^3$$

$$= 17.7 \text{ in}^3$$

21.

$$A_{old} = 2\pi rh$$

$$A_{old} = 2\pi \left(\frac{2.78 \text{ in.}}{2}\right)(5.50 \text{ in.})$$

$$= 48.0 \text{ in}^2$$

$$A_{new} = 2\pi \left(\frac{2.86 \text{ in.}}{2}\right)(5.50 \text{ in.})$$

$$= 49.4 \text{ in}^2$$

$$A_{increase} = 49.4 \text{ in}^2 - 48.0 \text{ in}^2$$

$$= 1.40 \text{ in}^2$$

23.

$$A = 2\pi rh$$

$$A = 2\pi \left(\frac{15 \text{ ft}}{2}\right)(26 \text{ ft})$$

$$= 1225 \text{ ft}^2$$

$$\frac{1225 \text{ ft}^2}{200 \text{ ft}^2} = 6.1 \text{ gal}$$

25.

$$A_{lateral} = 2\pi rh$$

$$A_{lateral} = 2\pi\left(\frac{2.38 \text{ cm}}{2}\right)(7.22 \text{ cm})$$

$$= 54.0 \text{ cm}^2$$

$$A_{ends} = 2 \times \pi r^2$$

$$A_{ends} = 2 \times \pi\left(\frac{2.38 \text{ cm}}{2}\right)^2$$

$$= 8.90 \text{ cm}^2$$

$$A_{can} = 54.0 \text{ cm}^2 + 8.90 \text{ cm}^2$$

$$= 62.9 \text{ cm}^2$$

$$Weight = 2,700,000\left(62.9 \text{ cm}^2\right)\left(\frac{0.000147 \text{ g}}{1 \text{ cm}^2}\right)$$

$$= 25,000 \text{ g} = 25 \text{ kg}$$

27. a.

$$r = 37.0 \text{ in.} - 3.00 \text{ in.} = 34.0 \text{ in.}$$

$$d = 2(34.0 \text{ in.}) = 68.0 \text{ in.}$$

b.

$$V = \pi r^2 h$$

$$V_{outer} = \pi\left(37.0 \text{ in.} \times \frac{1 \text{ ft}}{12 \text{ in.}}\right)^2 (20.0 \text{ ft})$$

$$= 597 \text{ ft}^3$$

$$V_{inner} = \pi\left(34.0 \text{ in.} \times \frac{1 \text{ ft}}{12 \text{ in.}}\right)^2 (20.0 \text{ ft})$$

$$= 504 \text{ ft}^3$$

$$V_{concrete} = 597 \text{ ft}^3 - 504 \text{ ft}^3$$

$$= 93.0 \text{ ft}^3$$

$$W_{concrete} = 93.0 \text{ ft}^3 \times \frac{148 \text{ lb}}{1 \text{ ft}^3} = 13,800 \text{ lb}$$

c.

$$A = 2\pi rh$$

$$A = 2\pi\left(37.0 \text{ in.} \times \frac{1 \text{ ft}}{12 \text{ in.}}\right)(20.0 \text{ ft})$$

$$= 387 \text{ ft}^2$$

$$Cost = 387 \text{ ft}^2 \times \frac{1 \text{ gal}}{186 \text{ ft}^2}$$

$$= 2.08 \text{ gal}$$

d.

$$W_{concrete} = 597 \text{ ft}^3 \times \frac{148 \text{ lb}}{1 \text{ ft}^3}$$

$$= 88,400 \text{ lb}$$

29. The volume of one tank is

$$\frac{1}{2}(22,020 \text{ gal}) \times \frac{231 \text{ in}^3}{1 \text{ gal}} \times \left(\frac{1 \text{ ft}}{12 \text{ in.}}\right)^3$$

$$= 1472 \text{ ft}^3$$

$$V = \pi r^2 h$$

$$h = \frac{V}{\pi r^2}$$

$$h = \frac{1472 \text{ ft}^3}{\pi\left(\dfrac{11.0 \text{ ft}}{2}\right)^2}$$

$$= 15.5 \text{ ft}$$

31.

$$V = \pi r^2 h$$

$$V = \pi\left(\frac{20.0 \text{ ft}}{2}\right)^2 (40.0 \text{ ft})$$

$$= 12,600 \text{ ft}^3$$

$$V = 12,600 \text{ ft}^3 \times \frac{1 \text{ bu}}{1.2445 \text{ ft}^3}$$

$$= 10,100 \text{ bu}$$

33.

$$h = 20.0 \text{ in.} - 2.0 \text{ in.} = 18.0 \text{ in.}$$

$$V = \pi r^2 h$$

$$V = \pi \left(\frac{18.0 \text{ in.}}{2}\right)^2 (18.0 \text{ in.})$$

$$= 4580 \text{ in}^3$$

$$V = 4580 \text{ in}^3 \times \frac{1 \text{ cup}}{14.4 \text{ in}^3}$$

$$= 318 \text{ cups}$$

35. Each tier will be $\frac{2}{3}$ full of batter before baking.

$$V_{cake} = V_{top} + V_{middle} + V_{bottom}$$

$$V_{cake} = \pi \left(\frac{8 \text{ in.}}{2}\right)^2 (3 \text{ in.})(2)\left(\frac{2}{3}\right) + \pi \left(\frac{12 \text{ in.}}{2}\right)^2 (3 \text{ in.})(2)\left(\frac{2}{3}\right) + \pi \left(\frac{16 \text{ in.}}{2}\right)^2 (3 \text{ in.})(2)\left(\frac{2}{3}\right)$$

$$= 1460 \text{ in}^3$$

$$V_{cake} = 1460 \text{ in}^3 \times \frac{1 \text{ cup}}{14.4 \text{ in}^3} = 101 \text{ cups}$$

Section 12.9: Pyramids and Cones

1.

$$B = (8.10 \text{ in.})^2 = 65.6 \text{ in}^2$$

$$V = \frac{1}{3}Bh$$

$$V = \frac{1}{3}(65.6 \text{ in}^2)(6.70 \text{ in})$$

$$= 147 \text{ in}^3$$

3.

$$B = (10.8 \text{ m})(18.8 \text{ m})$$

$$= 203 \text{ m}^2$$

$$V = \frac{1}{3}Bh$$

$$V = \frac{1}{3}(203 \text{ m}^2)(16.2 \text{ m})$$

$$= 1100 \text{ m}^3$$

5.

$$B = \frac{1}{2}(101 \text{ mm})(115 \text{ mm})$$

$$= 5810 \text{ mm}^2$$

$$V = \frac{1}{3}Bh$$

$$V = \frac{1}{3}(5810 \text{ mm}^2)(36.0 \text{ mm})$$

$$= 69,700 \text{ mm}^3$$

7.

$$s = \frac{1}{2}(15.4 \text{ mm} + 24.6 \text{ mm} + 19.2 \text{ mm}) = 29.6 \text{ mm}$$

$$B = \sqrt{(29.6 \text{ mm})(29.6 \text{ mm} - 15.4 \text{ mm})(29.6 \text{ mm} - 24.6 \text{ mm})(29.6 \text{ mm} - 19.2 \text{ mm})}$$

$$= 148 \text{ mm}^2$$

$$V = \frac{1}{3}Bh$$

$$V = \frac{1}{3}(148 \text{ mm}^2)(29.2 \text{ mm})$$

$$= 1440 \text{ mm}^3$$

9.

$$B = (6.00 \text{ ft})(6.00 \text{ ft}) = 36.0 \text{ ft}^2$$

$$V_{prism} = Bh = (36.0 \text{ ft}^2)(6.00 \text{ ft})$$

$$= 216 \text{ ft}^2$$

$$V_{pyramid} = \frac{1}{3}Bh$$

$$V_{pyramid} = \frac{1}{3}(36.0 \text{ ft}^2)(8.00 \text{ ft})$$

$$= 96.0 \text{ ft}^3$$

$$V = 216 \text{ ft}^2 + 96.0 \text{ ft}^3 = 312 \text{ ft}^3$$

11. a.

$$r = \frac{1}{2}(16.0 \text{ cm}) = 8.00 \text{ cm}$$

$$V = \frac{1}{3}\pi r^2 h$$

$$V = \frac{1}{3}\pi (8.00 \text{ cm})^2 (15.0 \text{ cm})$$

$$= 1010 \text{ cm}^3$$

b.

$$s = \sqrt{(8.00 \text{ cm})^2 + (15.0 \text{ cm})^2}$$

$$= 17.0 \text{ cm}$$

$$A = \pi rs$$

$$A = \pi (8.00 \text{ cm})(17.0 \text{ cm})$$

$$= 427 \text{ cm}^2$$

13.

$$V = \frac{1}{3}\pi r^2 h$$

$$V = \frac{1}{3}\pi (10.0 \text{ ft})^2 (18.0 \text{ ft}) \times \frac{0.804 \text{ bu}}{1 \text{ ft}^3}$$

$$= 1520 \text{ bu}$$

15.

$$A = \pi rs$$

$$A = \pi (5.50 \text{ in.})(13.0 \text{ in.})$$

$$= 225 \text{ in}^2$$

17.

$$C = 2\pi r$$

$$224 \text{ ft} = 2\pi r$$

$$r = \frac{224 \text{ ft}}{2\pi} = 35.7 \text{ ft}$$

$$h = \sqrt{(45 \text{ ft})^2 - (35.7 \text{ ft})^2} = 27.4 \text{ ft}$$

$$V = \frac{1}{3}\pi r^2 h$$

$$V = \frac{1}{3}\pi (35.7 \text{ ft})^2 (27.4 \text{ ft})$$

$$= 37,000 \text{ ft}^3$$

$$W = 37,000 \text{ ft}^3 \times \left(\frac{1 \text{ yd}}{3 \text{ ft}}\right)^3 \times \frac{3200 \text{ lb}}{1 \text{ ft}^3}$$

$$= 4,390,000 \text{ lb}$$

$$22 \text{ tons} \times \frac{2000 \text{ lb}}{1 \text{ ton}} = 44,000 \text{ lb}$$

$$\frac{4,390,000 \text{ lb}}{44,000 \text{ lb}} = \overline{1}00 \text{ truckloads}$$

19.

$$B_1 = (1.32 \text{ m})(0.750 \text{ m}) = 0.990 \text{ m}^2$$

$$B_2 = (2.35 \text{ m})(4.15 \text{ m}) = 9.75 \text{ m}^2$$

$$V = \frac{1}{3}h\left(B_1 + B_2 + \sqrt{B_1 B_2}\right)$$

$$V = \frac{1}{3}(3.25 \text{ m})\left(0.990 \text{ m}^2 + 9.75 \text{ m}^2 + \sqrt{\left(0.990 \text{ m}^2\right)\left(9.75 \text{ m}^2\right)}\right)$$

$$= 15.0 \text{ m}^3$$

21.

$$B_1 = \pi(15.0 \text{ ft})^2 = 707 \text{ ft}^2$$

$$B_2 = \pi(25.0 \text{ ft})^2 = 1960 \text{ ft}^2$$

$$V_{frustum} = \frac{1}{3}h\left(B_1 + B_2 + \sqrt{B_1 B_2}\right)$$

$$V_{frustum} = \frac{1}{3}(22.9 \text{ ft})\left(707 \text{ ft}^2 + 1960 \text{ ft}^2 + \sqrt{\left(707 \text{ ft}^2\right)\left(1960 \text{ ft}^2\right)}\right)$$

$$= 29,300 \text{ ft}^3$$

$$V_{cylinder} = \pi r^2 h$$

$$V_{cylinder} = \pi r^2 h$$

$$V_{cylinder} = \pi(25.0 \text{ ft})^2(40.0 \text{ ft})$$

$$= 78,500 \text{ ft}^3$$

$$V = 29,300 \text{ ft}^3 + 78,500 \text{ ft}^3 = 108,000 \text{ ft}^3$$

$$A_{frustum} = \pi s(r_1 + r_2)$$

$$A_{frustum} = \pi(25.0 \text{ ft})(15.0 \text{ ft} + 25.0 \text{ ft})$$

$$= 3140 \text{ ft}^2$$

$$A_{cylinder} = 2\pi rh$$

$$A_{cylinder} = 2\pi(25.0 \text{ ft})(40.0 \text{ ft})$$

$$= 6280 \text{ ft}^2$$

$$A_{bin} = 3140 \text{ ft}^2 + 6280 \text{ ft}^2$$

$$= 9420 \text{ ft}^2$$

23.

$$5.00 \text{ ft}^3 \times \left(\frac{12 \text{ in.}}{1 \text{ ft}}\right)^3 = 8640 \text{ in}^3$$

$$V_{cylinder} = \pi r^2 h$$

$$V_{cylinder} = \pi \left(\frac{18.0 \text{ in.}}{2}\right)^2 h$$

$$= \left(81\pi \text{ in}^2\right) h$$

$$B_1 = \pi \left(\frac{18.0 \text{ in.}}{2}\right)^2 = 254 \text{ in}^2$$

$$B_2 = \pi \left(\frac{3.00 \text{ in.}}{2}\right)^2 = 7.07 \text{ in}^2$$

$$V_{frustum} = \frac{1}{3} h \left(B_1 + B_2 + \sqrt{B_1 B_2}\right)$$

$$V_{frustum} = \frac{1}{3}(15.0 \text{ in.}) \left(254 \text{ in}^2 + 7.07 \text{ in}^2 + \sqrt{\left(254 \text{ in}^2\right)\left(7.07 \text{ in}^2\right)}\right)$$

$$= 1520 \text{ in}^3$$

$$V_{bin} = V_{cylinder} + V_{frustum}$$

$$8640 \text{ in}^3 = \left(81\pi \text{ in}^2\right) h + 1520 \text{ in}^3$$

$$h = 28.0 \text{ in}$$

25. 33.9 ft^3

27.

$$A_{old} = \pi s \left(r_1 + r_2\right)$$

$$A_{old} = \pi (15.0 \text{ in.}) \left(\frac{15.0 \text{ in.}}{2} + \frac{13.5.0 \text{ in.}}{2}\right)$$

$$= 672 \text{ in}^2$$

$$A_{new} = \pi s \left(r_1 + r_2\right)$$

$$A_{new} = \pi (14.0 \text{ in.}) \left(\frac{14.0 \text{ in.}}{2} + \frac{12.5.0 \text{ in.}}{2}\right)$$

$$= 583 \text{ ft}^2$$

$$Change = \frac{672 \text{ in}^2 - 583 \text{ ft}^2}{672 \text{ in}^2} = 13.2\%$$

Yes, 13.2% was saved, so at least 10% was saved.

29.

$$B = (26.0 \text{ in.})^2 = 676 \text{ in}^2$$

$$V = \frac{1}{3} B h$$

$$V = \frac{1}{3} \left(676 \text{ in}^2\right)(128 \text{ in.}) \times \frac{1 \text{ gal}}{231 \text{ in}^3}$$

$$= 125 \text{ gal}$$

Section 12.10: Spheres

1. a.

$$A = 4\pi r^2$$
$$A = 4\pi (8.00 \text{ in.})^2$$
$$= 804 \text{ in}^2$$

b.

$$V = \frac{4}{3}\pi r^3$$
$$V = \frac{4}{3}\pi (8.00 \text{ in.})^3$$
$$= 2140 \text{ in}^3$$

3. a.

$$r = \frac{1}{2}d = \frac{1}{2}(36.2 \text{ in.}) = 18.1 \text{ in.}$$
$$A = 4\pi r^2$$
$$A = 4\pi (18.1 \text{ in.})^2$$
$$= 4120 \text{ in}^2$$

b.

$$V = \frac{4}{3}\pi r^3$$
$$V = \frac{4}{3}\pi (18.1 \text{ in.})^3$$
$$= 24,800 \text{ in}^3$$

5.

$$V = \frac{4}{3}\pi r^3$$
$$V = \frac{4}{3}\pi (30.1 \text{ m})^3$$
$$= 114,000 \text{ m}^3$$

7.

$$V = \frac{1}{2} \times \frac{4}{3}\pi r^3$$
$$V = \frac{1}{2} \times \frac{4}{3}\pi (9.00 \text{ in.})^3$$
$$= 1530 \text{ in}^3$$

9.

$$r = \frac{1}{2}d = \frac{1}{2}(6.00 \text{ in.})$$
$$= 3.00 \text{ in.}$$
$$V_{old} = \frac{4}{3}\pi r^3$$
$$V_{old} = \frac{4}{3}\pi (3.00 \text{ in.})^3$$
$$= 113 \text{ in}^3$$
$$V_{new} = 2 \times 113 \text{ in}^3 = 226 \text{ in}^3$$
$$V_{new} = \frac{4}{3}\pi r^3$$
$$226 \text{ in}^3 = \frac{4}{3}\pi r^3$$
$$\frac{3 \cdot 226 \text{ in}^3}{4\pi} = r^3$$
$$\sqrt[3]{\frac{3 \cdot 226 \text{ in}^3}{4\pi}} = r$$
$$3.78 \text{ in.} = r$$
$$d = 2r = 2(3.78 \text{ in.}) = 7.56 \text{ in.}$$

11.

$$r = \frac{1}{2}d = \frac{1}{2}(62.0 \text{ ft}) = 31.0 \text{ ft}$$
$$V = \frac{4}{3}\pi r^3$$
$$V = \frac{4}{3}\pi (31.0 \text{ ft})^3 \times \frac{7.48 \text{ gal}}{1 \text{ ft}^3}$$
$$= 933,000 \text{ gal}$$

13. a.

$$r = \frac{1}{2}d = \frac{1}{2}(16.0 \text{ in.}) = 8.00 \text{ in.}$$

$$\frac{A}{V} = \frac{4\pi(8.00 \text{ in.})^2}{\frac{4}{3}\pi(8.00 \text{ in.})^3} = 0.375 = \frac{3}{8}$$

b.

$$r = \frac{1}{2}d = \frac{1}{2}(24.0 \text{ in.}) = 12.0 \text{ in.}$$

$$\frac{A}{V} = \frac{4\pi(12.0 \text{ in.})^2}{\frac{4}{3}\pi(12.0 \text{ in.})^3} = 0.250 = \frac{1}{4}$$

c. $\dfrac{A}{V} = \dfrac{4\pi r^2}{\frac{4}{3}\pi r^3} = \dfrac{3}{r}$

17.

$$r = \frac{1}{2}d = \frac{1}{2}(9.0 \text{ ft}) = 4.5 \text{ ft}$$

$$V = \frac{4}{3}\pi r^3$$

$$V = \frac{4}{3}\pi(4.5 \text{ ft})^3$$

$$= 380 \text{ ft}^3$$

15. a.

$$A = \frac{1}{2} \times 4\pi r^2$$

$$A = \frac{1}{2} \times 4\pi\left(\frac{40.0 \text{ ft}}{2}\right)^2$$

$$= 2510 \text{ ft}^2$$

$$A_{stucco} = 85\% \times 2510 \text{ ft}^2 = 2140 \text{ ft}^2$$

b.

$$V_{hemisphere} = \frac{1}{2} \times \frac{4}{3}\pi r^3$$

$$V_{hemisphere} = \frac{1}{2} \times \frac{4}{3}\pi\left(\frac{40.0 \text{ ft}}{2}\right)^3$$

$$= 16{,}800 \text{ ft}^3$$

c.

$$V_{outer} = \frac{1}{2} \times \frac{4}{3}\pi\left(\frac{40.0 \text{ ft}}{2}\right)^3$$

$$V_{inner} = \frac{1}{2} \times \frac{4}{3}\pi\left(\frac{38.0 \text{ ft}}{2}\right)^3$$

$$V_{concrete} = \frac{2}{3}\pi\left((20.0 \text{ ft})^3 - (19.0 \text{ ft})^3\right)$$

$$Weight = 0.85 \times V_{concrete} \times \frac{148}{1 \text{ ft}^3}$$

$$= 301{,}000 \text{ lb}$$

d. $2\left(\dfrac{2140 \text{ ft}^2}{110 \text{ ft}^2}\right) = 39 \text{ gal}$

Chapter 12 Review

1. acute angle

2. obtuse angle

3.

$$\angle 2 = \angle 5 = 121°$$

$$\angle 4 = 180° - 121° = 59°$$

$$\angle 3 = \angle 4 = 59°$$

$$\angle 1 = \angle 3 = 59°$$

4. adjacent or supplementary

5.

$$\angle 1 + \angle 2 = 180°$$

$$(4x+5) + (2x+55) = 180$$

$$6x + 60 = 180$$

$$6x = 120$$

$$x = 20°$$

6. a. quadrilateral

b. pentagon

c. hexagon

d. triangle

e. octagon

7.

$$P = 2(a+b)$$

$$P = 2(6.00\text{cm} + 12.00 \text{ cm})$$

$$= 36 \text{ cm}$$

$$A = bh$$

$$A = (12.00 \text{ cm})(5.00\text{cm})$$

$$= 60 \text{ cm}^2$$

8.

$P = 2(a+b)$

$P = 2(79.2 \text{ m} + 101.0 \text{ m})$

$\quad = 360 \text{ m}$

$A = bh$

$A = (101.0 \text{ m})(79.2 \text{ m})$

$\quad = 80\overline{0}0 \text{ m}^2$

9.

$P = a+b+c+d$

$P = 10.21 \text{ cm} + 15.63 \text{ cm} + 7.91 \text{ cm} + 9.59 \text{ cm}$

$\quad = 43.34 \text{ cm}$

$A = \left(\dfrac{a+b}{2}\right)h$

$A = \left(\dfrac{10.21 \text{ cm} + 15.63 \text{ cm}}{2}\right)(7.91 \text{ cm})$

$\quad = 102 \text{ cm}^2$

10.

$A = lw$

$w = \dfrac{A}{l}$

$w = \dfrac{79.60 \text{ cm}^2}{10.3 \text{ cm}}$

$\quad = 7.73 \text{ cm}$

11.

$A = bh$

$h = \dfrac{A}{b}$

$h = \dfrac{2.53 \text{ cm}^2}{1.76 \text{ cm}}$

$\quad = 1.44 \text{ cm}$

12.

$P = a+b+c$

$P = (16.5 \text{ m}) + (21.9 \text{ m}) + (20.3 \text{ m})$

$\quad = 58.7 \text{ m}$

$A = \dfrac{1}{2}bh$

$A = \dfrac{1}{2}(20.3 \text{ m})(15.7 \text{ m})$

$\quad = 159 \text{ m}^2$

13.

$P = a+b+c$

$P = (46.0 \text{ m}) + (73.9 \text{ m}) + (39.6 \text{ m})$

$\quad = 159.5 \text{ m}$

$A = \dfrac{1}{2}bh$

$A = \dfrac{1}{2}(39.6 \text{ m})(40.2 \text{ m})$

$\quad = 796 \text{ m}^2$

14.

$P = a+b+c$

$P = (2.42 \text{ cm}) + (4.60 \text{ cm}) + (3.91 \text{ cm})$

$\quad = 10.93 \text{ cm}$

$A = \dfrac{1}{2}bh$

$A = \dfrac{1}{2}(3.91 \text{ cm})(2.42 \text{ cm})$

$\quad = 4.73 \text{ cm}^2$

15.

$c = \sqrt{a^2 + b^2}$

$c = \sqrt{(29.1 \text{ m})^2 + (30.2 \text{ m})^2}$

$\quad = 41.9 \text{ m}$

16.

$c = \sqrt{a^2 + b^2}$

$c = \sqrt{(1.72 \text{ cm})^2 + (3.44 \text{ cm})^2}$

$\quad = 3.85 \text{ cm}$

17. $180° - 58° - 36° = 86°$

18.

$\dfrac{BC}{DE} = \dfrac{AB}{AE}$

$\dfrac{BC}{4.0} = \dfrac{8.0}{4.0}$

$4.0(BC) = (4.0)(8.0)$

$BC = \dfrac{(4.0)(8.0)}{4.0} = 8.0 \text{ m}$

19.

$$r = \frac{1}{2}(23.2 \text{ cm}) = 11.6 \text{ cm}$$

$$A = \pi r^2$$

$$A = \pi (11.6 \text{ cm})^2 = 423 \text{ cm}^2$$

$$C = 2\pi r$$

$$C = 2\pi (11.6 \text{ cm}) = 72.9 \text{ cm}$$

20.

$$A = \pi r^2$$

$$462 \text{ cm}^2 = \pi r^2$$

$$\sqrt{\frac{462 \text{ cm}^2}{\pi}} = r$$

$$12.1 \text{ cm} = r$$

21. $\dfrac{3}{5} \times 360° = 216°$

22. $24.0° \times \dfrac{\pi \text{ rad}}{180°} = \dfrac{2\pi}{15} \text{ rad} = 0.419 \text{ rad}$

23. $\dfrac{\pi}{18} \text{ rad} \times \dfrac{180°}{\pi \text{ rad}} = 10°$

24.

$$s = r\theta$$

$$s = (75.3 \text{ cm})(0.561 \text{ rad})$$

$$= 42.2 \text{ cm}$$

25.

$$r = \frac{1}{2}(25.8 \text{ cm}) = 12.9 \text{ cm}$$

$$s = r\theta$$

$$\theta = \frac{s}{r}$$

$$s = \frac{20.0 \text{ cm}}{12.9 \text{ cm}}$$

$$= 1.55 \text{ rad}$$

26.

$$1028° \times \frac{\pi \text{ rad}}{180°} = \frac{257\pi}{45} \text{ rad}$$

$$s = r\theta$$

$$s = (16.2 \text{ cm})\left(\frac{257\pi}{45}\right)$$

$$= 291 \text{ cm}$$

27. a.

$$A_{front} = (18.0 \text{ m})(32.0 \text{ m}) = 576 \text{ m}^2$$

$$A_{side} = (32.0 \text{ m})(10.0 \text{ m}) = 320 \text{ m}^2$$

$$A = 2 \times \left(A_{front} + A_{side}\right)$$

$$A = 2 \times \left(576 \text{ m}^2 + 320 \text{ m}^2\right)$$

$$= 1790 \text{ m}^2$$

b.

$$A_{bottom} = (18.0 \text{ m})(10.0 \text{ m}) = 18\overline{0} \text{ m}^2$$

$$A = 2 \times \left(A_{front} + A_{side} + A_{bottom}\right)$$

$$A = 2 \times \left(576 \text{ m}^2 + 320 \text{ m}^2 + 18\overline{0} \text{ m}^2\right)$$

$$= 2150 \text{ m}^2$$

28.

$$B = (18.0 \text{ m})(10.0 \text{ m}) = 18\overline{0} \text{ m}^2$$

$$V = Bh$$

$$V = \left(18\overline{0} \text{ m}^2\right)(32.0 \text{ m})$$

$$= 5760 \text{ m}^3$$

29.

$$V_{cylinder} = \pi r^2 h$$

$$V_{cylinder} = \pi (5.62 \text{ cm})^2 (12.7 \text{ cm})$$

$$= 1260 \text{ cm}^3$$

30.

$$r = \frac{1}{2}(4.00 \text{ cm}) = 2.00 \text{ cm}$$

$$V_{cylinder} = \pi r^2 h$$

$$V_{cylinder} = \pi (2.00 \text{ cm})^2 (5.60 \text{ cm})$$

$$= 70.4 \text{ cm}^3$$

31. a.
$$r = \frac{1}{2}(4.00 \text{ cm}) = 2.00 \text{ cm}$$

$$A_{lateral} = 2\pi rh$$

$$A_{lateral} = 2\pi(2.00 \text{ cm})(5.60 \text{ cm})$$

$$= 70.4 \text{ cm}^2$$

b.
$$A_{base} = \pi r^2$$

$$A_{base} = \pi(2.00 \text{ cm})^2$$

$$= 12.7 \text{ cm}^2$$

$$A_{total} = 2 \times A_{base} + A_{lateral}$$

$$= 2(12.7 \text{ cm}^2) + 70.4 \text{ cm}^2$$

$$= 70.4 \text{ cm}^2$$

32.
$$B = \frac{1}{2}(8.47 \text{ m})(6.71 \text{ m}) = 28.4 \text{ m}^2$$

$$V = \frac{1}{3}Bh$$

$$V = \frac{1}{3}(28.4 \text{ m}^2)(13.8 \text{ m}) = 131 \text{ m}^3$$

33. a.
$$V = \frac{1}{3}\pi r^2 h$$

$$V = \frac{1}{3}\pi(37.6 \text{ m})^2(38.7 \text{ m})$$

$$= 57,300 \text{ m}^3$$

b.
$$s = \sqrt{(37.6 \text{ m})^2 + (38.7 \text{ m})^2}$$

$$= 54.0 \text{ m}$$

$$A_{cone} = \pi rs$$

$$A_{cone} = \pi(37.6 \text{ m})(54.0 \text{ m})$$

$$= 6380 \text{ m}^2$$

34. a.
$$V = \frac{4}{3}\pi r^3$$

$$V = \frac{4}{3}\pi(5.92 \text{ m})^3$$

$$= 869 \text{ m}^3$$

b.
$$A = 4\pi r^2$$

$$A = 4\pi(5.92 \text{ m})^2$$

$$= 44\overline{0} \text{ m}^2$$

35.
$$B_1 = \pi(18.0 \text{ in.})^2 = 1020 \text{ ft}^2$$

$$B_2 = \pi(25.0 \text{ in.})^2 = 1960 \text{ in}^2$$

$$V_{frustum} = \frac{1}{3}h\left(B_1 + B_2 + \sqrt{B_1 B_2}\right)$$

$$V_{frustum} = \frac{1}{3}(45.0 \text{ in.})\left(1020 \text{ in}^2 + 1960 \text{ in}^2 + \sqrt{(1020 \text{ in}^2)(1960 \text{ in}^2)}\right)$$

$$= 65,900 \text{ in}^3$$

$$s = \sqrt{(45.0 \text{ in.})^2 + (7.00 \text{ in.})^2} = 45.5 \text{ in.}$$

$$A = \pi s(r_1 + r_2)$$

$$A = \pi(45.5 \text{ in.})(18.0 \text{ in.} + 25.0 \text{ in.})$$

$$= 6150 \text{ in}^2$$

36. a.

$$A_{front/back} = 4(6.00 \text{ ft})(2.25 \text{ ft}) = 54.0 \text{ ft}^2$$

$$A_{sides} = 4(4.00 \text{ ft})(2.25 \text{ ft}) = 36.0 \text{ ft}^2$$

$$A_{bottom} = 2(6.00 \text{ ft})(4.00 \text{ ft}) = 48.0 \text{ ft}^2$$

$$A = 54.0 \text{ ft}^2 + 36.0 \text{ ft}^2 + 48.0 \text{ ft}^2 = 138 \text{ ft}^2$$

$$\frac{138 \text{ ft}^2}{11.0 \text{ ft}^2} = 13 \text{ whole gallon containers}$$

$$13 \times \$40 = \$520$$

b.

$$h = \frac{4}{5}(2.25) = 1.80 \text{ ft}$$

$$V = Bh$$

$$V = (24.0 \text{ ft}^2)(1.80 \text{ ft})$$

$$= 43.2 \text{ ft}^3$$

$$43.2 \text{ ft}^3 \times \frac{62.4 \text{ lb}}{1 \text{ ft}^3} = 2\overline{7}00 \text{ lb}$$

Chapter 12 Test

1.

$$A = bh$$

$$A = (18.0 \text{ ft})(6.00 \text{ ft})$$

$$= 108 \text{ ft}^2$$

3.

$$A = \left(\frac{a+b}{2}\right)h$$

$$A = \left(\frac{8.00 \text{ cm} + 15.0 \text{ cm}}{2}\right)(5.00 \text{ cm})$$

$$= 57.5 \text{ cm}^2$$

5.

$$a = \sqrt{c^2 - b^2}$$

$$a = \sqrt{(30.0 \text{ km})^2 - (24.0 \text{ km})^2}$$

$$= 18.0 \text{ km}$$

7.

$$C = 2\pi r$$

$$C = 2\pi(20.0 \text{ cm})$$

$$= 126 \text{ cm}$$

9. $\dfrac{7\pi}{4} \text{ rad} \times \dfrac{180°}{\pi \text{ rad}} = 315°$

11.

$$A_{front} = (12.0 \text{ ft})(8.00 \text{ ft}) = 96.0 \text{ ft}^2$$

$$A_{side} = (12.0 \text{ ft})(9.00 \text{ ft}) = 108 \text{ ft}^2$$

$$A_{bottom} = (8.00 \text{ ft})(9.00 \text{ ft}) = 72.0 \text{ ft}^2$$

$$A = 2 \times (A_{front} + A_{side} + A_{bottom})$$

$$A = 2 \times (96.0 \text{ ft}^2 + 108 \text{ ft}^2 + 72.0 \text{ ft}^2)$$

$$= 552 \text{ ft}^2$$

13. a.

$$r = \frac{1}{2}(8.00 \text{ ft}) = 4.00 \text{ ft}$$

$$V_{cone} = \frac{1}{3}\pi r^2 h$$

$$V_{cone} = \frac{1}{3}\pi(4.00 \text{ ft})^2(6.00 \text{ ft})$$

$$= 101 \text{ ft}^3$$

$$V_{cylinder} = \pi r^2 h$$

$$V_{cylinder} = \pi(4.00 \text{ ft})^2(12.0 \text{ ft})$$

$$= 603 \text{ ft}^3$$

$$V_{bin} = 101 \text{ ft}^3 + 603 \text{ ft}^3$$

$$= 704 \text{ ft}^3$$

13. (continued)

b.

$$s = \sqrt{(4.00 \text{ ft})^2 + (6.00 \text{ ft})^2}$$
$$= 7.21 \text{ ft}$$
$$A_{cone} = \pi r s$$
$$A_{cone} = \pi (4.00 \text{ ft})(7.21 \text{ ft})$$
$$= 90.1 \text{ ft}^2$$
$$A_{cylinder} = 2\pi r h$$
$$A_{cylinder} = 2\pi (4.00 \text{ ft})(12.0 \text{ ft})$$
$$= 302 \text{ ft}^2$$
$$A_{bin} = 90.1 \text{ ft}^3 + 302 \text{ ft}^2$$
$$= 322 \text{ ft}^2$$

15.

a.
$$B = \left(\frac{18.0 \text{ in.} + 24.0 \text{ in.}}{2}\right)(12.0 \text{ in.})$$
$$= 252 \text{ in}^2$$
$$V = Bh$$
$$V = \left(252 \text{ in}^2\right)(72 \text{ in.}) \times \left(\frac{1 \text{ ft}}{12 \text{ in.}}\right)^3$$
$$= 10.5 \text{ ft}^3$$

b.
$$y = \sqrt{h^2 + x^2}, \text{ where}$$
$$x = \frac{24.0 \text{ in.} - 18.0 \text{ in.}}{2} = 3.0 \text{ in.}$$
$$y = \sqrt{(12.0 \text{ in.})^2 + (3.0 \text{ in.})^2}$$
$$= 12.4 \text{ in.}$$

Cumulative Review Chapters 1-12

1. -4

2.
$$P = 2(l + w)$$
$$P = 2\left(4\frac{1}{8} \text{ in.} + 2\frac{3}{4} \text{ in.}\right)$$
$$= 2\left(\frac{33}{8} \text{ in.} + \frac{11}{4} \text{ in.}\right)$$
$$= 2\left(\frac{33}{8} \text{ in.} + \frac{22}{8} \text{ in.}\right)$$
$$= 2\left(\frac{55}{8} \text{ in.}\right)$$
$$= \frac{55}{4} \text{ in.} = 13\frac{3}{4} \text{ in.}$$

3. b

4. a. 100 L

b. $\dfrac{100 \text{ L}}{2} = 50 \text{ L}$

5.
$$\frac{14x^3 - 56x^2 - 28x}{7x}$$
$$= \frac{7x\left(2x^2 - 8x - 4\right)}{7x}$$
$$= 2x^2 - 8x - 4$$

6.
$$\begin{array}{r} -2x^2 + 7x - 3 \\ 4x + 5 \\ \hline -8x^3 + 28x^2 - 12x \\ -10x^2 + 35x - 15 \\ \hline -8x^3 + 18x^2 + 23x - 15 \end{array}$$

7.

$$4 - 2x = 18$$
$$4 - 2x - 4 = 18 - 4$$
$$-2x = 14$$
$$\frac{-2x}{-2} = \frac{14}{-2}$$
$$x = -7$$

8.

$$E = mv^2$$
$$m = \frac{E}{v^2}$$
$$m = \frac{952}{(7.00)^2} = 19.4$$

9. $\dfrac{3.4 \text{ tons}}{200 \text{ bu}} \times \dfrac{2000 \text{ lb}}{1 \text{ ton}} = 34 \text{ lb/bu}$

10.

$$F_1 d_1 = F_2 d_2$$
$$(x)(80.0 \text{ in.}) = (160 \text{ lb})(28.0 \text{ in.})$$
$$x = \frac{(160 \text{ lb})(28.0 \text{ in.})}{80.0 \text{ in.}}$$
$$x = 56 \text{ lb}$$

11.

$$5x - 8y = 10$$
$$5x - 8y - 5x = 10 - 5x$$
$$-8y = 10 - 5x$$
$$\frac{-8y}{-8} = \frac{10 - 5x}{-8}$$
$$y = \frac{10 - 5x}{-8}$$
$$y = \frac{(10 - 5x)(-1)}{(-8)(-1)}$$
$$y = \frac{5x - 10}{8} \text{ or } y = \frac{5}{8}x - \frac{5}{4}$$

12. $m = \dfrac{y_2 - y_1}{x_2 - x_1} = \dfrac{(-8) - (-1)}{(5) - (2)} = -\dfrac{7}{3}$

13.

$$5x - y = 12$$
$$y = 2x$$

$$5x - y = 12$$
$$5x - (2x) = 12$$
$$3x = 12$$
$$x = 4$$

$$y = 2x$$
$$y = 2(4)$$
$$y = 8$$

The solution is $(4, 8)$.

14.

let x = larger resistor

y = smaller resistor

$$x + y = 1300$$
$$x = 3y$$

$$x + y = 1300$$
$$(3y) + y = 1300$$
$$4y = 1300$$
$$y = 325$$

$$x = 3y$$
$$x = 3(325)$$
$$x = 975$$

The larger resistor is $975\ \Omega$ and the smaller resistor is $325\ \Omega$.

15.

$$(3x - 5)(2x + 7)$$
$$= 6x^2 + (21x - 10x) - 35$$
$$= 6x^2 + 11x - 35$$

16. $(4x - 3)^2 = 16x^2 - 24x + 9$

17. $5x^3 - 15x = 5x(x^2 - 3)$

18. $x^2 - 3x - 28 = (x - 7)(x + 4)$

19.

$$10x^2 - 5x = 105$$
$$10x^2 - 5x - 105 = 0$$
$$5(x+3)(2x-7) = 0$$
$$x = -3, \frac{7}{2}$$

20.

$$2x^2 - x = 3$$
$$2x^2 - x - 3 = 0$$
$$(x+1)(2x-3) = 0$$
$$x = -1, \frac{3}{2}$$

21.

$$x = \frac{-b \pm \sqrt{b^2 - 4ac}}{2a}$$
$$a = 5, \ b = 13, \ c = -6$$

So $x = \dfrac{-(13) \pm \sqrt{(13)^2 - 4(5)(-6)}}{2(5)}$

$$= \frac{-13 \pm \sqrt{289}}{10}$$
$$= \frac{-13 \pm 17}{10}$$
$$= \frac{-13 - 17}{10} \quad \text{or} \quad \frac{-13 + 17}{10}$$
$$= -3 \quad \text{or} \quad \frac{2}{5}$$
$$= -3 \quad \text{or} \quad 0.4$$

22.

$$x = \frac{-b \pm \sqrt{b^2 - 4ac}}{2a}$$
$$a = 4, \ b = -14, \ c = 53$$

So $x = \dfrac{-(-10) \pm \sqrt{(-10)^2 - 4(4)(-29)}}{2(4)}$

$$= \frac{10 \pm \sqrt{564}}{8}$$
$$= \frac{10 - \sqrt{564}}{8} \quad \text{or} \quad \frac{10 + \sqrt{564}}{8}$$
$$= -1.72 \quad \text{or} \quad 4.22$$

23.

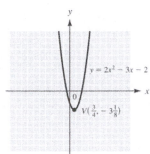

24.

$$\angle 1 = \angle 4 = 82°$$
$$\angle 2 = \angle 1 = 82°$$
$$\angle 3 = 180° - 82° = 98°$$
$$\angle 5 = \angle 3 = 98°$$

25.

$$\angle 4 + \angle 5 = 180°$$
$$(2x - 3) + (x + 6) = 180$$
$$3x + 3 = 180$$
$$3x = 177$$
$$x = 59°$$

26.

$$P = a + b + c$$
$$P = (22.2 \text{ cm}) + (24.3 \text{ cm}) + (29.8 \text{ cm})$$
$$= 76.3 \text{ cm}$$
$$A = \frac{1}{2}bh$$
$$A = \frac{1}{2}(29.8 \text{ cm})(17.8 \text{ cm})$$
$$= 265 \text{ cm}^2$$

27.

$$c = \sqrt{a^2 + b^2}$$
$$c = \sqrt{(1.82 \text{ m})^2 + (2.04 \text{ m})^2}$$
$$= 2.73 \text{ m}$$
$$P = a + b + c$$
$$P = (1.82 \text{ m}) + (2.04 \text{ m}) + (2.73 \text{ m})$$
$$= 6.59 \text{ m}$$
$$A = \frac{1}{2}bh$$
$$A = \frac{1}{2}(2.04 \text{ m})(1.82 \text{ m})$$
$$= 1.86 \text{ cm}^2$$

28.

$$r = \frac{1}{2}(15.6 \text{ cm}) = 7.80 \text{ cm}$$

$$A = \pi r^2$$

$$A = \pi (7.80 \text{ cm})^2$$

$$= 191 \text{ cm}^2$$

$$C = 2\pi r$$

$$C = 2\pi (7.80 \text{ cm})$$

$$= 49.0 \text{ cm}$$

29.

$$A = \pi r^2$$

$$168 \text{ cm}^2 = \pi r^2$$

$$\sqrt{\frac{168 \text{ cm}^2}{\pi}} = r$$

$$7.31 \text{ cm} = r$$

30.

$$r = \frac{1}{2}(8.92 \text{ cm}) = 4.46 \text{ cm}$$

$$V_{cylinder} = \pi r^2 h$$

$$V_{cylinder} = \pi (4.46 \text{ cm})^2 (15.8 \text{ cm})$$

$$= 987 \text{ cm}^3$$

$$A_{lateral} = 2\pi r h$$

$$A_{lateral} = 2\pi (4.46 \text{ cm})(15.8 \text{ cm})$$

$$= 443 \text{ cm}^2$$

$$A_{base} = \pi r^2$$

$$A_{base} = \pi (4.46 \text{ cm})^2$$

$$= 62.5 \text{ cm}^2$$

$$A_{total} = 2 \times A_{base} + A_{lateral}$$

$$= 2(62.5 \text{ cm}^2) + 443 \text{ cm}^2$$

$$= 568 \text{ cm}^2$$

Chapter 13: Right Triangle Trigonometry

Section 13.1: Trigonometric Ratios

1. a

3. c

5. a

7. B

9. B

11. $b = \sqrt{(75.0 \text{ m})^2 - (45.0 \text{ m})^2} = 60.0 \text{ m}$

13. $c = \sqrt{(29.0 \text{ mi})^2 + (47.0 \text{ mi})^2} = 55.2 \text{ mi}$

15. $b = \sqrt{(18.9 \text{ cm})^2 - (6.71 \text{ cm})^2} = 17.7 \text{ cm}$

17. $c = \sqrt{(171 \text{ ft})^2 + (203 \text{ ft})^2} = 265 \text{ ft}$

19. $b = \sqrt{(404 \text{ m})^2 - (202 \text{ m})^2} = 35\overline{0} \text{ m}$

21. $a = \sqrt{(2160 \text{ m})^2 - (1520 \text{ m})^2} = 1530 \text{ m}$

23.

$c = \sqrt{(45,800 \text{ m})^2 + (38,600 \text{ m})^2}$

$\quad = 59,900 \text{ m}$

25. $\sin A = \dfrac{31.0 \text{ m}}{62.0 \text{ m}} = 0.5000$

27. $\tan A = \dfrac{31.0 \text{ m}}{54.0 \text{ m}} = 0.5741$

29. $\cos B = \dfrac{31.0 \text{ m}}{62.0 \text{ m}} = 0.5000$

31. $\sin 49.6° = 0.7615$

33. $\tan 65.3° = 2.174$

35. $\cos 29.7° = 0.8686$

37. $\sin 31.64° = 0.5246$

39. $\cos 75.31° = 0.2534$

41. $\tan 3.05° = 0.05328$

43. $\sin 37.62° = 0.6104$

45. $\tan 21.45° = 0.3929$

47. $\cos 47.16° = 0.6800$

49. $A = \sin^{-1} 0.7941 = 52.6°$

51. $B = \cos^{-1} 0.4602 = 62.6°$

53. $B = \tan^{-1} 1.386 = 54.2°$

55. $B = \sin^{-1} 0.1592 = 9.2°$

57. $A = \cos^{-1} 0.8592 = 30.8°$

59. $A = \tan^{-1} 0.8644 = 40.8°$

61. $A = \tan^{-1} 0.1941 = 10.98°$

63. $B = \cos^{-1} 0.3572 = 69.07°$

65. $A = \sin^{-1} 0.1506 = 8.66°$

67. $B = \tan^{-1} 3.806 = 75.28°$

69. $B = \cos^{-1} 0.7311 = 43.02°$

71. $B = \sin^{-1} 0.3441 = 20.13°$

73. $\sin A$ and $\cos B$; $\cos A$ and $\sin B$

Section 13.2: Using Trigonometric Ratios to Find Angles

1.

$\tan A = \dfrac{a}{b}$

$\tan A = \dfrac{36.0 \text{ m}}{50.9 \text{ m}} = 0.7073$

$\quad A = 35.3°$

$\quad B = 90° - 35.3° = 54.7°$

3.

$\cos A = \dfrac{b}{c}$

$\cos A = \dfrac{39.7 \text{ cm}}{43.6 \text{ cm}} = 0.9106$

$\quad A = 24.4°$

$\quad B = 90° - 24.4° = 65.6°$

5.

$$\cos A = \frac{b}{c}$$

$$\cos A = \frac{13.6 \text{ m}}{18.7 \text{ m}} = 0.7273$$

$$A = 43.3°$$

$$B = 90° - 43.3° = 46.7°$$

7.

$$\sin A = \frac{a}{c}$$

$$\sin A = \frac{29.7 \text{ mm}}{42.0 \text{ mm}} = 0.7071$$

$$A = 45.0°$$

$$B = 90° - 45.0° = 45.0°$$

9.

$$\tan A = \frac{a}{b}$$

$$\tan A = \frac{2902 \text{ km}}{1412 \text{ km}} = 2.055$$

$$A = 64.05°$$

$$B = 90° - 64.05° = 25.95°$$

11.

$$\sin A = \frac{a}{c}$$

$$\sin A = \frac{0.6341 \text{ cm}}{0.7982 \text{ cm}} = 0.7944$$

$$A = 52.60°$$

$$B = 90° - 52.60° = 37.40°$$

13.

$$\cos A = \frac{b}{c}$$

$$\cos A = \frac{1455 \text{ ft}}{1895 \text{ ft}} = 0.7678$$

$$A = 39.84°$$

$$B = 90° - 39.84° = 50.16°$$

15.

$$\tan A = \frac{a}{b}$$

$$\tan A = \frac{243.2 \text{ km}}{271.5 \text{ km}} = 0.8958$$

$$A = 41.85°$$

$$B = 90° - 41.85° = 48.15°$$

17.

$$\sin A = \frac{a}{c}$$

$$\sin A = \frac{16.7 \text{ m}}{81.4 \text{ m}} = 0.2052$$

$$A = 11.8°$$

$$B = 90° - 11.8° = 78.2°$$

19.

$$\cos A = \frac{b}{c}$$

$$\cos A = \frac{1185 \text{ ft}}{1384 \text{ ft}} = 0.8562$$

$$A = 31.11°$$

$$B = 90° - 31.11° = 58.89°$$

21.

$$\tan A = \frac{a}{b}$$

$$\tan A = \frac{845 \text{ km}}{2960 \text{ km}} = 0.2855$$

$$A = 15.9°$$

$$B = 90° - 15.9° = 74.1°$$

23.

$$\sin A = \frac{a}{c}$$

$$\sin A = \frac{8897 \text{ m}}{9845 \text{ m}} = 0.9037$$

$$A = 64.65°$$

$$B = 90° - 64.65° = 24.35°$$

Section 13.3: Using Trigonometric Ratios to Find Sides

1.

$$\tan A = \frac{a}{b}$$

$$\tan 42.1° = \frac{36.7 \text{ m}}{b}$$

$$b\left(\tan 42.1°\right) = 36.7 \text{ m}$$

$$b = \frac{36.7 \text{ m}}{\tan 42.1°} = 40.6 \text{ m}$$

$$\sin A = \frac{a}{c}$$

$$\sin 42.1° = \frac{36.7 \text{ m}}{c}$$

$$c\left(\sin 42.1°\right) = 36.7 \text{ m}$$

$$c = \frac{36.7 \text{ m}}{\sin 42.1°} = 54.7 \text{ m}$$

3.

$$\tan B = \frac{b}{a}$$

$$\tan 49.7° = \frac{b}{236 \text{ km}}$$

$$b = \left(\tan 49.7°\right)\left(236 \text{ km}\right) = 278 \text{ km}$$

$$\cos B = \frac{a}{c}$$

$$\cos 49.7° = \frac{236 \text{ km}}{c}$$

$$c\left(\cos 49.7°\right) = 236 \text{ km}$$

$$c = \frac{236 \text{ km}}{\cos 49.7°} = 365 \text{ km}$$

5.

$$\sin A = \frac{a}{c}$$

$$\sin 36.7° = \frac{a}{49.1 \text{ cm}}$$

$$a = \left(\sin 36.7°\right)\left(49.1 \text{ cm}\right) = 29.3 \text{ cm}$$

$$\cos A = \frac{b}{c}$$

$$\cos 36.7° = \frac{b}{49.1 \text{ cm}}$$

$$b = \left(\cos 36.7°\right)\left(49.1 \text{ cm}\right) = 39.4 \text{ cm}$$

7.

$$\tan A = \frac{a}{b}$$

$$\tan 23.7° = \frac{a}{23.7 \text{ cm}}$$

$$a = \left(\tan 23.7°\right)\left(23.7 \text{ cm}\right)$$

$$= 10.4 \text{ cm}$$

$$\sin B = \frac{b}{c}$$

$$\cos 23.7° = \frac{23.7 \text{ cm}}{c}$$

$$c\left(\cos 23.7°\right) = 23.7 \text{ cm}$$

$$c = \frac{23.7 \text{ cm}}{\cos 23.7°} = 25.9 \text{ cm}$$

9.

$$\tan A = \frac{a}{b}$$

$$\tan 12.9° = \frac{a}{29,200 \text{ km}}$$

$$a = \left(\tan 12.9°\right)\left(29,200 \text{ km}\right)$$

$$= 6690 \text{ km}$$

$$\cos A = \frac{b}{c}$$

$$\cos 12.9° = \frac{29,200 \text{ km}}{c}$$

$$c\left(\cos 12.9°\right) = 29,200 \text{ km}$$

$$c = \frac{29,200 \text{ km}}{\cos 12.9°} = 30,\overline{0}00 \text{ km}$$

11.

$$\tan A = \frac{a}{b}$$

$$\tan 19.75° = \frac{19.72 \text{ m}}{b}$$

$$b\left(\tan 19.75°\right) = 19.72 \text{ m}$$

$$b = \frac{19.72 \text{ m}}{\tan 19.75°} = 54.92 \text{ m}$$

$$\sin A = \frac{a}{c}$$

$$\sin 19.75° = \frac{19.72 \text{ m}}{c}$$

$$c\left(\sin 19.75°\right) = 19.72 \text{ m}$$

$$c = \frac{19.72 \text{ m}}{\sin 19.75°} = 58.36 \text{ m}$$

13.

$$\sin A = \frac{a}{c}$$

$$\sin 39.25° = \frac{a}{255.6 \text{ mi}}$$

$$a = \left(\sin 39.25°\right)\left(255.6 \text{ mi}\right)$$

$$= 161.7 \text{ mi}$$

$$\cos A = \frac{b}{c}$$

$$\cos 39.25° = \frac{b}{255.6 \text{ mi}}$$

$$b = \left(\cos 39.25°\right)\left(255.6 \text{ mi}\right)$$

$$= 197.9 \text{ mi}$$

15.

$$\tan B = \frac{b}{a}$$

$$\tan 69.72° = \frac{12{,}350 \text{ m}}{a}$$

$$a\left(\tan 69.72°\right) = 12{,}350 \text{ m}$$

$$a = \frac{12{,}350 \text{ m}}{\tan 69.72°}$$

$$= 4564 \text{ m}$$

$$\sin B = \frac{b}{c}$$

$$\sin 69.72° = \frac{12{,}350 \text{ m}}{c}$$

$$c\left(\sin 69.72°\right) = 12{,}350 \text{ m}$$

$$c = \frac{12{,}350 \text{ m}}{\sin 69.72°}$$

$$= 13{,}170 \text{ m}$$

17.

$$\tan A = \frac{a}{b}$$

$$\tan 18.4° = \frac{1980 \text{ m}}{b}$$

$$b\left(\tan 18.4°\right) = 1980 \text{ m}$$

$$b = \frac{1980 \text{ m}}{\tan 18.4°}$$

$$= 5950 \text{ m}$$

$$\sin A = \frac{a}{c}$$

$$\sin 18.4° = \frac{1980 \text{ m}}{c}$$

$$c\left(\sin 18.4°\right) = 1980 \text{ m}$$

$$c = \frac{1980 \text{ m}}{\sin 18.4°}$$

$$= 6270 \text{ m}$$

19.

$$\tan A = \frac{a}{b}$$

$$\tan 18.91° = \frac{a}{841.6 \text{ km}}$$

$$a = \left(\tan 18.91°\right)\left(841.6 \text{ km}\right)$$

$$= 288.3 \text{ cm}$$

$$\cos A = \frac{b}{c}$$

$$\cos 18.91° = \frac{841.6 \text{ km}}{c}$$

$$c\left(\cos 18.91°\right) = 841.6 \text{ km}$$

$$c = \frac{841.6 \text{ km}}{\cos 18.91°}$$

$$= 889.6 \text{ km}$$

21.

$$\sin B = \frac{b}{c}$$

$$\sin 61.45° = \frac{b}{185.6 \text{ m}}$$

$$b = \left(\sin 61.45°\right)\left(185.6 \text{ m}\right)$$

$$= 163.0 \text{ cm}$$

$$\cos B = \frac{a}{c}$$

$$\cos 61.45° = \frac{a}{185.6 \text{ m}}$$

$$a = \left(\cos 61.45°\right)\left(185.6 \text{ m}\right)$$

$$= 88.70 \text{ cm}$$

23.

$$\sin A = \frac{a}{c}$$

$$\sin 25.6° = \frac{a}{256 \text{ cm}}$$

$$a = \left(\sin 25.6°\right)\left(256 \text{ cm}\right) = 111 \text{ cm}$$

$$\cos A = \frac{b}{c}$$

$$\cos 25.6° = \frac{b}{256 \text{ cm}}$$

$$b = \left(\cos 25.6°\right)\left(256 \text{ cm}\right) = 231 \text{ cm}$$

Section 13.4: Solving Right Triangles

1.

$$B = 90° - 50.6° = 39.4°$$

$$\sin A = \frac{a}{c}$$

$$\sin 50.6° = \frac{a}{49.0 \text{ m}}$$

$$a = \left(\sin 50.6°\right)\left(49.0 \text{ m}\right) = 37.9 \text{ m}$$

$$\sin B = \frac{b}{c}$$

$$\sin 39.4° = \frac{b}{49.0 \text{ m}}$$

$$b = \left(\sin 39.4°\right)\left(49.0 \text{ m}\right) = 31.1 \text{ m}$$

3.

$$A = 90° - 41.2° = 48.8°$$

$$\tan B = \frac{b}{a}$$

$$\tan 41.2° = \frac{b}{267 \text{ ft}}$$

$$b = \left(\tan 41.2°\right)\left(267 \text{ ft}\right) = 234 \text{ ft}$$

$$\cos B = \frac{a}{c}$$

$$\cos 41.2° = \frac{267 \text{ ft}}{c}$$

$$c = \frac{267 \text{ ft}}{\cos 41.2°} = 355 \text{ ft}$$

5.

$$a = \sqrt{c^2 - b^2}$$

$$a = \sqrt{(78.0 \text{ mi})^2 - (72.0 \text{ mi})^2}$$

$$= 30.0 \text{ mi}$$

$$\cos A = \frac{b}{c}$$

$$\cos A = \frac{72.0 \text{ mi}}{78.0 \text{ mi}} = 0.9231$$

$$A = 22.6°$$

$$B = 90° - 22.6° = 67.4°$$

7.

$$B = 90° - 52.1° = 37.9°$$

$$\tan B = \frac{b}{a}$$

$$\tan 37.9° = \frac{b}{72.0 \text{ mm}}$$

$$b = (\tan 37.9°)(72.0 \text{ mm})$$

$$= 56.1 \text{ mm}$$

$$\sin A = \frac{a}{c}$$

$$\sin 52.1° = \frac{72.0 \text{ mm}}{c}$$

$$c = \frac{72.0 \text{ mm}}{\sin 52.1°}$$

$$= 91.2 \text{ mm}$$

9.

$$B = 90° - 68.8° = 21.2°$$

$$\sin A = \frac{a}{c}$$

$$\sin 68.8° = \frac{a}{39.4 \text{ m}}$$

$$a = (\sin 68.8°)(39.4 \text{ m})$$

$$= 36.7 \text{ m}$$

$$\sin B = \frac{b}{c}$$

$$\sin 21.2° = \frac{b}{39.4 \text{ m}}$$

$$b = (\sin 21.2°)(39.4 \text{ m})$$

$$= 14.2 \text{ m}$$

11.

$$c = \sqrt{a^2 + b^2}$$

$$c = \sqrt{(12.00 \text{ m})^2 + (24.55 \text{ m})^2}$$

$$= 27.33 \text{ m}$$

$$\tan A = \frac{a}{b}$$

$$\tan A = \frac{12.00 \text{ m}}{24.55 \text{ m}} = 0.4888$$

$$A = 26.05°$$

$$B = 90° - 26.05° = 63.95°$$

13.

$$B = 90° - 29.19° = 60.81°$$

$$\sin A = \frac{a}{c}$$

$$\sin 29.19° = \frac{a}{2975 \text{ ft}}$$

$$a = (\sin 29.19°)(2975 \text{ ft})$$

$$= 1451 \text{ ft}$$

$$\sin B = \frac{b}{c}$$

$$\sin 60.81° = \frac{b}{2975 \text{ ft}}$$

$$b = (\sin 60.81°)(2975 \text{ ft})$$

$$= 2597 \text{ ft}$$

15.

$$c = \sqrt{a^2 + b^2}$$

$$c = \sqrt{(46.72 \text{ m})^2 + (19.26 \text{ m})^2}$$

$$= 50.53 \text{ m}$$

$$\tan A = \frac{a}{b}$$

$$\tan A = \frac{46.72 \text{ m}}{19.26 \text{ m}} = 2.426$$

$$A = 67.60°$$

$$B = 90° - 67.60° = 22.40°$$

17.

$$B = 90° - 41.1° = 48.9°$$

$$\sin A = \frac{a}{c}$$

$$\sin 41.1° = \frac{a}{485 \text{ m}}$$

$$a = \left(\sin 41.1°\right)\left(485 \text{ m}\right)$$

$$= 319 \text{ m}$$

$$\sin B = \frac{b}{c}$$

$$\sin 48.9° = \frac{b}{485 \text{ m}}$$

$$b = \left(\sin 48.9°\right)\left(485 \text{ m}\right)$$

$$= 365 \text{ m}$$

19.

$$A = 90° - 9.45° = 80.55°$$

$$\tan A = \frac{a}{b}$$

$$\tan 80.55° = \frac{1585 \text{ ft}}{b}$$

$$b = \frac{1585 \text{ ft}}{\tan 80.55°}$$

$$= 263.8 \text{ ft}$$

$$\cos B = \frac{a}{c}$$

$$\cos 9.45° = \frac{1585 \text{ ft}}{c}$$

$$c = \frac{1585 \text{ ft}}{\cos 9.45°}$$

$$= 1607 \text{ ft}$$

21.

$$a = \sqrt{c^2 - b^2}$$

$$a = \sqrt{\left(380.5 \text{ m}\right)^2 - \left(269.5 \text{ m}\right)^2}$$

$$= 268.6 \text{ m}$$

$$\cos A = \frac{b}{c}$$

$$\cos A = \frac{269.5 \text{ m}}{380.5 \text{ m}} = 0.7083$$

$$A = 44.90°$$

$$B = 90° - 44.90° = 45.10°$$

23.

$$A = 90° - 81.5° = 8.5°$$

$$\tan A = \frac{a}{b}$$

$$\tan 8.5° = \frac{a}{9370 \text{ ft}}$$

$$a = \left(\tan 8.5°\right)\left(9370 \text{ ft}\right)$$

$$= 1400 \text{ ft}$$

$$\sin B = \frac{b}{c}$$

$$\sin 81.5° = \frac{9370 \text{ ft}}{c}$$

$$c = \frac{9370 \text{ ft}}{\sin 81.5°}$$

$$= 9470 \text{ ft}$$

Section 13.5: Applications Involving Trigonometric Ratios

1. Let $x =$ length of conveyor.

$$\sin 35.8° = \frac{11.0 \text{ m}}{x}$$

$$x = \frac{11.0 \text{ m}}{\sin 35.8°}$$

$$= 18.8 \text{ m}$$

3. Let $x =$ distance from wall.

$$\tan 12° = \frac{\left(2.3 \text{ m} - 1.2 \text{ m}\right)}{x}$$

$$\tan 12° = \frac{1.1 \text{ m}}{x}$$

$$x = \frac{1.1 \text{ m}}{\tan 12°}$$

$$= 5.2 \text{ m}$$

5.

$$x = \sqrt{(38.0 \text{ ft})^2 - (20.0 \text{ ft})^2}$$
$$= 32.3 \text{ ft}$$
$$\cos A = \frac{20.0 \text{ ft}}{38.0 \text{ ft}} = 0.5263$$
$$A = 58.2°$$

9. Let $x =$ difference in height.

$$\tan 1.0° = \frac{x}{5280 \text{ ft}}$$
$$x = \left(\tan 1.0°\right)\left(5280 \text{ ft}\right)$$
$$= 92.2 \text{ ft}$$

13.

$$\cos 60.0° = \frac{BC}{11.4 \text{ cm}}$$
$$BC = \left(\cos 60.0°\right)\left(11.4 \text{ cm}\right)$$
$$= 5.70 \text{ cm}$$

17. a.

$$\cos \theta = \frac{R}{Z}$$
$$\cos 35° = \frac{350 \ \Omega}{Z}$$
$$Z = \frac{350 \ \Omega}{\cos 35°}$$
$$= 430 \ \Omega$$

b.

$$\cos \theta = \frac{R}{Z}$$
$$\cos \theta = \frac{550 \ \Omega}{\overline{7}00 \ \Omega} = 0.7857$$
$$\theta = 38°$$

c.

$$Z = \sqrt{X^2 + R^2}$$
$$Z = \sqrt{(182 \ \Omega)^2 + (240 \ \Omega)^2}$$
$$Z = 3\overline{0}0 \ \Omega$$
$$\tan \theta = \frac{X}{R}$$
$$\tan \theta = \frac{182 \ \Omega}{240 \ \Omega} = 0.7583$$
$$\theta = 37°$$

7. Let $A =$ angle of inclination.

$$\tan A = \frac{220 \text{ ft}}{2300 \text{ ft}} = 0.09565$$
$$A = 5°$$

11.

$$\cos \theta = \frac{15.0 \text{ ft}}{17.0 \text{ ft}} = 0.8824$$
$$A = 28.1°$$

15. Let $x =$ tower height.

$$\sin 60.0° = \frac{x}{1\overline{1}0 \text{ m}}$$
$$x = \left(\sin 60.0°\right)\left(1\overline{1}0 \text{ m}\right)$$
$$= 95.3 \text{ m}$$

19. a.

Voltage applied

$$\sqrt{(40.2 \text{ V})^2 + (35.6 \text{ V})^2} = 53.7 \text{ V}$$

b.

Voltage across resistance

$$\sqrt{(378 \text{ V})^2 - (268 \text{ V})^2} = 267 \text{ V}$$

c.

Voltage across coil

$$\sqrt{(448 \text{ V})^2 - (381 \text{ V})^2} = 236 \text{ V}$$

21. $\dfrac{360°}{12} = 30°$, so the angle opposite x is

$$\frac{30°}{2} = 15°.$$
$$\sin 15° = \frac{x}{14.500 \text{ in.}}$$
$$x = \left(\sin 15°\right)\left(14.500 \text{ in.}\right)$$
$$= 3.7529 \text{ in.}$$

23. Let X be the point where the dotted line intersects $\overline{CD}$.

$$DX = \frac{3.00 \text{ in.} - 2.00 \text{ in.}}{2}$$
$$= 0.500 \text{ in.}$$

$$\tan \angle DAX = \frac{0.500 \text{ in.}}{6.00 \text{ in.}} = 0.08333$$

$$\angle DAX = 4.76°$$
$$\theta = 2(\angle DAX)$$
$$\theta = 2(4.76°) = 9.5°$$

25. a.

$$\cos 51.0° = \frac{x}{5.50 \text{ in.}}$$
$$x = (\cos 51.0°)(5.50 \text{ in.})$$
$$= 3.46 \text{ in.}$$

b.

$$BD = BC - x$$
$$BD = 5.50 \text{ in.} - 3.46 \text{ in.}$$
$$= 2.04 \text{ in.}$$

31. The angle between the 33.0 ft side and side a is $180° - 112.0° = 68.0°$

$$\cos 68.0° = \frac{a}{33.0 \text{ ft}}$$
$$a = (\cos 68.0°)(33.0 \text{ ft})$$
$$= 12.4 \text{ ft}$$

$$\sin 68.0° = \frac{b}{33.0 \text{ ft}}$$
$$a = (\sin 68.0°)(33.0 \text{ ft})$$
$$= 30.6 \text{ ft}$$

Chapter 13 Review

1. 29.7 m

2. B

27.

$$\tan A = \frac{18.7 \text{ ft}}{55.0 \text{ ft} - 42.0 \text{ ft}}$$
$$= \frac{18.7 \text{ ft}}{13.0 \text{ ft}}$$
$$= 1.438$$
$$A = 55.2°$$
$$x = \sqrt{(18.7 \text{ ft})^2 + (13.0 \text{ ft})^2}$$
$$= 22.8 \text{ ft}$$

29. If the hexagon is inscribed in a circle, the arc of one side would be $60.0°$, so the angle between x and y is $\frac{60.0°}{2} = 30.0°$.

$$\sin 30.0° = \frac{2.25 \text{ cm}}{x}$$
$$x = \frac{2.25 \text{ cm}}{\sin 30.0°}$$
$$= 4.50$$

$$\tan 30.0° = \frac{2.25 \text{ cm}}{y}$$
$$y = \frac{2.25 \text{ cm}}{\tan 30.0°}$$
$$= 3.90 \text{ cm}$$

33. The angle of elevation is
$$0.90(42.44° + 29°) = 67.2°$$

a. Let $x =$ height above roof.

$$\sin 67.2° = \frac{x}{8.0 \text{ ft}}$$
$$x = (\sin 67.2°)(8.0 \text{ ft})$$
$$= 7.4 \text{ ft}$$

b.

$$\cos 67.2° = \frac{x}{8.0 \text{ ft}}$$
$$x = (\cos 67.2°)(8.0 \text{ ft})$$
$$= 3.1 \text{ ft}$$

3. Hypotenuse

4.
$$c = \sqrt{a^2 + b^2}$$
$$c = \sqrt{(29.7 \text{ m})^2 + (32.2 \text{ m})^2}$$
$$= 43.8 \text{ m}$$

5. $\sin A$

6. Length of side adjacent to $\angle A$

7. $\dfrac{\text{length of side opposite } \angle B}{\text{length of side adjacent to } \angle B}$

8. $\cos 36.2° = 0.8070$

9. $\tan 48.7° = 1.138$

10. $\sin 23.72° = 0.4023$

11. $A = \sin^{-1} 0.7136 = 45.5°$

12. $B = \tan^{-1} 0.1835 = 10.4°$

13. $A = \cos^{-1} 0.4104 = 65.8°$

14.
$$\tan A = \frac{2.62 \text{ m}}{3.21 \text{ m}} = 0.8162$$
$$B = 39.2°$$

15.
$$\tan B = \frac{3.21 \text{ m}}{2.62 \text{ m}} = 1.225$$
$$B = 50.8°$$

16.
$$\tan A = \frac{a}{b}$$
$$\tan 29.7° = \frac{4.09 \text{ m}}{b}$$
$$b = \frac{4.09 \text{ m}}{\tan 29.7°}$$
$$= 7.17 \text{ m}$$

17.
$$\sin A = \frac{a}{c}$$
$$\sin 29.7° = \frac{4.09 \text{ m}}{c}$$
$$c = \frac{4.09 \text{ m}}{\sin 29.7°}$$
$$= 8.25 \text{ m}$$

18.
$$A = 90° - 49.6° = 40.4°$$
$$\sin A = \frac{a}{c}$$
$$\sin 40.4° = \frac{a}{28.7 \text{ m}}$$
$$a = (\sin 40.4°)(28.7 \text{ m})$$
$$= 18.6 \text{ m}$$
$$\sin B = \frac{b}{c}$$
$$\sin 49.6° = \frac{b}{28.7 \text{ m}}$$
$$b = (\sin 49.6°)(28.7 \text{ m})$$
$$= 21.9 \text{ m}$$

19.
$$B = 90° - 31.2° = 58.8°$$
$$\tan B = \frac{b}{a}$$
$$\tan 58.8° = \frac{b}{61.7 \text{ m}}$$
$$b = (\tan 58.8°)(61.7 \text{ m})$$
$$= 102 \text{ m}$$
$$\sin A = \frac{a}{c}$$
$$\sin 31.2° = \frac{61.7 \text{ m}}{c}$$
$$c = \frac{61.7 \text{ m}}{\sin 31.2°}$$
$$= 119 \text{ m}$$

20.
$$b = \sqrt{c^2 - a^2}$$
$$b = \sqrt{(136 \text{ mi})^2 - (68.0 \text{ mi})^2}$$
$$= 118 \text{ mi}$$
$$\sin A = \frac{68.0 \text{ mi}}{136 \text{ mi}} = 0.5000$$
$$A = 30.0°$$
$$B = 90° - 30.0° = 60.0°$$

21. Let $x =$ height of satellite.

$$\tan 68.0° = \frac{x}{2000 \text{ m}}$$

$$x = \left(\tan 68.0°\right)\left(2000 \text{ m}\right)$$

$$= 5940 \text{ m}$$

22. Let $x =$ ground distance.

$$\tan 3.0° = \frac{275 \text{ ft}}{x}$$

$$x = \frac{275 \text{ ft}}{\tan 3.0°}$$

$$= 5250 \text{ ft}$$

Chapter 13 Test

1. $\sin 35.5° = 0.5807$

3. $\tan 57.1° = 1.546$

9.

$$\cos 34.2° = \frac{15.8 \text{ cm}}{c}$$

$$c = \frac{15.8 \text{ cm}}{\cos 34.2°}$$

$$= 19.1 \text{ cm}$$

11.

$$\cos B = \frac{245 \text{ ft}}{305 \text{ ft}} = 0.8033$$

$$B = 36.6°$$

23. Let $x =$ angle of slope.

$$\tan x = \frac{12.0 \text{ ft}}{48.0 \text{ ft}} = 0.2500$$

$$x = 14.0°$$

5. $B = \tan^{-1} 0.8888 = 41.6°$

7. $B = 90° - 34.2° = 55.8°$

13. Let $x =$ length of guy wire.

$$x = \sqrt{\left(50.0 \text{ ft}\right)^2 + \left(15.0 \text{ ft}\right)^2}$$

$$= 52.2 \text{ ft}$$

15. Let $B =$ the angle from side x to the horizontal.

$$\tan B = \frac{20.0 \text{ ft}}{24.0 \text{ ft}} = 0.8333$$

$$B = 39.8°$$

$$A = 180° - 39.8° = 140.2°$$

Chapter 14: Trigonometry with Any Angle

Section 14.1: Sine and Cosine Graphs

1. $\sin\left(137°\right) = 0.6820$

3. $\cos\left(246°\right) = -0.4067$

5. $\sin\left(205.8°\right) = -0.4352$

7. $\cos\left(166.5°\right) = -0.9724$

9. $\tan\left(217.6°\right) = 0.7701$

11. $\tan\left(156.3°\right) = -0.4390$

13.

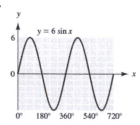

15.

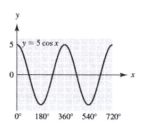

17.

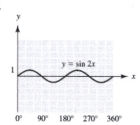

19.

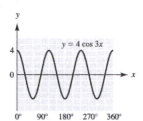

21.

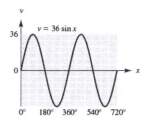

23.

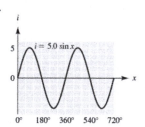

25. 25 V; -33 V

27. 3.5 A; -3.5 A

29.

$$f = \frac{1}{T}$$

$$T = \frac{1}{f}$$

$$T = \frac{1}{5.0 \text{ kHz}}$$

$$= \frac{1}{5.0 \times 10^3 \text{ Hz}}$$

$$= \frac{1}{5.0 \times 10^3} \cdot \frac{1}{\text{Hz}}$$

$$= 2.0 \times 10^{-4} \text{ s}$$

31.

$$f = \frac{1}{T}$$

$$= \frac{1}{0.56 \text{ s}}$$

$$= 1.8/\text{s}$$

$$= 1.8 \text{ Hz}$$

33.

$$v = \lambda f$$

$$f = \frac{v}{\lambda}$$

$$f = \frac{3.0 \times 10^8 \text{ m/s}}{3.4 \text{ cm}}$$

$$= \frac{3.0 \times 10^8 \text{ m/s}}{0.034 \text{ m}}$$

$$= 8.8 \times 10^9 \text{/s}$$

$$= 8.8 \times 10^9 \text{ Hz}$$

$$= 8.8 \text{ GHz}$$

35.

$$v = \lambda f$$

$$v = (0.500 \text{ m/s})(4.50\text{/s})$$

$$= 2.25 \text{ m/s}$$

37.

$$v = \lambda f$$

$$\lambda = \frac{v}{f}$$

$$\lambda = \frac{3.0 \times 10^8 \text{ m/s}}{98.0 \text{ MHz}}$$

$$= \frac{3.0 \times 10^8 \text{ m/s}}{98.0 \times 10^6 \text{ Hz}}$$

$$= \frac{3.0 \times 10^8 \text{ m/s}}{98.0 \times 10^6 \text{ waves/s}}$$

$$= 3.06 \text{ m}$$

$$L = \frac{1}{2}(3.06 \text{ m}) = 1.53 \text{ m}$$

Section 14.2: Period and Phase Shift

1. $P = \dfrac{360°}{3} = 120°; A = 3$

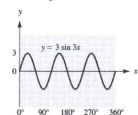

3. $P = \dfrac{360°}{6} = 60°; A = 8$

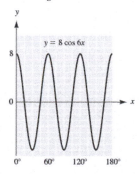

5. $P = \dfrac{360°}{9} = 40°; A = 10$

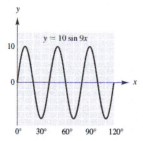

7. $P = \dfrac{360°}{1/2} = 720°; A = 6$

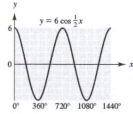

9. $P = \dfrac{360°}{2/3} = 540°; A = 3.5$

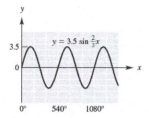

11. $P = \dfrac{360°}{5/2} = 144°; A = 4$

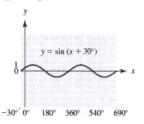

13.

$$P = \dfrac{360°}{1} = 360°; A = 1$$

$$\dfrac{C}{B} = \dfrac{30°}{1} = 30° \text{ or } 30° \text{ left}$$

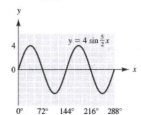

15.

$$P = \dfrac{360°}{1} = 360°; A = 2$$

$$\dfrac{C}{B} = \dfrac{-60°}{1} = -60° \text{ or } 60° \text{ right}$$

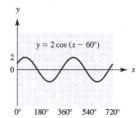

17.

$$P = \dfrac{360°}{3} = 120°; A = 4$$

$$\dfrac{C}{B} = \dfrac{180°}{3} = 60° \text{ or } 60° \text{ left}$$

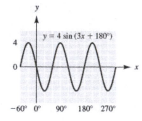

19.

$$P = \dfrac{360°}{4} = 90°; A = 10$$

$$\dfrac{C}{B} = \dfrac{-120°}{4} = -30° \text{ or } 30° \text{ right}$$

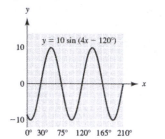

21.

$$P = \dfrac{360°}{1/2} = 720°; A = 5$$

$$\dfrac{C}{B} = \dfrac{90°}{1/2} = 180° \text{ or } 180° \text{ left}$$

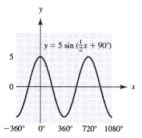

23.

$$P = \dfrac{360°}{1/4} = 1140°; A = 10$$

$$\dfrac{C}{B} = \dfrac{180°}{1/4} = 720° \text{ or } 720° \text{ left}$$

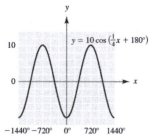

Section 14.3: Solving Oblique Triangles: Law of Sines

1.

$$\frac{\sin B}{b} = \frac{\sin A}{a}$$

$$\frac{\sin B}{17.5 \text{ m}} = \frac{\sin 68.0°}{24.5 \text{ m}}$$

$$\sin B = \frac{(17.5 \text{ m})(\sin 68.0°)}{24.5 \text{ m}}$$

$$B = 41.5°$$

$$C = 180° - 68.0° - 41.5°$$

$$= 70.5°$$

$$\frac{c}{\sin C} = \frac{a}{\sin A}$$

$$\frac{c}{\sin 70.5°} = \frac{24.5 \text{ m}}{\sin 68.0°}$$

$$c = \frac{(\sin 70.5°)(24.5 \text{ m})}{\sin 68.0°}$$

$$= 24.9 \text{ m}$$

3.

$$\frac{a}{\sin A} = \frac{b}{\sin B}$$

$$\frac{a}{\sin 61.5°} = \frac{255 \text{ ft}}{\sin 75.6°}$$

$$a = \frac{(\sin 61.5°)(255 \text{ ft})}{\sin 75.6°}$$

$$= 231 \text{ ft}$$

$$C = 180° - 61.5° - 75.6°$$

$$= 42.9°$$

$$\frac{c}{\sin C} = \frac{b}{\sin B}$$

$$\frac{c}{\sin 42.9°} = \frac{255 \text{ ft}}{\sin 75.6°}$$

$$c = \frac{(\sin 42.9°)(255 \text{ ft})}{\sin 75.6°}$$

$$= 179 \text{ ft}$$

5.

$$C = 180° - 14.6° - 35.1°$$

$$= 130.3°$$

$$\frac{a}{\sin A} = \frac{c}{\sin C}$$

$$\frac{a}{\sin 14.6°} = \frac{43.7 \text{ cm}}{\sin 130.3°}$$

$$a = \frac{(\sin 14.6°)(43.7 \text{ cm})}{\sin 130.3°}$$

$$= 14.4 \text{ cm}$$

$$\frac{b}{\sin B} = \frac{c}{\sin C}$$

$$\frac{b}{\sin 35.1°} = \frac{43.7 \text{ cm}}{\sin 130.3°}$$

$$b = \frac{(\sin 35.1°)(43.7 \text{ cm})}{\sin 130.3°}$$

$$= 32.9 \text{ cm}$$

7.

$$B = 180° - 54.0° - 43.1°$$

$$= 82.9°$$

$$\frac{b}{\sin B} = \frac{a}{\sin A}$$

$$\frac{b}{\sin 82.9°} = \frac{26.5 \text{ m}}{\sin 54.0°}$$

$$b = \frac{(\sin 82.9°)(26.5 \text{ m})}{\sin 54.0°}$$

$$= 32.5 \text{ m}$$

$$\frac{c}{\sin C} = \frac{a}{\sin A}$$

$$\frac{c}{\sin 43.1°} = \frac{26.5 \text{ m}}{\sin 54.0°}$$

$$c = \frac{(\sin 43.1°)(26.5 \text{ m})}{\sin 54.0°}$$

$$= 22.4 \text{ m}$$

9.

$$\frac{\sin C}{c} = \frac{\sin A}{a}$$

$$\frac{\sin C}{35.6 \text{ mi}} = \frac{\sin 20.1°}{47.5 \text{ mi}}$$

$$\sin C = \frac{(35.6 \text{ mi})(\sin 20.1°)}{47.5 \text{ mi}}$$

$$C = 14.9°$$

$$B = 180° - 20.1° - 14.9°$$

$$= 145.0°$$

$$\frac{b}{\sin B} = \frac{a}{\sin A}$$

$$\frac{b}{\sin 145.0°} = \frac{47.5 \text{ mi}}{\sin 20.1°}$$

$$b = \frac{(\sin 145.0°)(47.5 \text{ mi})}{\sin 20.1°}$$

$$= 79.3 \text{ mi}$$

11.

$$A = 180° - 56.4° - 48.7°$$

$$= 74.9°$$

$$\frac{a}{\sin A} = \frac{b}{\sin B}$$

$$\frac{a}{\sin 74.9°} = \frac{5960 \text{ m}}{\sin 56.4°}$$

$$a = \frac{(\sin 74.9°)(5960 \text{ m})}{\sin 56.4°}$$

$$= 6910 \text{ m}$$

$$\frac{c}{\sin C} = \frac{b}{\sin B}$$

$$\frac{c}{\sin 48.7°} = \frac{5960 \text{ m}}{\sin 56.4°}$$

$$c = \frac{(\sin 48.7°)(5960 \text{ m})}{\sin 56.4°}$$

$$= 5380 \text{ m}$$

13.

$$\frac{\sin C}{c} = \frac{\sin B}{b}$$

$$\frac{\sin C}{11.3 \text{ km}} = \frac{\sin 105.5°}{31.4 \text{ km}}$$

$$\sin C = \frac{(11.3 \text{ km})(\sin 105.5°)}{31.4 \text{ km}}$$

$$C = 20.3°$$

$$A = 180° - 105.5° - 20.3°$$

$$= 54.2°$$

$$\frac{a}{\sin A} = \frac{b}{\sin B}$$

$$\frac{a}{\sin 54.2°} = \frac{31.4 \text{ km}}{\sin 105.5°}$$

$$a = \frac{(\sin 54.2°)(31.4 \text{ km})}{\sin 105.5°}$$

$$= 26.4 \text{ km}$$

15.

$$\frac{\sin B}{b} = \frac{\sin A}{a}$$

$$\frac{\sin B}{189 \text{ ft}} = \frac{\sin 16.5°}{206 \text{ ft}}$$

$$\sin B = \frac{(189 \text{ ft})(\sin 16.5°)}{206 \text{ ft}}$$

$$B = 15.1°$$

$$C = 180° - 16.5° - 15.1°$$

$$= 148.4°$$

$$\frac{c}{\sin C} = \frac{a}{\sin A}$$

$$\frac{c}{\sin 148.4°} = \frac{206 \text{ ft}}{\sin 16.5°}$$

$$c = \frac{(\sin 148.4°)(206 \text{ ft})}{\sin 16.5°}$$

$$= 380 \text{ ft}$$

17.

$$C = 180° - 121.0° - 24.0°$$
$$= 35.0°$$

$$\frac{AC}{\sin 24.0°} = \frac{89.5 \text{ m}}{\sin 35.0°}$$

$$AC = \frac{\left(\sin 24.0°\right)\left(89.5 \text{ m}\right)}{\sin 35.0°}$$
$$= 63.5 \text{ m}$$

19.

$$C = 180° - 24.0° - 122.0°$$
$$= 34.0°$$

$$\frac{AB}{\sin 34.0°} = \frac{5.30 \text{ km}}{\sin 24.0°}$$

$$AB = \frac{\left(\sin 34.0°\right)\left(5.30 \text{ km}\right)}{\sin 24.0°}$$
$$= 7.29 \text{ km}$$

21. The obtuse angle in the triangle is $180° - 24.0° = 156.0°$ and the third angle of the triangle is $180° - 12.5° - 156.0° = 11.5°$. Let $x =$ distance from street.

$$\frac{x}{\sin 156.0°} = \frac{105 \text{ ft}}{\sin 11.5°}$$

$$x = \frac{\left(\sin 156.0°\right)\left(105 \text{ ft}\right)}{\sin 11.5°}$$
$$= 214 \text{ ft}$$

Section 14.4: Law of Sines: The Ambiguous Case

1. a. $h = \left(32.5 \text{ m}\right)\left(\sin 38.0°\right) = 20.0 \text{ m}$; $20.0 \text{ m} < 32.5 \text{ m} < 42.3 \text{ m}$; one triangle

b.

$$\frac{\sin B}{b} = \frac{\sin A}{a}$$

$$\frac{\sin B}{32.5 \text{ m}} = \frac{\left(\sin 38.0°\right)}{42.3 \text{ m}}$$

$$\sin B = \frac{\left(32.5 \text{ m}\right)\left(\sin 38.0°\right)}{42.3 \text{ m}}$$

$$B = 28.2°$$

$$C = 180° - 38.0° - 28.2°$$
$$= 113.8°$$

$$\frac{c}{\sin C} = \frac{a}{\sin A}$$

$$\frac{c}{\sin 113.8°} = \frac{42.3 \text{ m}}{\sin 38.0°}$$

$$c = \frac{\left(\sin 113.9°\right)\left(42.3 \text{ m}\right)}{\sin 38.0°}$$
$$= 62.9 \text{ mi}$$

3. a. $h = (275 \text{ m})(\sin 25.6°) = 119 \text{ m}$; $119 \text{ m} < 275 \text{ m} < 306 \text{ m}$; two triangles

   b.

   first triangle

$$\frac{\sin B}{b} = \frac{\sin A}{a}$$

$$\frac{\sin B}{306 \text{ m}} = \frac{\sin 25.6°}{275 \text{ m}}$$

$$\sin B = \frac{(306 \text{ m})(\sin 25.6°)}{275 \text{ m}}$$

$$B = 28.7°$$

$$C = 180° - 28.7° - 25.6°$$

$$= 125.7°$$

$$\frac{c}{\sin C} = \frac{a}{\sin A}$$

$$\frac{c}{\sin 125.7°} = \frac{275 \text{ m}}{\sin 25.6°}$$

$$c = \frac{(\sin 125.7°)(275 \text{ m})}{\sin 25.6°}$$

$$= 517 \text{ m}$$

   second triangle

$$B = 180° - 28.7° = 151.3°$$

$$C = 180° - 25.6° - 151.3°$$

$$= 3.1°$$

$$\frac{c}{\sin C} = \frac{a}{\sin A}$$

$$\frac{c}{\sin 3.1°} = \frac{275 \text{ m}}{\sin 25.6°}$$

$$c = \frac{(\sin 3.1°)(275 \text{ m})}{\sin 25.6°}$$

$$= 34.4 \text{ m}$$

5. a. $h = (48.5 \text{ m})(\sin 71.6°) = 46.0 \text{ m}$; $15.7 \text{ m} < 46.0 \text{ m}$; no triangle

7. a. $h = (245 \text{ cm})(\sin 71.2°) = 232 \text{ cm}$; $232 \text{ cm} < 238 \text{ cm} < 245 \text{ cm}$; two triangles

   b.

   first triangle

$$\frac{\sin A}{a} = \frac{\sin C}{c}$$

$$\frac{\sin A}{245 \text{ cm}} = \frac{\sin 71.2°}{238 \text{ cm}}$$

$$\sin A = \frac{(245 \text{ cm})(\sin 71.2°)}{238 \text{ cm}}$$

$$A = 77.0°$$

$$B = 180° - 77.0° - 71.2°$$

$$= 31.8°$$

$$\frac{b}{\sin B} = \frac{c}{\sin C}$$

$$\frac{b}{\sin 31.8°} = \frac{238 \text{ cm}}{\sin 71.2°}$$

$$b = \frac{(\sin 31.8°)(238 \text{ cm})}{\sin 71.2°}$$

$$= 132 \text{ ft}$$

7. (continued)

second triangle

$$\frac{b}{\sin B} = \frac{c}{\sin C}$$

$A = 180° - 77.0° = 103.0°$

$B = 180° - 103.0° - 71.2°$

$= 5.8°$

$$\frac{b}{\sin 5.8°} = \frac{238 \text{ cm}}{\sin 71.2°}$$

$$b = \frac{(\sin 5.8°)(238 \text{ cm})}{\sin 71.2°}$$

$$= 25.4 \text{ cm}$$

9. a. 24.0 mi < 33.0 mi; one triangle

b.

$$\frac{\sin A}{a} = \frac{\sin B}{b}$$

$$\frac{\sin A}{24.0 \text{ mi}} = \frac{\sin 105.0°}{33.0 \text{ mi}}$$

$$\sin A = \frac{(24.0 \text{ mi})(\sin 105.0°)}{33.0 \text{ mi}}$$

$$A = 44.6°$$

$$C = 180° - 44.6° - 105.0° = 30.4°$$

$$\frac{c}{\sin C} = \frac{b}{\sin B}$$

$$\frac{c}{\sin 30.4°} = \frac{33.0 \text{ mi}}{\sin 105.0°}$$

$$c = \frac{(\sin 30.4°)(33.0 \text{ mi})}{\sin 105.0°}$$

$$= 17.3 \text{ mi}$$

11. a. $h = (406 \text{ m})(\sin 31.5°) = 212 \text{ m}$; 212 m < 376 m < 406 m; two triangles

b.

first triangle

$$\frac{\sin C}{c} = \frac{\sin A}{a}$$

$$\frac{\sin C}{406 \text{ m}} = \frac{\sin 31.5°}{376 \text{ m}}$$

$$\sin C = \frac{(406 \text{ m})(\sin 31.5°)}{376 \text{ m}}$$

$$C = 34.3°$$

$$B = 180° - 31.5° - 34.3° = 114.2°$$

$$\frac{b}{\sin B} = \frac{a}{\sin A}$$

$$\frac{b}{\sin 114.2°} = \frac{376 \text{ m}}{\sin 31.5°}$$

$$b = \frac{(\sin 114.2°)(376 \text{ m})}{\sin 31.5°}$$

$$= 656 \text{ m}$$

second triangle

$$\frac{b}{\sin B} = \frac{a}{\sin A}$$

$C = 180° - 34.3° = 145.7°$

$B = 180° - 31.5° - 145.7°$

$= 2.8°$

$$\frac{b}{\sin 2.8°} = \frac{376 \text{ m}}{\sin 31.5°}$$

$$b = \frac{(\sin 2.8°)(376 \text{ m})}{\sin 31.5°}$$

$$= 35.2 \text{ m}$$

13. a. $h = (181 \text{ m})(\sin 60.0°) = 157 \text{ m}$; 151 m < 157 m; no triangle

15. a. $h = (855 \text{ m})(\sin 8.0°) = 119 \text{ m}; \ 119 \text{ m} < 451 \text{ m} < 855 \text{ m};$ two triangles

b.

first triangle

$$\frac{\sin C}{c} = \frac{\sin B}{b}$$

$$\frac{\sin C}{855 \text{ m}} = \frac{\sin 8.0°}{451 \text{ m}}$$

$$\sin C = \frac{(855 \text{ m})(\sin 8.0°)}{451 \text{ m}}$$

$$C = 15.3°$$

$$A = 180° - 15.3° - 8.0°$$

$$= 156.7°$$

$$\frac{a}{\sin A} = \frac{b}{\sin B}$$

$$\frac{a}{\sin 156.7°} = \frac{451 \text{ m}}{\sin 8.0°}$$

$$a = \frac{(\sin 156.7°)(451 \text{ m})}{\sin 8.0°}$$

$$= 1280 \text{ m}$$

second triangle

$$C = 180° - 15.3° = 164.7°$$

$$A = 180° - 164.7° - 8.0°$$

$$= 7.3°$$

$$\frac{a}{\sin A} = \frac{b}{\sin B}$$

$$\frac{a}{\sin 7.3°} = \frac{451 \text{ m}}{\sin 8.0°}$$

$$a = \frac{(\sin 7.3°)(451 \text{ m})}{\sin 8.0°}$$

$$= 412 \text{ m}$$

17. $h = (295 \text{ ft})(\sin 36.0°) = 173 \text{ ft}; \ 173 \text{ ft} < 295 \text{ ft} < 355 \text{ ft};$ two triangles

first triangle

$$\frac{\sin C}{AB} = \frac{\sin A}{BC}$$

$$\frac{\sin C}{355 \text{ ft}} = \frac{\sin 36.0°}{295 \text{ ft}}$$

$$\sin C = \frac{(355 \text{ ft})(\sin 36.0°)}{295 \text{ ft}}$$

$$C = 45.0°$$

$$B = 180° - 36.0° - 45.0°$$

$$= 99.0°$$

$$\frac{AC}{\sin B} = \frac{BC}{\sin A}$$

$$\frac{AC}{\sin 99.0°} = \frac{295 \text{ ft}}{\sin 36.0°}$$

$$AC = \frac{(\sin 99.0°)(295 \text{ ft})}{\sin 36.0°}$$

$$= 496 \text{ ft}$$

second triangle

$$C = 180° - 45.0° = 135.0°$$

$$B = 180° - 36.0° - 135.0°$$

$$= 9.0°$$

$$\frac{AC}{\sin B} = \frac{BC}{\sin A}$$

$$\frac{AC}{\sin 9.0°} = \frac{295 \text{ ft}}{\sin 36.0°}$$

$$AC = \frac{(\sin 9.0°)(295 \text{ ft})}{\sin 36.0°}$$

$$= 78.5 \text{ ft}$$

19.

Let x = length of lower support

θ = angle between support and building

A = angle between supports

$$\frac{\sin A}{20.00 \text{ ft}} = \frac{\sin 80.0°}{20.75 \text{ ft}}$$

$$\sin A = \frac{(20.00 \text{ ft})(\sin 80.0°)}{20.75 \text{ ft}}$$

$$A = 71.7°$$

$$\theta = 180° - 71.7° - 80.0°$$

$$= 28.3°$$

$$\frac{x}{\sin 28.3°} = \frac{20.75 \text{ ft}}{\sin 80.0°}$$

$$x = \frac{(\sin 28.3°)(20.75 \text{ ft})}{\sin 80.0°}$$

$$= 9.99 \text{ ft}$$

Section 14.5: Solving Oblique Triangles: Law of Cosines

1.

$$a^2 = b^2 + c^2 - 2bc \cos A$$

$$a^2 = (20.2 \text{ m})^2 + (25.0 \text{ m})^2 - 2(20.2 \text{ m})(25.0 \text{ m})\cos 55.0°$$

$$a = 21.3 \text{ m}$$

$$\frac{\sin B}{b} = \frac{\sin A}{a}$$

$$\frac{\sin B}{20.2 \text{ m}} = \frac{\sin 55.0°}{21.3 \text{ m}}$$

$$\sin B = \frac{(20.2 \text{ m})(\sin 55.0°)}{21.3 \text{ m}}$$

$$B = 51.0°$$

$$C = 180° - 51.0° - 55.0° = 74.0°$$

3.

$$c^2 = b^2 + c^2 - 2bc \cos C$$

$$c^2 = (247 \text{ ft})^2 + (316 \text{ ft})^2 - 2(247 \text{ ft})(316 \text{ ft})\cos 115.0°$$

$$c = 476 \text{ ft}$$

$$\frac{\sin A}{a} = \frac{\sin C}{c}$$

$$\frac{\sin A}{247 \text{ ft}} = \frac{\sin 115.0°}{476 \text{ ft}}$$

$$\sin A = \frac{(247 \text{ ft})(\sin 115.0°)}{476 \text{ ft}}$$

$$A = 28.1°$$

$$B = 180° - 28.1° - 115.0° = 36.9°$$

5.

$$b^2 = a^2 + c^2 - 2ac\cos B$$

$$(67,500\text{ mi})^2 = (38,500\text{ mi})^2 + (47,200\text{ mi})^2 - 2(38,500\text{ mi})(47,200\text{ mi})\cos B$$

$$\cos B = \frac{(67,500\text{ mi})^2 - (38,500\text{ mi})^2 - (47,200\text{ mi})^2}{-2(38,500\text{ mi})(47,200\text{ mi})}$$

$$B = 103.5°$$

$$\frac{\sin C}{c} = \frac{\sin B}{b}$$

$$\frac{\sin C}{47,200\text{ mi}} = \frac{\sin 103.5°}{67,500\text{ mi}}$$

$$\sin C = \frac{(47,200\text{ mi})(\sin 103.5°)}{67,500\text{ mi}}$$

$$C = 42.8°$$

$$A = 180° - 103.5° - 42.8° = 33.7°$$

7.

$$b^2 = a^2 + c^2 - 2ac\cos B$$

$$b^2 = (4820\text{ ft})^2 + (1930\text{ ft})^2 - 2(4820\text{ ft})(1930\text{ ft})\cos 19.3°$$

$$b = 3070\text{ ft}$$

$$\frac{\sin C}{c} = \frac{\sin B}{b}$$

$$\frac{\sin C}{1930\text{ ft}} = \frac{\sin 19.3°}{3070\text{ ft}}$$

$$\sin C = \frac{(1930\text{ ft})(\sin 19.3°)}{3070\text{ ft}}$$

$$C = 12.0°$$

$$A = 180° - 19.3° - 12.0° = 148.7°$$

9.

$$b^2 = a^2 + c^2 - 2ac\cos B$$

$$(36.5\text{ m})^2 = (19.5\text{ m})^2 + (25.6\text{ m})^2 - 2(19.5\text{ m})(25.6\text{ m})\cos B$$

$$\cos B = \frac{(36.5\text{ m})^2 - (19.5\text{ m})^2 - (25.6\text{ m})^2}{-2(19.5\text{ m})(25.6\text{ m})}$$

$$B = 107.3°$$

9. (continued)

$$\frac{\sin A}{a} = \frac{\sin B}{b}$$

$$\frac{\sin A}{19.5 \text{ m}} = \frac{\sin 107.3^\circ}{36.5 \text{ m}}$$

$$\sin A = \frac{(19.5 \text{ m})(\sin 107.3^\circ)}{36.5 \text{ m}}$$

$$A = 30.7^\circ$$

$$C = 180^\circ - 30.7^\circ - 107.3^\circ = 42.0^\circ$$

11.

$$a^2 = (105 \text{ m})^2 + (125 \text{ m})^2 - 2(105 \text{ m})(125 \text{ m})\cos 31.5^\circ$$

$$a = 65.3 \text{ m}$$

13.

$$(5.80 \text{ m})^2 = (4.50 \text{ m})^2 + (8.40 \text{ m})^2 - 2(4.50 \text{ m})(8.40 \text{ m})\cos A$$

$$\cos A = \frac{(5.80 \text{ m})^2 - (4.50 \text{ m})^2 - (8.40 \text{ m})^2}{-2(4.50 \text{ m})(8.40 \text{ m})}$$

$$A = 40.9^\circ$$

$$\frac{\sin C}{4.50 \text{ m}} = \frac{\sin 40.9^\circ}{5.80 \text{ m}}$$

$$\sin C = \frac{(4.50 \text{ m})(\sin 40.9^\circ)}{5.80 \text{ m}}$$

$$C = 30.5^\circ$$

15.

Let A = angle between 24.0 in. and 12.0 in. side

B = angle between 24.0 in. and 21.0 in. side

C = angle between 12.0 in. and 21.0 in. side

$$(24.0 \text{ in.})^2 = (12.0 \text{ in.})^2 + (21.0 \text{ in.})^2 - 2(12.0 \text{ ft})(8.40 \text{ m})\cos C$$

$$\cos C = \frac{(24.0 \text{ in.})^2 - (12.0 \text{ in.})^2 - (21.0 \text{ in.})^2}{-2(12.0 \text{ in.})(21.0 \text{ in.})}$$

$$C = 89.0^\circ$$

$$\frac{\sin B}{12.0 \text{ in.}} = \frac{\sin 89.0^\circ}{24.0 \text{ in.}}$$

$$\sin B = \frac{(12.0 \text{ in.})(\sin 89.0^\circ)}{24.0 \text{ in.}}$$

$$B = 30.0^\circ$$

$$A = 180^\circ - 30.0^\circ - 89.0^\circ = 61.0^\circ$$

17. The angle between the legs of the flight is $90.0° + 12.0° = 110.0°$.

Let d = distance to airport

$$d^2 = (70.0 \text{ mi})^2 + (90.0 \text{ mi})^2 - 2(70.0 \text{ mi})(90.0 \text{ mi})\cos 110.0°$$

$$d = 132 \text{ mi}$$

19.

Let d = distance from front tire to back tire

$$d^2 = (120.0 \text{ in.})^2 + (54.0 \text{ in.})^2 - 2(120.0 \text{ in.})(54.0 \text{ in.})\cos 110.0°$$

$$d = 147 \text{ in.}$$

21. Label the vertices between the 20.0 ft and 28.0 ft walls A and C, the vertices between the 20.0 ft walls B, and the vertices between the 28.0 ft walls D.

$$BD^2 = AB^2 + AD^2 - 2(AB)(AD)\cos \angle BAD$$

$$BD^2 = (20.0 \text{ ft})^2 + (28.0 \text{ ft})^2 - 2(20.0 \text{ ft})(28.0 \text{ ft})\cos 130.0°$$

$$BD = 43.6 \text{ ft}$$

$$\frac{\sin \angle ABD}{AD} = \frac{\sin \angle BAD}{BD}$$

$$\frac{\sin \angle ABD}{28.0 \text{ ft}} = \frac{\sin 130.0°}{43.6 \text{ ft}}$$

$$\sin \angle ABD = \frac{(28.0 \text{ ft})(\sin 130.0°)}{43.6 \text{ ft}}$$

$$\angle ABD = 29.5°$$

$$\angle ABC = 2(29.5°) = 59.0°$$

$$AC^2 = AB^2 + BC^2 - 2(AB)(BC)\cos \angle ABC$$

$$AC^2 = (20.0 \text{ ft})^2 + (20.0 \text{ ft})^2 - 2(20.0 \text{ ft})(20.0 \text{ ft})\cos 59.0°$$

$$AC = 19.7 \text{ ft}$$

23. $\angle EBA = \angle DBC = \dfrac{180° - 80.0°}{2} = 50.0°$

a.

$$\frac{\sin \angle BEA}{AB} = \frac{\sin \angle ABE}{AE}$$

$$\frac{\sin \angle BEA}{4.50 \text{ m}} = \frac{\sin 50.0°}{4.00 \text{ m}}$$

$$\sin \angle BEA = \frac{(4.50 \text{ m})(\sin 50.0°)}{4.00 \text{ m}}$$

$$\angle BEA = 59.5°$$

b. $\angle A = 180° - 50.0° - 59.5° = 70.5°$

23. (continued)

c.

$$BE^2 = AE^2 + AB^2 - 2(AE)(AB)\cos A$$

$$BE^2 = (4.00 \text{ m})^2 + (4.50 \text{ m})^2 - 2(4.00 \text{ m})(4.50 \text{ m})\cos 70.5°$$

$$BE = 4.92 \text{ m}$$

d.

$$DE^2 = BD^2 + BE^2 - 2(BD)(BE)\cos \angle DBE$$

$$DE^2 = (4.92 \text{ m})^2 + (4.92 \text{ m})^2 - 2(4.92 \text{ m})(4.92 \text{ m})\cos 80.0°$$

$$DE = 6.33 \text{ m}$$

25. The largest angle is opposite the longest side.

Let θ = largest angle

$$(2860 \text{ ft})^2 = (1580 \text{ ft})^2 + (1820 \text{ ft})^2 - 2(1580 \text{ ft})(1820 \text{ ft})\cos\theta$$

$$\cos\theta = \frac{(2860 \text{ ft})^2 - (1580 \text{ ft})^2 - (1820 \text{ ft})^2}{-2(1580 \text{ ft})(1820 \text{ ft})} = -0.4122$$

$$\theta = 114.3°$$

27.

$$AC^2 = (4\bar{0} \text{ yd})^2 + (5\bar{0} \text{ yd})^2 - 2(4\bar{0} \text{ yd})(5\bar{0} \text{ yd})\cos 45°$$

$$AC = 36 \text{ yd}$$

29. Let x = distance from horizontal top of side of house to peak of roof.

$$\frac{x}{30.0 \text{ ft}} = \frac{1}{12}$$

$$12x = (30.0 \text{ ft})$$

$$x = 2.50 \text{ ft}$$

$$x = \sqrt{(30.0 \text{ ft})^2 + (9.50 \text{ ft})^2} = 31.5 \text{ft}$$

31.

Let d = distance from person to kite

$$d^2 = (125 \text{ m})^2 + (90.0 \text{ m})^2 - 2(125 \text{ m})(90.0 \text{ m})\cos 75.0°$$

$$d = 134 \text{ m}$$

Chapter 14 Review

1. $\tan 143° = -0.7536$

2. $\sin 209.8° = -0.4970$

3. $\cos 317.4° = 0.7361$

4.

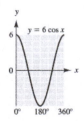

5.

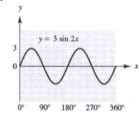

6. $P = \dfrac{360^\circ}{3} = 120^\circ$; $A = 5$

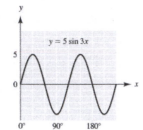

7. $P = \dfrac{360^\circ}{4} = 90^\circ$; $A = 3$

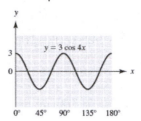

8.

$$P = \frac{360^\circ}{1} = 360^\circ; \ A = 4$$

$$\frac{C}{B} = \frac{60^\circ}{1} = 60^\circ \ \text{or} \ 60^\circ \ \text{left}$$

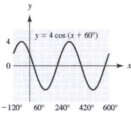

9.

$$P = \frac{360^\circ}{2} = 180^\circ; \ A = 6$$

$$\frac{C}{B} = \frac{-180^\circ}{2} = -90^\circ \ \text{or} \ 90^\circ \ \text{right}$$

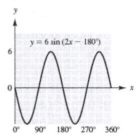

10. $h = (175 \text{ m})(\sin 52.7^\circ) = 139 \text{ m}$; $139 \text{ m} < 175 \text{ m} < 206 \text{ m}$; one triangle

$$\frac{\sin A}{a} = \frac{\sin B}{b}$$

$$\frac{\sin A}{175 \text{ m}} = \frac{\sin 52.7^\circ}{206 \text{ m}}$$

$$\sin A = \frac{(175 \text{ m})(\sin 52.7^\circ)}{206 \text{ m}}$$

$$A = 42.5^\circ$$

$$C = 180^\circ - 42.5^\circ - 52.7^\circ$$

$$= 84.8^\circ$$

$$\frac{c}{\sin C} = \frac{b}{\sin B}$$

$$\frac{c}{\sin 84.8^\circ} = \frac{206 \text{ m}}{\sin 52.7^\circ}$$

$$c = \frac{(\sin 84.8^\circ)(206 \text{ m})}{\sin 52.7^\circ}$$

$$= 258 \text{ m}$$

11.

$$B = 180^\circ - 61.2^\circ - 75.6^\circ$$

$$= 43.2^\circ$$

$$\frac{a}{\sin A} = \frac{c}{\sin C}$$

$$\frac{a}{\sin 61.2^\circ} = \frac{88.0 \text{ m}}{\sin 75.6^\circ}$$

$$a = \frac{(\sin 61.2^\circ)(88.0 \text{ m})}{\sin 75.6^\circ}$$

$$= 79.6 \text{ m}$$

$$\frac{b}{\sin B} = \frac{c}{\sin C}$$

$$\frac{a}{\sin 43.2^\circ} = \frac{88.0 \text{ m}}{\sin 75.6^\circ}$$

$$a = \frac{(\sin 43.2^\circ)(88.0 \text{ m})}{\sin 75.6^\circ}$$

$$= 62.2 \text{ m}$$

12.

$$b^2 = a^2 + c^2 - 2ac\cos B$$

$$b^2 = (345 \text{ m})^2 + (405 \text{ m})^2 - 2(345 \text{ m})(405 \text{ m})\cos 17.5°$$

$$b = 129 \text{ m}$$

$$\frac{\sin A}{a} = \frac{\sin B}{b}$$

$$\frac{\sin A}{(345 \text{ m})} = \frac{\sin 17.5°}{129 \text{ m}}$$

$$\sin A = \frac{(345 \text{ m})(\sin 17.5°)}{129 \text{ m}}$$

$$A = 53.5°$$

$$C = 180° - 53.5° - 17.5° = 109.0°$$

13.

$$c^2 = a^2 + b^2 - 2ab\cos C$$

$$(51.5 \text{ cm})^2 = (48.6 \text{ cm})^2 + (31.2 \text{ cm})^2 - 2(48.6 \text{ cm})(31.2 \text{ cm})\cos C$$

$$\cos C = \frac{(51.5 \text{ cm})^2 - (48.6 \text{ cm})^2 - (31.2 \text{ cm})^2}{-2(48.6 \text{ cm})(31.2 \text{ cm})}$$

$$C = 77.0°$$

$$\frac{\sin B}{b} = \frac{\sin C}{c}$$

$$\frac{\sin B}{31.2 \text{ cm}} = \frac{\sin 77.0°}{51.5 \text{ cm}}$$

$$\sin B = \frac{(31.2 \text{ cm})(\sin 77.0°)}{51.5 \text{ cm}}$$

$$B = 36.2°$$

$$A = 180° - 36.2° - 77.0° = 66.8°$$

14. $h = (20.5 \text{ m})(\sin 29.5°) = 10.1 \text{ m}; \ 10.1 \text{ m} < 18.5 \text{ m} < 20.5 \text{ m}; \text{ two triangles}$

first triangle

$$\frac{\sin B}{b} = \frac{\sin A}{a}$$

$$\frac{\sin B}{20.5 \text{ m}} = \frac{\sin 29.5°}{18.5 \text{ m}}$$

$$\sin B = \frac{(20.5 \text{ m})(\sin 29.5°)}{18.5 \text{ m}}$$

$$B = 33.1°$$

$$C = 180° - 29.5° - 33.1°$$

$$= 117.4°$$

$$\frac{c}{\sin C} = \frac{b}{\sin B}$$

$$\frac{c}{\sin 117.4°} = \frac{20.5 \text{ m}}{\sin 33.1°}$$

$$c = \frac{(\sin 117.4°)(20.5 \text{ m})}{\sin 33.1°}$$

$$= 33.4 \text{ m}$$

14. (continued)

second triangle

$$B = 180° - 33.1° = 146.9°$$

$$C = 180° - 29.5° - 146.9°$$

$$= 3.6°$$

$$\frac{c}{\sin C} = \frac{b}{\sin B}$$

$$\frac{c}{\sin 3.6°} = \frac{20.5 \text{ m}}{\sin 146.9°°}$$

$$c = \frac{\left(\sin 3.6°\right)\left(20.5 \text{ m}\right)}{\sin 146.9°}$$

$$= 2.36 \text{ m}$$

15. $h = (1680 \text{ m})\left(\sin 18.5°\right) = 553 \text{ m};\ 553 \text{ m} < 1520 \text{ m} < 1680 \text{ m};$ two triangles

first triangle

$$\frac{\sin A}{a} = \frac{\sin B}{b}$$

$$\frac{\sin A}{1680 \text{ m}} = \frac{\sin 18.5°}{1520 \text{ m}}$$

$$\sin A = \frac{(1680 \text{ m})\left(\sin 18.5°\right)}{1520 \text{ m}}$$

$$A = 20.5°$$

$$C = 180° - 20.5° - 18.5°$$

$$= 141.0°$$

$$\frac{c}{\sin C} = \frac{b}{\sin B}$$

$$\frac{c}{\sin 141.0°} = \frac{1520 \text{ m}}{\sin 18.5°}$$

$$c = \frac{\left(\sin 141.0°\right)\left(1520 \text{ m}\right)}{\sin 18.5°}$$

$$= 3010 \text{ m}$$

second triangle

$$A = 180° - 20.5° = 159.5°$$

$$C = 180° - 159.5° - 18.5°$$

$$= 2.0°$$

$$\frac{c}{\sin C} = \frac{b}{\sin B}$$

$$\frac{c}{\sin 2.0°} = \frac{1520 \text{ m}}{\sin 18.5°}$$

$$c = \frac{\left(\sin 2.0°\right)\left(1520 \text{ m}\right)}{\sin 18.5°}$$

$$= 167 \text{ m}$$

16.

$$(1250 \text{ ft})^2 = (575 \text{ ft})^2 + (1080 \text{ ft})^2 - 2(575 \text{ ft})(1080 \text{ ft})\cos C$$

$$\cos C = \frac{(1250 \text{ ft})^2 - (575 \text{ ft})^2 - (1080 \text{ ft})^2}{-2(575 \text{ ft})(1080 \text{ ft})}$$

$$C = 93.0°$$

$$\frac{\sin A}{575 \text{ ft}} = \frac{\sin 93.0°}{1250 \text{ ft}}$$

$$\sin A = \frac{(575 \text{ ft})\left(\sin 93.0°\right)}{1250 \text{ ft}}$$

$$A = 27.3°$$

$$B = 180° - 27.3° - 93.0° = 59.7°$$

17. $h = (81.2 \text{ ft})(\sin 73.5°) = 77.9 \text{ ft}; \ 58.2 \text{ ft} < 77.9 \text{ ft}; \text{ no triangle}$

18. a. $AB = 12.5 \text{ m} + 88.0 \text{ m} = 100.5 \text{ m}$

$$b^2 = a^2 + c^2 - 2ac \cos B$$

$$(60.0 \text{ m})^2 = (60.0 \text{ m})^2 + (100.5 \text{ m})^2 - 2(100.5 \text{ m})(77.0 \text{ m})\cos B$$

$$\cos B = \frac{(60.0 \text{ m})^2 - (100.5 \text{ m})^2 - (77.0 \text{ m})^2}{-2(100.5 \text{ m})(77.0 \text{ m})}$$

$$B = 36.6°$$

b.

$$b^2 = a^2 + c^2 - 2ac \cos B$$

$$x^2 = (88.0 \text{ m})^2 + (77.0 \text{ m})^2 - 2(88.0 \text{ m})(77.0 \text{ m})\cos 36.6°$$

$$x = 52.8 \text{ m}$$

19. $\dfrac{360°}{5} = 72°; \ r = \dfrac{1}{2}(16.0 \text{ in.}) = 8.00 \text{ in.}$

Let d = distance between holes

$$d^2 = (8.00 \text{ in.})^2 + (8.00 \text{ in.})^2 - 2(8.00 \text{ in.})(8.00 \text{ in.})\cos 72°$$

$$d = 9.4 \text{ in.}$$

20. $AC = \dfrac{1}{2}AE = \dfrac{1}{2}(36.0 \text{ m}) = (18.0 \text{ m})$

a.

$$\cos A = \frac{AC}{AF}$$

$$\cos 40.0° = \frac{18.0 \text{ m}}{AF}$$

$$AF = \frac{18.0 \text{ m}}{\cos 40.0°}$$

$$= 23.5 \text{ m}$$

b.

$$\angle ABF = 180° - 65.0° = 115.0°$$

$$\frac{BF}{\sin A} = \frac{AF}{\sin \angle ABF}$$

$$\frac{BF}{\sin 40.0°} = \frac{23.5 \text{ m}}{\sin 115.0°}$$

$$BF = \frac{(\sin 40.0°)(23.5 \text{ m})}{\sin 115.0°}$$

$$= 16.7 \text{ m}$$

c.

$$\tan A = \frac{CF}{AC}$$

$$\tan 40.0° = \frac{CF}{18.0 \text{ m}}$$

$$CF = (\tan 40.0°)(18.0 \text{ m})$$

$$= 15.1 \text{ m}$$

d.

$$\tan \angle FBC = \frac{CF}{BC}$$

$$\tan 65.0° = \frac{15.1 \text{ m}}{BC}$$

$$BC = \frac{15.1 \text{ m}}{\tan 65.0°}$$

$$= 7.04 \text{ m}$$

Chapter 14 Test

1. $\cos 182.9° = -0.9988$

5. $C = 180° - 68.3° - 34.0° = 77.7°$

3.

$$P = \frac{360°}{3} = 120°; A = 2$$

$$\frac{C}{B} = \frac{45°}{3} = 15° \text{ or } 15° \text{ left}$$

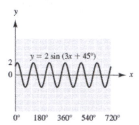

7.

$$(20.2 \text{ cm})^2 = (19.6 \text{ cm})^2 + (17.8 \text{ cm})^2 - 2(19.6 \text{ cm})(17.8 \text{ cm})\cos C$$

$$\cos C = \frac{(20.2 \text{ cm})^2 - (19.6 \text{ cm})^2 - (17.8 \text{ cm})^2}{-2(19.6 \text{ cm})(17.8 \text{ cm})}$$

$$C = 65.2°$$

9.

$$(17.8 \text{ cm})^2 = (19.6 \text{ cm})^2 + (20.2 \text{ cm})^2 - 2(19.6 \text{ cm})(20.2 \text{ cm})\cos A$$

$$\cos A = \frac{(17.8 \text{ cm})^2 - (19.6 \text{ cm})^2 - (20.2 \text{ cm})^2}{-2(19.6 \text{ cm})(20.2 \text{ cm})}$$

$$A = 53.1°$$

11.

$$\frac{\sin A}{11.7 \text{ft}} = \frac{\sin 63.0°}{21.4 \text{ ft}}$$

$$\sin A = \frac{(11.7 \text{ ft})(\sin 63.0°)}{21.4 \text{ ft}}$$

$$A = 29.2°$$

Cumulative Review Chapters 1-14

1.

$$R = \frac{Vt}{I}$$

$$R = \frac{(32)(5)}{20} = 8$$

2.

$$-\frac{5}{6} - \left(-\frac{3}{5}\right)$$

$$= -\frac{25}{30} - \left(-\frac{18}{30}\right)$$

$$= -\frac{25}{30} + \frac{18}{30}$$

$$= -\frac{7}{30}$$

3. 4.18×10^4

4. $90 \text{ kg} \times \dfrac{2.205 \text{ lb}}{1 \text{ kg}} = 198 \text{ lb}$

5. $13 \, \Omega$

6. a. 0.0001 m

b. $\dfrac{0.0001 \text{ m}}{2} = 0.00005 \text{ m}$

c. $\dfrac{0.00005 \text{ m}}{0.0018 \text{ m}} = 0.0278$

d. $0.0278 = 2.78\%$

7.

$$\left(2a^2 - 5a + 3\right) + \left(4a^2 + 3a - 1\right)$$
$$= \left(2a^2 + 4a^2\right) + \left(-5a + 3a\right) + \left(3 - 1\right)$$
$$= 6a^2 - 2a + 2$$

8.

$$6 + 3(x - 2) = 24$$
$$6 + 3x - 6 = 24$$
$$3x = 24$$
$$\dfrac{3x}{3} = \dfrac{24}{3}$$
$$x = 8$$

9.

$$\dfrac{x}{2} - \dfrac{2}{7} = \dfrac{1}{3}$$
$$42\left(\dfrac{x}{2} - \dfrac{2}{7}\right) = \left(\dfrac{1}{3}\right)42$$
$$21x - 12 = 14$$
$$21x - 12 + 12 = 14 + 12$$
$$21x = 26$$
$$\dfrac{21x}{21} = \dfrac{26}{21}$$
$$x = \dfrac{26}{21}$$

10.

$$\dfrac{A_1}{l_1} = \dfrac{A_2}{l_2}$$
$$\dfrac{A_1}{42.5 \text{ m}} = \dfrac{30.8 \text{ m}^2}{12.8 \text{ m}}$$
$$(12.8 \text{ m})\left(A_1\right) = (42.5 \text{ m})\left(30.8 \text{ m}^2\right)$$
$$A_1 = \dfrac{(42.5 \text{ m})\left(30.8 \text{ m}^2\right)}{12.8 \text{ m}}$$
$$= 102 \text{ m}^2$$

11.

$$-5x - 3y = -8$$
$$-5x - 3y + 5x = -8 + 5x$$
$$-3y = 5x - 8$$
$$\dfrac{-3y}{-3} = \dfrac{5x - 8}{-3}$$
$$y = \dfrac{8 - 5x}{3}$$

12.

$$7x - y = 4$$
$$14x - 2y = 8$$

$$(-2)(7x - y) = (4)(-2)$$
$$14x - 2y = 8$$

$$-14x + 2y = -8$$
$$14x - 2y = 8$$
$$0 = 0$$

The lines coincide, so there are many solutions.

13. $x^2 - 2x - 168 = (x - 14)(x + 12)$

$$3x^2 - 6x - 189$$

14. $= 3\left(x^2 - 2x - 63\right)$

$$= 3(x + 7)(x - 9)$$

15.

$$3x^2 - 13x = 10$$
$$3x^2 - 13x - 10 = 0$$
$$(3x + 2)(x - 5) = 0$$
$$3x + 2 = 0 \text{ or } x - 5 = 0$$
$$x = -\dfrac{2}{3} \text{ or } x = 5$$

16.

$$x = \frac{-b \pm \sqrt{b^2 - 4ac}}{2a}$$

$$a = 2, \ b = -1, \ c = -8$$

So $x = \dfrac{-(-1) \pm \sqrt{(-1)^2 - 4(2)(-8)}}{2(2)}$

$$= \frac{1 \pm \sqrt{1 + 64}}{4}$$

$$= \frac{1 \pm \sqrt{65}}{4}$$

$$= \frac{1 - \sqrt{65}}{4} \quad \text{or} \quad \frac{1 + \sqrt{65}}{4}$$

$$= -1.77 \quad \text{or} \quad 2.27$$

17. a.

$$A = \frac{1}{2}bh$$

$$A = \frac{1}{2}(20.1 \text{ ft})(51.2 \text{ ft}) = 515 \text{ ft}^2$$

b.

$$c = \sqrt{(20.1 \text{ ft})^2 + (51.2 \text{ ft})^2} = 55.0 \text{ ft}$$

$$P = a + b + c$$

$$P = 20.1 \text{ ft} + 51.2 \text{ ft} + 55.0 \text{ ft} = 126.3 \text{ ft}$$

18.

$$A = lw$$

$$w = \frac{A}{l}$$

$$w = \frac{307 \text{ ft}^2}{22.4 \text{ ft}}$$

$$= 13.7 \text{ ft}$$

19.

$$B_1 = (0.981 \text{ m})(0.827 \text{ m}) = 0.811 \text{ m}^2$$

$$B_2 = (1.92 \text{ m})(2.31 \text{ m}) = 4.44 \text{ m}^2$$

$$V = \frac{1}{3}h\left(B_1 + B_2 + \sqrt{B_1 B_2}\right)$$

$$V = \frac{1}{3}(2.68 \text{ m})\left(0.811 \text{ m}^2 + 4.44 \text{ m}^2 + \sqrt{(0.811 \text{ m}^2)(4.44 \text{ m}^2)}\right)$$

$$= 6.39 \text{ m}^3$$

20. $\tan 67.2° = 2.379$

21. $\cos^{-1} 0.6218 = 51.6°$

22. $\angle B = 90° - 34.2° = 55.8°$

23.

$$\tan A = \frac{a}{b}$$

$$\tan 34.2° = \frac{a}{15.8 \text{ cm}}$$

$$a = (15.8 \text{ cm})(\tan 34.2°)$$

$$= 10.7 \text{ cm}$$

24.

$$\cos A = \frac{b}{c}$$

$$\cos 34.2° = \frac{15.8 \text{ cm}}{c}$$

$$c = \frac{15.8 \text{ cm}}{\cos 34.2°}$$

$$= 19.1 \text{ cm}$$

25.

$$\text{pitch} = \frac{8.00 \text{ ft}}{24.0 \text{ ft}} = \frac{2}{3}$$

Let d = distance, eave to peak

$$d = \sqrt{(8.00 \text{ ft})^2 + (12.0 \text{ ft})^2}$$

$$= 14.4 \text{ ft}$$

$$d = 14.4 \text{ ft} \times \frac{12 \text{ in.}}{1 \text{ ft}} = 173 \text{ in.}$$

26. $\cos 191.13° = -0.9812$

27.

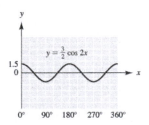

28. $P = \dfrac{360°}{1/2} = 720°; A = 2$

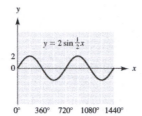

29.

$$(20.2 \text{ cm})^2 = (19.6 \text{ cm})^2 + (17.8 \text{ cm})^2 - 2(19.6 \text{ cm})(17.8 \text{ cm})\cos C$$

$$\cos C = \frac{(20.2 \text{ cm})^2 - (19.6 \text{ cm})^2 - (17.8 \text{ cm})^2}{-2(19.6 \text{ cm})(17.8 \text{ cm})}$$

$$C = 65.2°$$

30.

$$b^2 = (28.0 \text{ ft})^2 + (12.2 \text{ ft})^2 - 2(28.0 \text{ ft})(12.2 \text{ ft})\cos 60.0°$$

$$b = 24.3 \text{ ft}$$

$$\frac{\sin A}{12.2 \text{ ft}} = \frac{\sin 60.0°}{24.3 \text{ ft}}$$

$$\sin A = \frac{(12.2 \text{ ft})(\sin 60.0°)}{24.3 \text{ ft}}$$

$$A = 25.8°$$

Chapter 15: Basic Statistics

Section 15.1: Bar Graphs

1. $4400

3. $2100

5. $1600

7. $3700

9. $3200

11. 1.9 million barrels

13. United States

15. 3 million barrels

17. 19 million barrels

19. 2.2 million barrels

21.

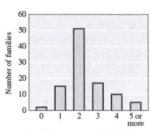

23.

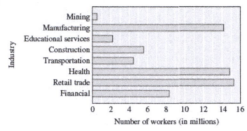

25.

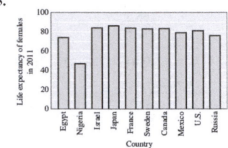

Section 15.2: Circle Graphs

1. $0.26 \times 360° = 94°$

3. $0.152 \times 360° = 55°$

5. $0.75 \times 360° = 270°$

7.
$$\frac{452}{744} = \frac{r}{100}$$
$$744r = 45,200$$
$$r = \frac{45,200}{744} = 60.8\%$$
$$0.608 \times 360° = 219°$$

9.
$$\frac{208}{5020} = \frac{r}{100}$$
$$5020r = 20,800$$
$$r = \frac{20,800}{5020} = 4.14\%$$
$$0.0414 \times 360° = 15°$$

11.
$$\frac{182,000 - 16,192}{182,000} = \frac{r}{100}$$
$$\frac{165,800}{182,000} = \frac{r}{100}$$
$$182,000r = 1,658,600$$
$$r = \frac{1,658,600}{182,000}$$
$$= 91.1\%$$
$$0.911 \times 360° = 328°$$

13.

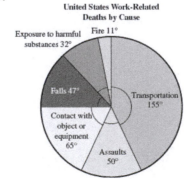

United States Work-Related Deaths by Cause

15.

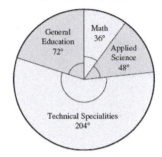

Industrial Technology
Credit Hour Requirements

17.

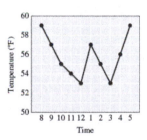

Total Forest by Continent in Year 2011
in Millions of Hectares

19.

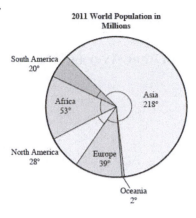

2011 World Population in
Millions

Section 15.3: Line Graphs

1.

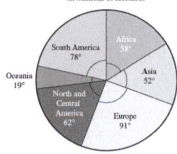

3.

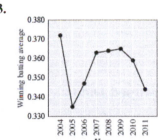

5. 29.52 in.; Fri. near midnight.

7. 29.34 in.

9. 98%

11. 90°

13. 95%

15. a. 75 ppmv

b. 1995

Section 15.4: Other Graphs

1. 900 mW; 14 dB

3. 85 mW

5. 30 dB; 32 dB

7. 16 dB

9. 14 dB; 16 dB

Section 15.5: Mean Measurement

1. 6911

3. 2020

5. 1.096

7. 51

9. 4.58

11. 47.89 cm

13. 0.2619 in.

15. 30,292 mi

17. 73°

19. 161 kW

21. 1101.2 million tons

Section 15.6: Other Average Measurements and Percentiles

1. 6185.5

3. 2023

5. 1.102

7. 53

9. 4.17

11. 47.87 cm

13. 0.2618 in.

15. 28,985 mi

17. 72.5°

19. 169 kW

21. 163 $\left(0.94 \times 50 = 47\right)$

23. 107 $\left(0.55 \times 50 = 28\right)$

25. 23 $\left(0.05 \times 50 = 3\right)$

27. 2.85 mm

29. 2 modes: 7291, 7285

31. 28

33. 6.2 in.

35. 0.024 in.

37. a. 9

b. 9

c. $\dfrac{7+11}{2} = 9$

Section 15.7: Range and Standard Deviation

1. 6061

3. 447

5. 0.208

7. 73

9. 8.54

11. 0.62 cm

13. 0.0010 in.

15. 19,393 mi

17. F 20° F

19. 101 kW

21. 1931

23. 150

25. 0.064

27. 25

29. 2.45

31. 0.26 cm

33. 0.0003 in.

35. 7221 mi

37. 7° F

39. 39 kW

Section 15.8: Grouped Data

1. 59

3. 7.7 months

5. 36 h

7. 42 lb

15. 58 players

9. 11

11. 4.2 days

13. 99 strokes

Interval	Midpoint	Frequency	Product
18.5–43.5	31	14	434
43.5–68.5	56	7	392
68.5–93.5	81	4	324
93.5–118.5	106	3	318
118.5–143.5	131	0	0
143.5–168.5	156	0	0
168.5–193.5	181	0	0
193.5–218.5	206	1	206
		29	1674
Mean = 58 players			

Section 15.9: Standard Deviation for Grouped Data

1. 9

3. 4 months

5. 5.2 h

7. 15 lb

9. 4

11. 3 days

13. 12 strokes

15. 38 players

17. a. 4 mm

b. 1.3 mm

Section 15.10: Statistical Process Control

1. a.

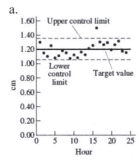

b. The process is out of control; in hour 16, the value was outside of limits.

3. a.

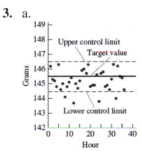

b. The process is out of control; at hour 12, the point is outside the control limits

5. a.

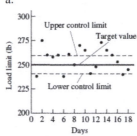

b. The process is out of control; days 2–4, 6, 7, 9–11, 13–15 all are outside the control limits.

Section 15.11: Other Graphs for Statistical Data

1.

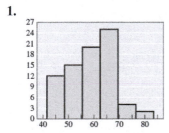

3.

5.

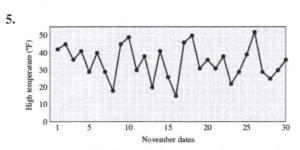

7.

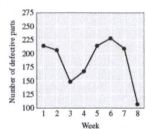

11. a.

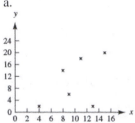

b. no linear correlation

9. a.

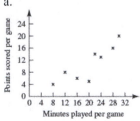

b. positive correlation

Section 15.12: Normal Distribution

1. $85 - 2(15) = 55$ and $85 + 2(15) = 115$, so approximately 95%.

3. a. 63

b. 6.9

c. The data does not form a normal distribution.

One standard deviation:

$$63 - 6.9 = 56.1 \text{ and } 63 + 6.9 = 69.9 \text{ ; } \frac{200 + 188}{508} = 76\%$$

Two standard deviations:

$$63 - 2(6.9) = 49.2 \text{ and } 63 + 2(6.9) = 76.8 \text{ ; } \frac{50 + 200 + 188 + 40}{508} = 94\%$$

Three standard deviations:

$$63 - 3(6.9) = 42.3 \text{ and } 63 + 3(6.9) = 83.7 \text{ ; } \frac{10 + 50 + 200 + 188 + 40 + 20}{508} = 100\%$$

5. a. 3544

b. 1033

c. The data does form a normal distribution

One standard deviation:

$$3544 - 1033 = 2511 \text{ and } 3544 + 1033 = 4577 \text{ ; } \frac{4100 + 2450}{9760} = 67\%$$

Two standard deviations:

$$3544 - 2(1033) = 1478 \text{ and } 3544 + 2(1033) = 5610 \text{ ; } \frac{2050 + 4100 + 2450 + 420}{9760} = 92\%$$

Three standard deviations:

$$3544 - 3(1033) = 445 \text{ and } 3544 + 3(1033) = 6643 \text{ ; } \frac{480 + 2050 + 4100 + 2450 + 420 + 155}{9760} = 99\%$$

7. One standard deviation:

$12 - 2.8 = 9.2$ and $12 + 2.8 = 14.8$

$0.68 \times 100,000 = 68,000$ blouses of sizes 10, 12, and 14.

Two standard deviations:

$12 - 2(2.8) = 6.4$ and $12 + 2(2.8) = 17.6$

$0.95 \times 100,000 = 95,000$ blouses of sizes 8, 10, 12, 14, and 16, so $95,000 - 68,000 = 27,000$ blouses of sizes 8 and 16.

Three standard deviations:

$12 - 3(2.8) = 3.6$ and $12 + 3(2.8) = 20.4$

$0.997 \times 100,000 = 99,700$ blouses of sizes 4, 6, 8, 10, 12, 14, 16, and 18, so $99,700 - 95,000 = 4700$ blouses of sizes 4, 6, and 18.

There will also be $100,000 - 99,700 = 300$ blouses of size 2.

Section 15.13: Probability

1. [A, 2, 3, 4, 5, 6, 7, 8, 9, 10, J, Q, K]

3. [J diamonds, Q diamonds, K diamonds, J hearts, Q hearts, K hearts]

5. [{red, red}, {red, white}, {red, white}]

7. $p = \dfrac{1}{13}$

9. $p = \dfrac{1}{6}$

11. $p = \dfrac{0}{8} = 0$

13. $p = \dfrac{6}{28} = \dfrac{3}{14}$

15. $p = \dfrac{3}{21} = \dfrac{1}{7}$

17. $p = \dfrac{52}{10,000} = \dfrac{13}{2500}$

Section 15.14: Independent Events

1. $p(\text{red and red}) = p(\text{red}) \cdot p(\text{red}) = \dfrac{1}{4} \cdot \dfrac{1}{4} = \dfrac{1}{16}$

3. $p(\text{green and red}) = p(\text{green}) \cdot p(\text{red}) = \dfrac{8}{16} \cdot \dfrac{5}{16} = \dfrac{5}{32}$

5. $p(\text{helmet and helmet}) = p(\text{helmet}) \cdot p(\text{helmet}) = \dfrac{6}{10} \cdot \dfrac{6}{10} = \dfrac{9}{25}$

7. $p(\text{H and 5}) = p(\text{H}) \cdot p(5) = \dfrac{1}{2} \cdot \dfrac{1}{6} = \dfrac{1}{12}$

9. $p(4 \text{ aces}) = p(\text{ace}) \cdot p(\text{ace}) \cdot p(\text{ace}) \cdot p(\text{ace}) = \dfrac{4}{52} \cdot \dfrac{4}{52} \cdot \dfrac{4}{52} \cdot \dfrac{4}{52} = \dfrac{1}{28,561}$

11. $p(3 \text{ and } 3 \text{ and } 3) = p(3) \cdot p(3) \cdot p(3) = \dfrac{1}{6} \cdot \dfrac{1}{6} \cdot \dfrac{1}{6} = \dfrac{1}{216}$

13. $p(6 \text{ and } 4) = p(6) \cdot p(4) = \dfrac{1}{7} \cdot \dfrac{1}{7} = \dfrac{1}{49}$

15. $p(\text{red and white and blue}) = p(\text{red}) \cdot p(\text{white}) \cdot p(\text{blue}) = \dfrac{7}{19} \cdot \dfrac{4}{19} \cdot \dfrac{3}{19} = \dfrac{84}{6859}$

Chapter 15 Review

1. $0.35 \times 360° = 126°$

2. $0.561 \times 360° = 202°$

3.

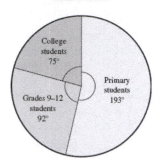

Students in 2000

4.

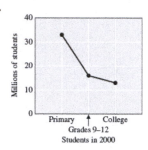

5. 80° F

6. 7.0036 mm

7. 7.0036 mm

8. 0.0001 mm

9. a. 49.9

   b. 18.3

10. 84.4

11. 83

12. 7.4

13. 19.2

14. 5.2

15. a. $[1, 2, 3, 4, 5]$

   b. $p(\text{odd}) = \dfrac{3}{5}$

16. $p(\text{red and red and black}) = p(\text{red}) \cdot p(\text{red}) \cdot p(\text{black}) = \dfrac{3}{18} \cdot \dfrac{3}{18} \cdot \dfrac{10}{18} = \dfrac{5}{324}$

Chapter 15 Test

1. Japan

3. $0.38 \times 360° = 137°$

5.

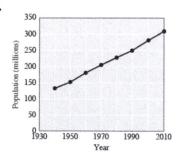

7. 50 mW

9.

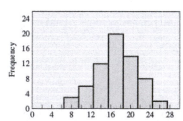

11. [ace of spades, two of spades, three of spades, ace of diamonds, two of diamonds, three of diamonds, ace of clubs, two of clubs, three of clubs, ace of hearts, two of hearts, three of hearts]

13. 3.0

15. $p(\text{spade}) = \dfrac{13}{52} = \dfrac{1}{4}$

Chapter 16: Binary and Hexadecimal Numbers

Section 16.1: Introduction to Binary Numbers

1. 3	**9.** 47	**17.** 140
3. 6	**11.** 78	**19.** 63
5. 9	**13.** 59	
7. 51	**15.** 156	

Section 16.2: Addition of Binary Numbers

1. 1000	**11.** 100100	**21.** 100110
3. 1011	**13.** 1000111	**23.** 1010001
5. 1100	**15.** 1000101	**25.** 1000000
7. 10000	**17.** 1110101	**27.** 1000011
9. 100001	**19.** 100100111	**29.** 111111

Section 16.3: Subtraction of Binary Numbers

1. 100	**13.** 10011	**25.** 10
3. 110	**15.** 10	**27.** −11
5. 10	**17.** 1001	**29.** −11101
7. 1	**19.** 1000111	**31.** 10111010
9. 1010	**21.** 10	**33.** −101110
11. 11100	**23.** −10	**35.** 110000010

Section 16.4: Multiplication of Binary Numbers

1. 1111	**9.** 100001	**17.** 1111110
3. 1100	**11.** 1001000	**19.** 100001001
5. 11001	**13.** 1010100	
7. 100100	**15.** 1001110011	

Section 16.5: Conversion from Decimal to Binary System

1. 1110	**9.** 11000	**17.** 1101111
3. 111111	**11.** 100000	**19.** 1110001
5. 1001000	**13.** 100101	
7. 10100	**15.** 1100100	

Section 16.6: Conversion from Binary to Decimal System

1. 2.5	**9.** 4.625	**17.** 4.8125
3. 2.25	**11.** 12.75	**19.** 3.8125
5. 5.75	**13.** 7.875	
7. 4.125	**15.** 26.5625	

Section 16.7: Hexadecimal System

1. 37	**7.** 197	**13.** 3010
3. 293	**9.** 1031	**15.** 2859
5. 30	**11.** 2594	**17.** 3294

19.	10,843	23.	3A	27.	15DA0
21.	EB	25.	D18	29.	8316

Section 16.8: Addition and Subtraction of Hexadecimal Numbers

1.	B	21.	CA51	41.	−91
3.	14	23.	10011	43.	−11F
5.	17	25.	B0EE	45.	−367
7.	A9	27.	E6CD	47.	−1EE
9.	13D	29.	17AEC	49.	−2B2
11.	169	31.	−C	51.	−A88
13.	77D	33.	−3	53.	C15
15.	1247	35.	−13	55.	−18AA
17.	1678	37.	−26	57.	1B8F
19.	CC3A	39.	−9	59.	−342F

Section 16.9: Binary to Hexadecimal Conversion

1.	6	17.	CE7	33.	101000110010
3.	A	19.	CCC	35.	11111100100
5.	17	21.	6891	37.	10011011101
7.	24	23.	F1CA	39.	101011001101
9.	44	25.	110	41.	100101000111011
11.	E6	27.	100100	43.	1011110010101111
13.	138	29.	101010		
15.	73D	31.	1001010001		

Chapter 16 Review

1.	13	11.	11110011	21.	83
2.	25	12.	100100	22.	293
3.	52	13.	11001101	23.	E63D
4.	22.75	14.	10000011010	24.	16
5.	100110	15.	225	25.	26
6.	10010110	16.	44	26.	173
7.	10100	17.	3102	27.	1001100
8.	1001	18.	138	28.	1101100101
9.	1101	19.	34	29.	101100101010
10.	10010	20.	1210	30.	100101010100001

Chapter 16 Test

1.	22	9.	10001110110	17.	2C
3.	10010000	11.	47	19.	1111111
5.	101001	13.	2C0		
7.	101011111	15.	4C8		

Cumulative Review Chapters 1-16

1. a. 2927.40

b. 2930

2. 37.0×10^{-9}

3.

$$C = \frac{5}{9}\left(F - 32°\right)$$

$$C = \frac{5}{9}\left(72° - 32°\right)$$

$$= \frac{5}{9}\left(40°\right) = 22.2° \text{ C}$$

4. 7890 km

5.

$$\left(2a^2 - 5a + 6\right) - \left(-2a^2 + 6a - 7\right) + \left(3a^2 - 7a - 2\right)$$

$$= \left(2a^2 + 2a^2 + 3a^2\right) + \left(-5a - 6a - 7a\right) + \left(6 + 7 - 2\right)$$

$$= 7a^2 - 18a + 11$$

6.

$$2\left(5y - 3\right) + 4\left(6y - 1\right) = 17\left(2y - 3\right) - 25y$$

$$10y - 6 + 24y - 4 = 34y - 51 - 25y$$

$$34y - 10 = 9y - 51$$

$$34y - 10 - 9y = 9y - 51 - 9y$$

$$25y - 10 = -51$$

$$25y - 10 + 10 = -51 + 10$$

$$25y = -41$$

$$\frac{25y}{25} = \frac{-41}{25}$$

$$y = -\frac{41}{25}$$

7.

$$\frac{x}{17} = \frac{14.6}{38.5}$$

$$38.5x = \left(17\right)\left(14.6\right)$$

$$x = \frac{\left(17\right)\left(14.6\right)}{38.5}$$

$$= 6.45$$

8.

$$y = mx + b$$

$$y = -\frac{2}{3}x + 4.5$$

$$y + \frac{2}{3}x = -\frac{2}{3}x + \frac{9}{2} + \frac{2}{3}x$$

$$\frac{2}{3}x + y = \frac{9}{2}$$

$$6\left(\frac{2}{3}x + y\right) = \left(\frac{9}{2}\right)6$$

$$4x + 6y = 27$$

9.

Let x = width

$x + 10.1$ = length

$$P = 2\left(l + w\right)$$

$$39.8 = 2\left(x + x + 10.1\right)$$

$$39.8 = 2\left(2x + 10.1\right)$$

$$39.8 = 4x + 20.2$$

$$19.6 = 4x$$

$$\frac{19.6}{4} = \frac{4x}{4}$$

$$x = 4.9$$

The width is 4.9 m and the length is $4.9 + 10.1 = 15.0$ m.

10.

$$30x^2 + 117x - 810$$

$$= 3\left(10x^2 + 39x - 270\right)$$

$$= 3\left(5x - 18\right)\left(2x + 15\right)$$

11.

$$3x^2 - 5x = 2$$

$$3x^2 - 5x - 2 = 0$$

$$\left(3x + 1\right)\left(x - 2\right) = 0$$

$$3x + 1 = 0 \text{ or } x - 2 = 0$$

$$x = -\frac{1}{3} \text{ or } x = 2$$

12.

$$x = \frac{-b \pm \sqrt{b^2 - 4ac}}{2a}$$

$$a = 7, \; b = 2, \; c = 15$$

$$\text{So } x = \frac{-(2) \pm \sqrt{(2)^2 - 4(7)(15)}}{2(7)}$$

$$= \frac{-2 \pm \sqrt{4 - 420}}{14}$$

$$= \frac{-2 \pm \sqrt{-416}}{14}$$

$$= \frac{-2 \pm 20.396\,j}{14}$$

$$= \frac{-2 - 20.396\,j}{14} \text{ or } \frac{-2 + 20.396\,j}{14}$$

$$= -0.143 - 1.46\,j \text{ or } -0.143 + 1.46\,j$$

13.

$$A = \left(\frac{a+b}{2}\right)h$$

$$A = \left(\frac{16.8 \text{ cm} + 22.4 \text{ cm}}{2}\right)(15.6 \text{ cm})$$

$$= 306 \text{ cm}^2$$

14.

$$V = \frac{4}{3}\pi r^3$$

$$V = \frac{4}{3}\pi (17.3 \text{ in.})^3$$

$$= 21,700 \text{ in}^3$$

15. $\cos^{-1} 0.9128 = 24.1°$

16.

$$\tan A = \frac{a}{b}$$

$$\tan 41.9° = \frac{a}{22.2 \text{ m}}$$

$$a = (22.2 \text{ m})(\tan 41.9°)$$

$$= 19.9 \text{ m}$$

17. $\cos(256°) = -0.2419$

18.

$$a^2 = b^2 + c^2 - 2bc \cos A$$

$$(18.2 \text{ cm})^2 = (20.5 \text{ cm})^2 + (26.1 \text{ cm})^2 - 2(20.5 \text{ cm})(26.1 \text{ cm}) \cos A$$

$$\cos A = \frac{(18.2 \text{ cm})^2 - (20.5 \text{ cm})^2 - (26.1 \text{ cm})^2}{-2(20.5 \text{ cm})(26.1 \text{ cm})}$$

$$A = 44.0°$$

19. 27.6

20. 28.5

21. 5.5

22.

Interval	Midpoint	Frequency
5.5–10.5	8	7
10.5–15.5	13	11
15.5–20.5	18	10
20.5–25.5	23	6
25.5–30.5	28	1
30.5–35.5	33	1

23. 16

24. 15.5

25. 10001000

26. 10001100111

27. 1001100100

28. 229

29. 10010101011

30. 1D02

Appendix B: Exponential Equations

1.

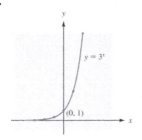

$y = 3^x$

$(0, 1)$

3.

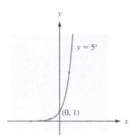

$y = 5^x$

$(0, 1)$

5.

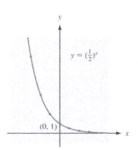

$y = (\frac{1}{2})^x$

$(0, 1)$

7.

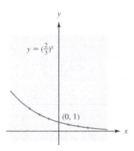

$y = (\frac{2}{3})^x$

$(0, 1)$

9.

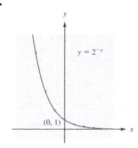

$y = 2^{-x}$

$(0, 1)$

11.

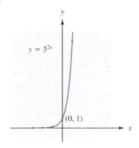

$y = 3^{2x}$

$(0, 1)$

13.

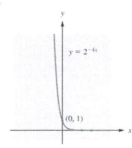

$y = 2^{-4x}$

$(0, 1)$

15.

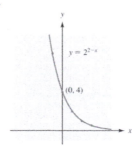

$y = 2^{2-x}$

$(0, 4)$

17.

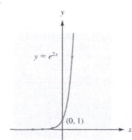

$y = e^{2x}$

$(0, 1)$

19.

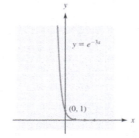

$y = e^{-3x}$

$(0, 1)$

261

21.

$$y = Ae^{rt}$$

$$y = (\$5000)e^{0.0375(2.5 \text{ yr})}$$

$$= \$5491.43$$

23.

$$y = Ae^{rt}$$

$$y = (65,000)e^{0.035(5 \text{ yr})}$$

$$= 77,400$$

25.

$$y = Ae^{-rt}$$

$$y = (\$75,000)e^{-0.05(10 \text{ yr})}$$

$$= \$45,490$$

27.

$$i = 0.025e^{-0.175t}$$

$$i = 0.025e^{-0.175(4.80 \text{ ms})}$$

$$= 0.0108 \text{ mA}$$

Appendix C: Simple Inequalities

1.

3.

5.

7.
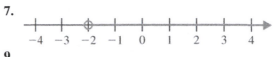

9.

11.
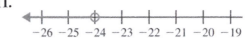

13. $x < 8$

15. $x \geq -1$

17. $x > 0$

19. $x \leq 3$

21. $x \geq -20$

23. $x \leq 16$

25.

$$5x < 15$$

$$\frac{5x}{5} < \frac{15}{5}$$

$$x < 3$$

27.

$$-4x \geq 20$$

$$\frac{-4x}{-4} \geq \frac{20}{-4}$$

$$x \leq -5$$

29.

$$16 \leq x + 7$$

$$16 - 7 \leq x + 7 - 7$$

$$9 \leq x$$

$$x \geq 9$$

31.

$$x - 12 \geq 8$$

$$x - 12 + 12 \geq 8 + 12$$

$$x \geq 20$$

33.

$$\frac{x}{2} > -4$$

$$2\left(\frac{x}{2}\right) > (-4)2$$

$$x > -8$$

35.

$$-10 \leq \frac{x}{5}$$

$$5(-10) \leq \left(\frac{x}{5}\right)5$$

$$-50 \leq x$$

$$x \geq -50$$

37.

$$3x + 5 < 29$$

$$3x + 5 - 5 < 29 - 5$$

$$3x < 24$$

$$\frac{3x}{3} < \frac{24}{3}$$

$$x < 8$$

39.

$$3 - 2x \geq -9$$
$$3 - 2x - 3 \geq -9 - 3$$
$$-2x \geq -12$$
$$\frac{-2x}{-2} \geq \frac{-12}{-2}$$
$$x \leq 6$$

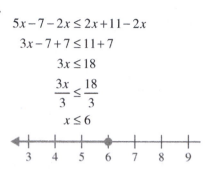

41.

$$5x - 7 - 2x \leq 2x + 11 - 2x$$
$$3x - 7 + 7 \leq 11 + 7$$
$$3x \leq 18$$
$$\frac{3x}{3} \leq \frac{18}{3}$$
$$x \leq 6$$

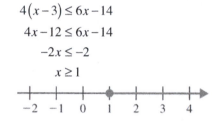

43.

$$4(x - 3) \leq 6x - 14$$
$$4x - 12 \leq 6x - 14$$
$$-2x \leq -2$$
$$x \geq 1$$

45.

$$3(2x + 4) < 2(x - 2)$$
$$6x + 12 < 2x - 4$$
$$4x < -16$$
$$x < -4$$

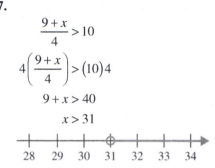

47.

$$\frac{9 + x}{4} > 10$$
$$4\left(\frac{9 + x}{4}\right) > (10)4$$
$$9 + x > 40$$
$$x > 31$$